W9-AXF-131

*Ebel/Bliefert/Russey*

# The Art of Scientific Writing

VCH

Distribution:

VCH Verlagsgesellschaft, Postfach 1260/1280, D-6940 Weinheim (Federal Republic of Germany)

USA and Canada:
VCH Publishers, Suite 909, 220 East 23rd Street, New York NY 10010-4606 (USA)

ISBN 3-527-26677-1 (VCH Verlagsgesellschaft)     ISBN 0-89573-645-4 (VCH Publishers)

Hans F. Ebel / Claus Bliefert / William E. Russey

# The Art of Scientific Writing

*From Student Reports to Professional Publications
in Chemistry and Related Fields*

**VCH**

Dr. Hans F. Ebel
Editor-in-Chief
VCH Verlagsgesellschaft mbH
Postfach 1260/1280
D-6940 Weinheim
Federal Republic of Germany

Prof. Dr. Claus Bliefert
Fachbereich Chemieingenieurwesen
Fachhochschule Münster
Stegerwaldstraße 39
D-4430 Steinfurt
Federal Republic of Germany

Dr. William E. Russey
Professor of Chemistry
Juniata College
Huntingdon PA 16652
USA

Production Manager: Heidi Lenz

Library of Congress Card Number 87-6165

Deutsche Bibliothek Cataloguing-in-Publication Data

Ebel, Hans F.:
The art of scientific writing: from student reports to professional publ. in chemistry and related fields / Hans F. Ebel; Claus Bliefert; William E. Russey. – Weinheim; New York: VCH, 1987.
  ISBN 3-527-26677-1 (Weinheim) kart.;
  ISBN 0-89573-645-4 (New York) kart.;
  ISBN 3-527-26469-8 (Weinheim) geb.;
  ISBN 0-89573-495-8 (New York) geb.
NE: Bliefert, Claus:; Russey, William E.:

Composition: Prof. Dr. Claus Bliefert, D-4437 Schöppingen
Printing: Zechnersche Buchdruckerei, D-6720 Speyer
Bookbinding: J. Schäffer oHG, D-6718 Grünstadt
Printed in the Federal Republic of Germany

APR 3 '90

*To Annie, Inge, Ruth, and Susan*

# *Preface*

This book is the product of an unusually close collaboration between three scientists from two countries. The three of us share varied backgrounds including teaching, research, and publishing in chemistry, but above all we share a deep interest in the role of the written word in science. Our book is especially dedicated to fellow chemists and aspiring chemists everywhere, but we hope it will be of service to others in the scientific community as well. Most of the examples in the book have been taken from chemistry, and some may be of interest only to chemists (such as the short excursion into chemical nomenclature in Chapter 6). Nevertheless, we have tried to present the basic ideas in such a way that they will be accessible to scientists in general. This is why our book has been given the title *The Art of Scientific Writing* and not "The Art of Chemical Writing" (which is not in any case an art in its own right).

Many books about writing already exist, particularly in English. However, we know of no book as broad in its coverage, as critical in its analysis of existing trends, and as international in its scope. Moreover, many of the books on the market do not supply the aspiring author with sufficient insight into the publishing process and the attitudes and expectations of referees, editors, and publishers. Nor are most books comprehensive enough to provide ready answers to the host of questions that inevitably arise during the communication process.

Our book is designed as both a reference manual and a basic text; a source of background information on causes and consequences of scientific writing, and a guide for the novice. We have therefore used an approach different from that of most other books on writing, an approach that is in itself scientific. Throughout the book we have tried to show the reader not only "how", but also "why" certain procedures are recommended. We also examine the incentives underlying scientific writing and some of the reasons scientists write as they do. *The Art of Scientific Writing* is thus more than a book on "How to Write".

All stages of the publication process, of which writing is a part, have been considered. Our starting point is the writing required of a future scientist in college or graduate school.* From there we proceed to more complicated material, in the end dealing with topics that may be of concern primarily to the mature author-scientist. In a way, the reader is led systematically from graduate student days to the heights of professional life. This idea is also expressed in the book's subtitle.

*The Art of Scientific Writing* consists of two parts. In Part I, basic types of scientific writing are discussed sequentially. Chapter 1 on reports is addressed primarily to science students, but also to those who write research reports and grant proposals in the academic world or industry. In it we describe what instructors and supervisors expect of an interim report, or even a properly maintained notebook. We hope that this chapter alone will substantially aid research directors in achieving their goals.

In Chapter 2 we accompany the graduate student through the horrifying maze of preparing a thesis, and the labor of achieving a degree. From here we move on to the more public types of writing, journal articles (Chapter 3). Finally, we offer the seasoned scientist advice applicable to the even more demanding—and far more rarely described—writing opportunities available later in a career: writing or editing a book (Chapter 4).

The technical problems entailed in scientific writing are almost independent of the kind of document to be produced. For this reason we felt it best to identify and isolate certain key subjects—the proper use of units and quantities and the preparation of figures and tables, for example—and treat them separately, one by one. The result is Part II of the book. Our objective here has been to share in a systematic way, and from a variety of perspectives, useful insights we have gained over the years. While primary attention has been directed to specific techniques that can lead to better scientific writing, there are also brief excursions into the aesthetic qualities of well-crafted

---

* A forthcoming supplement by the same authors, entitled *Study Guide to The Art of Scientific Writing—Problems and Exercises,* will provide the student with additional guidance.

technical prose and the ethics implicit in scientific writing. We have also enjoyed the challenge of sketching the peculiar demands and problems posed by the "game" of scientific writing and publishing.

Some may see *The Art of Scientific Writing* as a textbook to be introduced at the college or university level. Others will, hopefully, keep it close at hand as a reference manual. For the latter purpose we have been careful to provide a wide variety of specific examples, an extensive set of internal cross-references, and an especially thorough index; in addition, a set of appendixes incorporates supplementary information. The book should also be useful to science instructors called upon to stimulate and criticize student writing, and to editors, copyeditors, and others in the publishing business.

A word about language is perhaps in order. English has clearly become the "lingua franca", the language employed by scientists all over the world as the chief medium of scientific communication. Today, if a French chemist visits the laboratory of a German colleague, the discussions that ensue are likely to be held in English. Likewise, a lecture by a European, North American, Australian or foreign scientist of any nationality to colleagues in the People's Republic of China will probably be in English. English now holds the position once claimed by Latin. For this reason our thoughts on scientific writing have been presented in English, even though our target audience is by no means limited to native speakers of English; it encompasses all scientists who want or need to write about science. The stature of our publisher—VCH, with offices in both Europe and the United States—has been of particular help in gaining such an international perspective.

We are indebted to a number of people who helped us complete this book. Several chapters have been read carefully, and improved, by Dr. Christina Dyllick-Brenzinger and by Dr. David I. Loewus, member of the staff of *Angewandte Chemie International Edition in English*. Valuable guidance and advice has been given by Dr. Gerald Reed, Milwaukee, and by Dr. Steve Yamamoto of the editorial staff of *Ullmann's Encyclopedia of Industrial Chemistry*; by Gerlinde Kruse, nomenclature expert of *Angewandte Chemie*, and Prof. Gerhard Wenske, member of several DIN committees, with respect to Chapters 6, 7 and 8; and by Dr. Peter Hass, Mannheim, with respect to Appendix C. Special thanks for linguistic and other improvements are due to Dr. R.E. Reed and S. Scott Munshower. Both contributed many hours reading the entire manuscript. Technical advice in preparing the master copy for the print was received not only from our publisher but also from Eckehard Stümpel, Münster, and Christophe Villain, Paris.

We enjoyed writing this book and gathering experience from these and other friends and colleagues. Comments and criticism will be greatly appreciated from the readers to whom we have devoted so much effort in an attempt to help them in their scientific lives.

| | |
|---|---|
| Weinheim | H.F.E. |
| Steinfurt | C.B. |
| Huntingdon | W.E.R. |

March 1987

# Contents

XII  *Contents*

## Part II Scientific Writing: Materials, Tools, and Methods

*Appendixes*

# Part I
# Scientific Writing: Aims and Forms

# 1 Reports

## 1.1 Writing and Science

Picture the following situations, chosen because they illustrate the diverse nature of scientific activity:

- A nervous freshman carries out a simple titration for the first time.
- A graduate student considers how to analyze the final product in a synthetic sequence begun three long years ago.
- A technician re-calibrates a spectrometer for the sixth of a series of 20 routine quality-control measurements.
- A senior scientist examines data that seem to require a revolutionary modification of a long-held theory.

At first glance the four situations appear to have little in common, but one important characteristic they share: each endeavor will remain incomplete until its results have been *communicated*, or *reported*. Indeed, one could claim that the true significance of the work in each case lies in that *communication*, for in the absence of exchange of information there is no science. Leaver and Thomas (1974) phrase the same point somewhat differently in their perceptively written chapter on report writing (p 7): "The report is the end-product of research."

"Scientific" research is understood to be exploratory work conducted according to what has been characterized as the *scientific method*, and it almost invariably begins with the study of information acquired and made public by someone else. Another researcher's reported results attract a scientist's attention, they are subjected to close scrutiny, perhaps from a perspective never envisioned by their discoverer, some seem to demand re-examination—and suddenly a new research project is born. Previous work has become the basis for new experiments, and another chapter is opened in the record of mankind's quest for knowledge. It is in this sense that true research, true scientific work of any kind, is the antithesis of a purely "personal" experience.

If communication is obligatory in science, must it take any particular *form*? We generally regard the *spoken* word as our most important means of sharing information because it is most direct and immediate. Simple thoughts whose full import is easily grasped are indeed often best transmitted as spoken words:

> "O.K., I'll redo it."
> "The instrument is running."
> "Shut off the power!"

Scientific observations and deductions rarely fit in this category, however. In the first place, the truth of a scientific assertion is usually a function of the *precision* with which it is stated. Subtle details and equally subtle implications are what make scientific statements powerful. The spoken word is altogether too spontaneous, often too hastily formulated, and certainly too fleeting to be acceptable as the sole carrier of ideas of such precision and subtlety. It is no coincidence that oral presentations of research results at scientific meetings are considered less permanent and definitive than published papers.\*

There is a second important reason why the spoken word is of limited value for scientific and technical communication. One of life's biggest challenges is coping with a constant stream of information, and each day everyone's retention capacity is exceeded many times. If all scientific communication were verbal the vast majority of results would inevitably be ignored, forgotten, or, at best, housed in inaccessible corners of the mind. This would be intolerable. Scientific findings are obtained at substantial cost, and each is so unique and potentially valuable that it deserves special protection and faithful preservation.

Finally, there is a third factor to consider. For results to be regarded as truly scientific they must be made readily available for close, repeated inspection and *evaluation*—not just immediately and in the laboratory where they originate, or in the lecture hall where they are first disclosed, but also over time, and throughout the world. Isolated observations, experimental measurements, and test results are the foundations upon which far-reaching scientific conclusions are built. As long as each newly uncovered fact remains readily accessible, the potential exists for researchers unknown to the discoverer to ponder it, probe its significance, and attempt to fit it into a larger framework.

---

\* It is not our intent to dismiss oral presentations as of no value. On the contrary, they serve several important purposes. For example, a lecture can be an important stimulus: it can cause one to want to learn more about a subject (generally by reading!). Also, hearing someone talk about an idea can "personify" it, and members of an audience are often moved to try to establish closer contact with a speaker. Appendix A is devoted to a brief discussion of formal oral communication and techniques for making it more effective.

For thousands of years man's principal alternative to speech has been *writing*. Recent trends seem to suggest that this may someday cease to be the case, and it would be folly for us to attempt to foretell how communication among scientists will be structured in the future. No one knows what revolutionary developments will occur in information transfer, storage, and retrieval. Nonetheless, our primary concern here is with the present. It certainly remains true that the written word is the dominant medium for lasting communication and careful study. Today, in the world as we know it, written communication is indispensable, and scientists—including research chemists—have no choice but to come to terms with the fact.

It goes without saying that scientists need to be skillful *readers*. Extensive reading is the principal key to expanding one's knowledge and keeping up with developments in a discipline. The often ignored corollary to this assertion, however, is that scientists are also obliged to be skillful *writers*. Only the researcher who is competent in the art of *written* communication can play an active and effective role in *contributing* to science.

One of the aims of this book is to provide information and suggestions that will improve the average scientist's writing skills. We begin by considering what scientists write and how their writings are disseminated. Later we discuss in detail specific strategies and techniques that facilitate the preparation of high-quality documents.

## 1.2    The Purpose and Nature of a Report

### 1.2.1    A Matter of Definition

Virtually every productive scientist is regularly called upon to describe to others the progress of his or her work. Usually this entails producing some kind of *report*. The term "report" as we use it here is a very broad one, encompassing documents that differ considerably from one another in content, style, length, and importance. A few obvious examples are college and university *laboratory reports* (describing and explaining the results of assigned experiments), *project status reports* (destined for research supervisors or administrators), and the various reports related to research grants (including *grant proposals*).

Manufacturer's product descriptions, instrument manuals, information sheets, and patent applications are also "reports" in the broadest sense, as are theses, journal articles—even monographs. The latter more "advanced" literary species are treated in depth in subsequent chapters of our book. In this first chapter, however, we restrict ourselves to briefer writings. We begin with laboratory notebooks and those straightforward reports required of aspiring scientists during their studies. These writing exercises, properly done, can serve both as models for the more complex challenges a scientist encounters later in a professional career, and as preparation for "winning the games scientists play" (Sindermann, 1982; see also Hawkins and Sorgi, 1985).

The proper form for a given report depends on its purpose and intended recipients (Souther and White, 1984; see also Standard *ANSI Z39.18-1974*). Nevertheless, all reports do have certain features in common. We begin our exploration of that common ground by quoting from an official *standard*,* the *American National Standard for Bibliographic References* (Standard *ANSI Z39.29-1977*). This comprehensive (92-page) guide[†] actually deals with the proper form and content of citations, but within this context it defines a "report" as "a separately issued record of research results, research in progress, or other technical studies". It further notes that a report normally

---

* Standards, or "norms", are detailed technical definitions developed by various official bodies in an attempt to encourage uniformity among practitioners of a particular art or trade. Industrialized nations have long recognized the need for standardization within industry and commerce, resulting in adoption at the national level of standard weights, standard measurements—even standard colors and screw sizes. Gradually, such standards have taken on an increasingly international character, thanks largely to the efforts of the International Organization for Standardization (ISO). Coordination on the national level is encouraged by organizations like the American National Standards Institute (ANSI), the British Standards Institution (BSI), and the Deutsches Institut für Normung (DIN) in Germany. These are in turn supported and assisted in their efforts by professional organizations such as the American Chemical Society. ANSI, like its counterparts elsewhere, has for some time attempted to encourage a certain amount of "standardization" in intellectual activities, particularly with respect to the way scholars communicate. Over the years, ISO, ANSI, DIN, and others have published various norms having directly to do with writing, ranging from the way to write reports or abstracts, through recommended technical symbols, and descriptions of preferred book spine formats. These little-known recommendations summarize a wealth of wisdom and information, and we hope to encourage their wider study and utilization by making frequent reference to them in this book.

[†] Available from the American National Standards Institute, Inc., 1430 Broadway, New York NY 10018, USA.

concerns itself "only with a specific subject or a few closely related subjects". The ANSI definition concludes with the observation that reports usually are "not typeset, but reproduced from typing by a near-print process such as multilith, mimeograph, ditto, or photodirect".

For our purposes this final qualification is unfortunate, because it rules out such "report-like" technical writing as student laboratory reports. Even most research reports would fail to qualify, since they usually exist only as an original and at most one or two photocopies.

From the standpoint of the quoted ANSI standard a restrictive definition is logical, because the only reports with which it is concerned are those amenable to *citation* (inclusion in reference lists; cf. Sec. 11.3). This requires that a "report" be readily available to readers, hence the emphasis on multiple copies. Here we are only interested in the form and content of a report, so it is more appropriate that we adopt a broader definition, one that includes student laboratory reports, research reports, and grant proposals. These will in fact be the focus of our attention, because the lessons to be learned from them can easily be applied to reports in general.

## 1.2.2    Some Characteristics of a "Typical" Report

Therefore let us consider a report as *any document containing a systematic record of a particular technical study.* We further note that a report is something prepared primarily for the benefit of *others*, even though it may also be of value to its author.

What more can we say about reports in general? There are several far-reaching implications that follow from our definition, and it is these which point to a report's most important characteristics.

• The very act of preparing a report fulfills part of the report's assignment.

This assertion is not as paradoxical as it sounds, nor is it an inappropriate starting point for our discussion. In order to describe a piece of work in a comprehensive, meaningful, and useful way the report writer must first review the relevant facts and organize them in a manner that leads the reader naturally to the writer's own conclusions.

The process can be an enlightening one—for the writer! Even seasoned scientists are often surprised to discover the profoundly new levels of understanding they *themselves* reach as a result of writing a report of their own work. The reporting process clarifies the boundaries of their knowledge and it sometimes reveals unsuspected wider implications. Indeed, many find this "side-effect" ample reason in itself for preparing periodic progress reports. Nevertheless, reports are usually written for others. It follows that

● The expectations of intended *recipients* must be kept in mind throughout the process of report preparation.

Someone, perhaps the supervisor or sponsor of a project, is to be informed of a certain set of accomplishments (one's own or those of one's co-workers). The report will be the informant; if it is to be effective, its form, content, and style must be right for the background and interests of the recipient.

Often, the key to success in this regard is proper *balance*. For example, the amount of detail to be presented should be a subject of careful consideration. A very brief report stressing results and conclusions would be appropriate in some circumstances, but there are other times when methods or specific observations are what should be treated, and these at great length. Some readers may relish or even demand minute details, while others will only be confused by them. The successful writer must analyze what the intended reader is really interested in learning, and then strive to produce a report that will evoke a positive response.

It is especially important that the *value* of the reported work and the *effort* that went into it be accurately conveyed. A report that is overly wordy and dwells too long on the obvious or trivial will produce little more than boredom and irritation. Indeed, the recipient of such a report is likely to conclude that the writer was trying to disguise a pedestrian piece of work by "padding" the description of it. Too little information, on the other hand, can imply that little work was actually done.

Proper balance is difficult for a writer to achieve. Success depends almost entirely on the ability to analyze the reported work from the perspective of the intended reader.

● A report is a permanent, independent record, an entity complete in itself.

In a sense, a report is the author's personal representative, and proper attention must be directed to its *appearance* and the *words* in which it is framed. These can leave a deep and lasting (though sometimes unconscious) impression on a reader. Indeed, a report's *style* is almost as important as its *content*. As one is writing, it is wise to stop occasionally and ask what kind of an "ambassador" the emerging report is going to be. Are ideas being expressed with the same clarity, efficiency, precision, and accuracy that characterize one's scientific endeavors? Are purely mechanical matters such as *grammar* and *spelling* beyond reproach? (Specific suggestions for improvements in this regard can be found in Sec. 1.4 and Appendix B.) Is every reasonable precaution being taken to ensure *neatness*? These are crucial questions, because some readers are hypersensitive to just such "details".

The *independent* character of a report also has important implications. The writer will not be able to watch the reader's reactions and, at the proper moment, offer a defense such as: "But that's not what I meant! What I really wanted to say is ...". Also, the well-written report provokes no unforeseen questions. Writing will be of this caliber only if each sentence is carefully formulated and later re-examined within the context of the whole. Every sentence must say something significant, and the point each makes must be in the right place so the entire report will effectively convey its message.

● The preparation of any report relies upon the  broader *context* of a body of work previously undertaken by the author and others.

Establishing this context normally requires a large number of explicit *citations* of previous findings. The relevant background facts will have been reported at the time of their discovery, usually somewhere in what is known as "the literature" (cf. Sec. 3.1 and Appendix D). The purpose of a citation is to explain precisely *where* in the literature so that the interested reader can learn more. Literature citations are so crucial to scientific reports that standard conventions have been developed for their presentation, a subject we examine later in considerable detail (cf. Sec. 2.2.13 and especially Sec. 11.4 and Appendix J).

● Although reports are typically thought of as descriptions of previous accomplishments they also can have an influence on the *future*.

Nearly every report sets the stage for follow-up investigations of some kind. If industry is involved, a report may also be the key to far-reaching decisions, perhaps related to investments or to overall corporate strategy. Properly managing the forward-looking aspect of a report can be extremely important, even for the writer: a report is often the decisive factor in determining the extent to which one's efforts will continue to receive support. More will be said about this matter in Sec. 1.5.3.

● Finally, all reports have in common the need for proper *identification*.

A well-chosen *title*, for example, is essential for characterizing a report and providing some insight into its contents (cf. Sections 2.2.2, 3.3.1, and 4.5.2). With rare exceptions, the name(s) of the *author(s)* should be prominently displayed as part of a *heading*, which may also contain the *date* of writing or submission. If the report is likely to be read outside the writer's own department or institution, its *origin* needs to be apparent as well.

These are the minimal elements, but they may not suffice for an unusually formal report, or one expected to have wide distribution. An additional *identifier* may be necessary, such as a report number (assigned by the writer) or a contract or grant number. Numerical identifiers are common because of the help they provide in document classification and retrieval.*

We have now seen the principal features common to all reports. Characteristics of specific kinds of reports will emerge in subsequent sections of this chapter.

Not all of a scientist's "professional" writing is intended for the use of others. Some is more personal: weekly work plans, summaries of the current state of one's thinking, specific reminders, or the odd note designed as a "memory crutch". It might seem that writing of this kind is immune to generalizations, that it certainly need not be the subject of a book. Perhaps; but a word of caution is in order. We turn next to a discussion of the scientist's *notebook*, one of the most personal pieces of scientific writing, a kind of "diary" of one's professional work. We shall quickly find that a notebook cannot be treated strictly as an individual possession at the mercy of each writer's arbitrary whims. Indeed, a researcher's career success can be strongly influenced by habits developed in the process of notebook maintenance (Kanare, 1985).

## 1.3    The Laboratory Notebook

### 1.3.1    Is a Notebook Really That Important?

Before we discuss further the subject of report writing it is appropriate to consider the sources from which *experimental descriptions* are drawn. We

---

* Numbers carry the additional advantage that they sometimes elevate a report to membership in the body of information known as the "gray literature" (cf. Sec. 11.4.5).

begin here because most scientific reports are ultimately summaries of *experiments*. They may also incorporate theoretical arguments or speculative conclusions—indeed, it is often these parts that make reports memorable—but experimental evidence forms their basis.

One can write a meaningful account of an experiment only if one has a good set of *records*: on-the-spot descriptions, original spectra, computer outputs, and the like. Successful scientists are constantly alert and observant as they work, and they conscientiously take the time required to make clear, comprehensive notes of their actions, their observations, and their thoughts. They know that this is the only way of preserving the present for the future, and it allows them to write reports at a later time easily and without undue reliance on memory.

For example, every good scientist knows that the slightest deviation from a routine procedure is a "noteworthy" event, as is any unexpected phenomenon, no matter how irrelevant it may seem at the time. Details that a beginner might regard as trivial often turn out to be the most important facets of an entire experiment. The recollection of a peculiar color change on extracting an ethereal solution with aqueous alkali, finding a notation of a spur-of-the-moment decision to use $Na_2SO_4$ as a drying agent (because the $MgSO_4$ bottle was empty), being able to glance once more at the spectrum of a supposedly useless distillation residue—any of these might provide the key to a crucial insight.

Reliable and complete experimental records are thus indispensable to any truly scientific endeavor. Let us generalize further by saying that all science-related actions, observations, and measurements are important enough to be recorded. Most important, the writing must take place *immediately*, it must be *systematic*, and it must be in a *permanent* place. That place is a properly kept *laboratory notebook*, arguably a scientist's most important tool. Working in a laboratory—even pondering a complex set of ideas in one's office—without having a notebook open nearby should be unthinkable, unnatural. Someone once observed that a scientist without a notebook at hand is "out of uniform". This is no exaggeration. It is absolute foolishness to trust only one's memory, or to scribble isolated notes on loose pieces of paper or a blackboard. Memory is notoriously more fallible than one would wish or expect, and the best of intentions will not protect casual notes against loss—either of the notes themselves or of their significance.

The only acceptable place to record anything related to an experiment is in a permanent notebook, one which is firmly bound and has numbered, dated pages. A *bound, paginated, dated* format is always specified; single sheets are inappropriate, even if collected in a looseleaf binder, because they lack both permanence and a verifiable sequence.* The need for permanence also requires that notes be written exclusively in *ink*—of a kind that will not disappear if it gets wet!

## 1.3.2     What Should a Notebook Contain?

*Introduction*

Notebook records of one's findings must always be provided with a proper *context* if the *content* is to be meaningful. Each experimental write-up in a notebook should begin with an introduction containing:

- a *title* and an *identification number* (see below) for the experiment,
- the *date, time,* and (unless self-evident) *place* of the recorded events,
- a statement of the precise *goal* the experimenter had in mind, and
- annotated citations of all relevant *background literature.*

The introduction should also incorporate information about all materials used, as well as a record of experimental conditions. It should include, for example:

- the balanced equation for any chemical reaction involved;
- important instrumental parameters and measurement conditions;
- a record of significant external factors (such as the temperature and pressure in the laboratory);
- sketches of unusual or unfamiliar apparatus;
- molecular weights and physical properties of relevant chemicals (e.g., melting and boiling points of reactants and products, perhaps also expected solubilities, toxicities, etc.);

---

* The fact that a notebook is bound and intact, with numbered and dated pages, can be particularly important if it ever needs to be used as *legal evidence* (e.g., with respect to patent matters). Industrial laboratories usually have a rule that notebook pages must be *signed* and that all important results and records are to be *witnessed* ! Many companies provide their research scientists with special notebooks containing pre-printed page forms and a set of "Instruction" pages. For example, the instructions may inform the recipient that it is strictly forbidden to leave any pages empty; to do so would permit legal questions to arise concerning the chronology and documented authenticity of recorded events.

- identification of the supplier and grade of all commercial substances employed;
- sources of any non-commercial materials (e.g., by notebook cross-references, cf. Sec. 1.3.3);
- techniques used in purifying or testing any starting materials or reagents, together with the results.

It is also useful to outline here the experimental *procedure* one proposes to follow, but such an outline must never be allowed to substitute for a complete description of what was *actually* done and observed during the course of the experiment; the latter is the subject of the narrative.

## Narrative

The *narrative* is a description of the experiment proper. It must be a complete, self-contained, running account of all procedures and observations constituting the experiment. The distinction must be absolutely clear between, on the one hand, speculations and intentions (most of which should be part of the introduction), and on the other, observations and actions. Failing this, confusion will certainly develop later as to what is fact and what is wishful thinking. The inclusion of *subheadings* or other division markings (e.g., dividing lines), and careful, consistent use of *tense* are convenient devices for maintaining clarity. Thus, if the narrative were to take the form "I took ..." or "I added dropwise ...",* then the introduction would be constructed with phrases such as "The purpose of the experiment is ..." or "This should then ...".

As one is composing the experimental narrative (one or two sentences or phrases at a time, while the experiment is in progress), it is important to bear in mind that the resulting words will constitute the only reliable information available when the time comes to report the experiment. Ideally, notebook entries are written in such a way that they already approximate the form appropriate for the experimental section of a report, since this can substantially reduce the time required later for the "translation" process (cf. Sec. 1.4). In fact, the main difference between an ideal notebook entry and a description of the same experiment in a report is that the former is always considerably more *detailed*, since the information in a report will have been subjected to later editing.

---

* The narrative might also be written in the more impersonal passive form illustrated in Fig. 1-1a.

One other difference between the description of an experiment in a notebook and that in a report has to do with how *data* are presented. A proper notebook account always contains every *measurement* taken (weighings, tare values, buret readings, absorption intensities, etc.). These are accompanied by the *calculations* used to convert the raw numbers into more meaningful *derived data* (such as moles of starting materials, concentrations deduced from titrations, yields, etc.). Normally, only the latter would appear in a report, but all raw data, even though they are generally of no fundamental significance (and hence go unreported), need to be scrupulously preserved somewhere. The reason is simple: at a later time, logical fallacies may become apparent, requiring re-calculations based on original measurements. It is also not uncommon to discover that an instrument has for weeks given false information, but that results from it are nonetheless interpretable provided the data originally obtained can be recovered and suitably modified by a correction factor.

### 1.3.3    Organizational Considerations

Up to this point we have implicitly assumed that scientific work automatically divides itself into discrete *experiments,* each of which corresponds to a notebook entry. In doing so, however, we have ignored a fundamental question: just what *is* an experiment? When does one start and where does it end? The answers are not always obvious. Suppose, for example, that one has carried out a chemical reaction, and that it has led to a bewildering mixture of products. Chromatographic separation has produced six distinct fractions, and each fraction must now be studied in detail. Tests for uniformity will be required, as will attempts to characterize and identify the components. Are all these follow-up procedures still a part of the original experiment, or should the work on each fraction be regarded as a new experiment?

It is not our intent here to propose a general solution to the problem, nor even to answer the specific question raised by our hypothetical example. We only want to point out the kind of dilemma one often encounters in trying to arrange notebook entries. No matter what decision is reached in a particular case, it is inevitable that many individual experiments will ultimately be interrelated. The "product" of one experiment (a compound, a method, a theoretical equation, or a piece of data), regardless of the way "experiment" is defined, is almost always the "starting material" for another. Therefore, some system must be devised for notebook *cross-referencing.*

One good and often used device is a system of *experiment numbers,* each derived from the notebook page number on which the corresponding experiment begins.* Page number identifications should be augmented (usually preceded) by the experimenter's *initials,* particularly if one is working as part of a research team. Moreover, since notebooks—properly kept!—are quickly filled, inclusion of a notebook *volume* number is wise. Roman numerals are recommended, since volume numbers are thereby clearly distinguished from (Arabic) page numbers. A particular experiment might therefore be designated HJK-IV-127. Individual substances (or fractions) arising from this experiment could then be referred to by a code designation comprised of the experiment number and suitable subclassification numbers or letters. All coded items must of course be clearly identified in the notebook. Using such a scheme, chromatography fraction 5 from experiment HJK-IV-127 might be called HJK-IV-127-5. Codes or identifiers of this sort are convenient for a number of purposes, including labels on sample vials or spectra. They have the virtue of combining brevity with a minimal risk of confusion.

A scheme should also be devised for making notebook entries themselves readily accessible, not only to the researcher but also to others who may need to refer to them. Trying to locate information whose only organizing principle is its date of inception (i.e., the order in which it appears in a notebook) is frustratingly difficult. The simplest solution is an up-to-date *table of contents* for each notebook, preferably located at the front. Ample space should therefore be reserved for the purpose. Industrial notebooks often contain special provision for content records.

We began our discussion of research notebooks by emphasizing their importance. It seems appropriate to conclude by returning to this theme and, more particularly, to one of its corollaries: no research notebook should ever be discarded! Research groups usually insist that departing members leave their notebooks behind, because it is almost certain that other members of the team will someday need access to experimental details that appear here and nowhere else. Most research supervisors maintain whole shelves or even cabinets full of notebooks left by previous generations of co-workers. Industrial scientists

---

* This statement implies that a given experiment may extend over several pages. If this occurs, each page (except the last) should conclude with the words "continued p. xx". Free space remaining on the last page should be crossed out as soon as the experiment is complete and it is clear that nothing else need be written (cf. the first footnote in Sec. 1.3.2).

eventually regard returning their filled notebooks to a supervisor as a matter of routine, comparable (and related!) to receiving a salary.

The individual researcher is also likely someday to want access to the details of his or her past work, particularly since new research projects have a way of evolving from nagging questions put aside at an earlier time. For this reason it is a good idea to keep personal carbon copies (or, more likely, photocopies) of all notebook entries, assembled in loose-leaf binders.*,†

## 1.4    *Transforming Notebook Entries into a Report*

### 1.4.1    *Describing an Experiment*

Notebook entries prepared in the manner just described are easily re-worked into the experimental section of a report—or a thesis or journal article. The most difficult task is identifying what information and data should be reported. *Selectivity* is necessary since the writer should by this point possess more insight into the work than at the time the experiments were conducted. Certain observations recorded in the notebook will now be recognized as irrelevant or even erroneous, and some experiments will clearly have been misguided. The extent to which these false leads should be reported at all will depend on the role they played in subsequent work.

Being selective is a virtue, but *changing* information is a cardinal sin! No data should ever be altered for a report regardless of the temptation. Scientific reports, just like the work on which they are based, are assumed to be truthful. Puzzling results may prove awkward to deal with, but they class as *facts* and

---

* However, company policy usually prohibits *industrial* scientists from making copies of their notebooks.
† Much additional information regarding notebooks and their maintenance can be found in the book by Kanare (1985), which is illustrated by numerous examples.

deserve to be taken seriously. Ideally, one tries to explain or rationalize surprising findings, but even if that seems impossible it is wrong to alter or discard them.

Fig. 1-1a shows an excerpt from a typical notebook description of an experiment, written at the time the experiment was run. The example chosen describes the purification and characterization of an organic compound prepared according to a procedure (theoretical yield 1.85 g) described in the literature. The authors of the original paper report the compound's melting point as 85 °C. This particular researcher's notebook entry describes all observations and actions in a highly abbreviated way and relies heavily on key words and phrases rather than complete sentences.* Let us assume it is now to be elaborated for incorporation into a report.

A segment of a report dealing with this same information is reproduced as Fig. 1-1b. Notice the changes that have been made and the form in which the description has now been cast. The writing is quite formal. Full sentences are used, but an attempt is made to state everything as concisely as possible, though certainly not at the risk of ambiguity. All actions have been stated in the *past tense*, using the *passive voice* (i.e., without the use of "I" or any other identification of the person carrying out the work—who is *presumed* to have been the author).[†] Raw data have been converted to more meaningful terms [e.g., yield (in %), $pK_a$]. "Unimportant" or fairly obvious details have been omitted (type of funnel, drying agent, etc.), as have certain pieces of actual data—including some that might prove very helpful to someone trying to repeat the work.[‡] Omissions of this kind require judgments about the importance of the information, the needs of the reader, and the merits of brevity. Fortunately, however, even if important information was omitted from a report it could, in principle, be retrieved from the researcher's notebook, although with difficulty.

---

* Not everyone keeps such a terse notebook. Indeed, provided the result is clear and accurate, every scientist should feel free to establish his or her own notebook style.

† This grammatical idiosyncrasy, insistence upon passive constructions, is strictly a convention, but one that has traditionally been observed despite the fact that it often engenders awkward and even dull sentences. Recent books and articles on scientific writing encourage a more personal, simplified writing style (e.g., Schoenfeld, 1986; Day, 1983; O'Connor and Woodford, 1977). A corresponding change in practice is barely perceptible, however, and we concluded it would be wise to deal here with what *is* rather than what *might* be.

‡ A case in point is the volume of solvent employed for recrystallization.

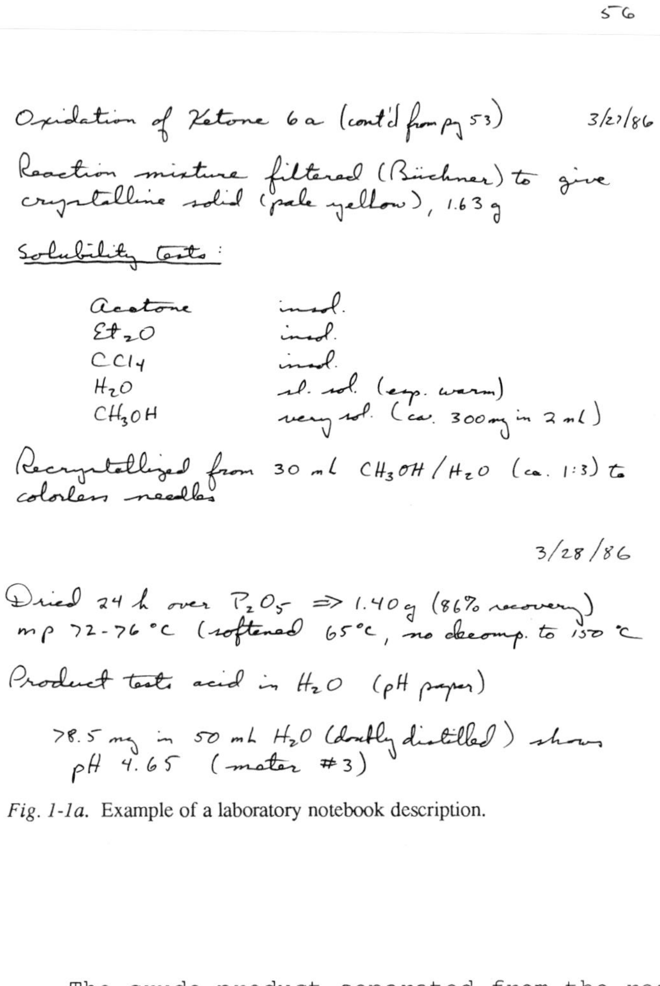

Oxidation of Ketone 6a (cont'd from pg 53)     3/27/86

Reaction mixture filtered (Büchner) to give crystalline solid (pale yellow), 1.63 g

<u>Solubility tests</u>:

| | |
|---|---|
| acetone | insol. |
| $Et_2O$ | insol. |
| $CCl_4$ | insol. |
| $H_2O$ | sl. sol. (esp. warm) |
| $CH_3OH$ | very sol. (ca. 300 mg in 2 ml) |

Recrystallized from 30 ml $CH_3OH/H_2O$ (ca. 1:3) to colorless needles

3/28/86

Dried 24 h over $P_2O_5$ ⟹ 1.40 g (86% recovery)
mp 72-76 °C (softened 65°C, no decomp. to 150 °C

Product tests acid in $H_2O$ (pH paper)

78.5 mg in 50 ml $H_2O$ (doubly distilled) shows
pH 4.65 (meter #3)

*Fig. 1-1a.* Example of a laboratory notebook description.

... The crude product separated from the reaction
mixture in crystalline form (slightly yellow due to
impurities). It was recrystallized from aqueous
methanol ($\varphi \approx 25\%$) to give a 75.7% yield of <u>3a</u> as
colorless needles, mp 72-76 °C [reported by Jones
(1980) to be 85 °C]. An aqueous solution of the
solid was found to be weakly acidic ($pK_a \approx 9$).

*Fig. 1-1b.* Information from Fig. 1-1a recast in the form appropriate for a report.

## 1.4.2    Preparing the Report

What follows is a collection of simple, pragmatic suggestions for report preparation. They are derived from experience, and while some may appear trivial or self-evident, others may help even experienced writers generate more effective (because more readable!) scientific prose.

First of all, before drafting a report, even a short one, an *outline* should be prepared, although this may consist of nothing more than an ordered list of key words. An outline is a device for organizing ideas so they flow smoothly and in a logical order. The object in preparing it is to be sure that no steps in an argument will be omitted, and that unnecessary or awkward backtracking will be avoided. Taking the time to prepare a proper outline actually can amount to a net *saving* of time, because it results in a framework that facilitates the subsequent writing process.

Lengthy reports should be divided into discrete *sections* bearing informative titles to guide the reader through the document. The longer a report becomes, the more nearly its sections begin to resemble those of a typical *thesis*, the subject of our next chapter (see especially Sec. 2.2 for information regarding thesis organization). Fig. 1-2 illustrates a sample outline for a report of a preparative organic experiment. It has been designed to meet the requirements typically imposed on a college-level report (cf. Sec. 1.5.1) and is organized and numbered according to the system described in Sec. 2.2.5.

Once a satisfactory outline exists, one that is complete, orderly, and sufficiently detailed, the next step is to compose a *first draft*. This preliminary version of the report is only a tentative, rough approximation of a finished document, but it is a major step beyond the outline. In the draft, ideas are expanded into complete *sentences* that should group themselves naturally into a series of *paragraphs*, each devoted to a single topic from the outline.

It is important to emphasize that paragraphs are not merely a cosmetic formality. On the contrary, they serve an important purpose. A new paragraph is a new text segment, shown by the fact that it begins on a new line. This "line-feed" is a signal to the reader that a new *idea* is about to be developed. A properly constructed paragraph nearly always begins with a *lead sentence* introducing the topic. Once the idea is established it is *elaborated* in the course of a few additional sentences, after which the paragraph is *closed*—without undue delay, and in a way that ends the discussion cleanly, while permitting a smooth, comfortable transition to the next idea (the next paragraph).

<u>Grignard Reaction of 2-Methylbutylmagnesium Bromide</u>
<u>with 3-Pentanone</u>

1       Introduction
1.1     Historical background
1.1.1   General overview of Grignard reaction with ketones
1.1.2   Chemical equation, mechanism, likely by-products
1.1.3   Beilstein:  Properties and alternative synthetic
        approaches to 3-ethyl-5-methyl-3-heptanol (**1**)
1.1.4   Recent journal articles discussing the synthesis, use,
        and physiological properties of **1**
1.2     Source of the specific procedure to be followed
        and reasons for its choice

2       Experimental procedure
2.1     Materials
2.1.1   Chemicals and their sources
2.1.2   Equipment
2.2     Description of the work
2.2.1   How the reaction was carried out
2.2.2   Work-up and product isolation
2.3     Product analysis
2.3.1   Chemical tests ($Br_2/CCl_4$, $KMnO_4$)
2.3.2   Physical data (b.p., $n_D^{20}$, density)
2.3.3   GC analysis
2.3.4   Spectroscopic data (IR, NMR, MS)

3       Results
3.1     Product yield, basis for its estimate
3.2     Purity (interpretation of GC and physical data)
3.3     Evidence for correctness of structural
        assignment (spectroscopic interpretation,
        literature comparisons)

4       Discussion
4.1     Comment regarding yield and purity, assessment
        of procedural flaws
4.2     Interesting and unexpected observations and
        attempts at interpreting them
4.3     Recommendations to others proposing to carry out
        this preparation

5       References

*Fig. 1-2.* Example of a typewritten outline for a laboratory report describing a preparative organic experiment.—The overview in Sec. 1.1.1 will probably rely heavily on textbook literature. The notebook is the source of virtually all data in Sections 2.1 through 2.3. References in Sec. 5 are to be given in standard format (cf. Appendix J).

More than one draft will be required before a manuscript emerges that is truly satisfactory—complete, accurate, and structurally sound. Even when this point is reached the writer's work is by no means finished; at best the text is ready to be *polished*. Polishing can be expected to take at least as much time and effort as the original writing, and probably more.

Regrettably, novice writers too often fail to be sufficiently critical at the polishing stage. Impatience leads them to submit inferior work for which they are penalized and from which they learn far less than they could. It requires considerable time and effort to produce a first-rate document. First, each individual sentence should be examined and carefully adjusted until it contains all the proper nuances. Next, clusters of sentences should be read to see how well they fit together. Finally, one must again focus attention on the whole report—which may show that further restructuring is necessary.

In the course of "fine-tuning" a manuscript one should remember that brief, pregnant sentences generally have more impact than complicated constructions. Unfortunately, overly complex sentences are a common outgrowth of the "passive voice" convention, but they can and should be avoided. Not all of one's sentences should be short, however. *Variety* in length is also useful, because it keeps the reader alert and interested. A monotonous rhythm is easily corrected: one simply chooses a few (appropriate!) sentences here and there to combine, and divides some that are too long.

Word choice should be reconsidered at this stage, taking into account both precision (the hallmark of science) and the "texture" of the writing. Reading a text aloud is a good way to become sensitive to the effect its "sounds" will have on a reader. Unnecessary repetition of words or phrases is especially detrimental because of its hypnotic influence. Nevertheless, the demand for accuracy in scientific writing occasionally makes frequent use of a given word inescapable, particularly if it is a technical term. Style guides usually point out that variation in wording is a sign of good prose, but in a scientific context this is often not permissible: the same entity must always be named in precisely the same way.

When the whole text finally seems "right" in every respect, a *final copy* is carefully typed. (For specific typing suggestions, see Chapter 5.) Thought should be given to details such as margin width and line spacing (cf. Sec. 5.3.1), since proper page layout makes a paper easier to read and more pleasing to the eye. If structures or diagrams are to be inserted by hand they should first be sketched on scratch paper. Nothing so completely destroys the appearance of a carefully typed report as careless, lackadaisical drawing.

Alleged lack of "artistic ability" is no excuse; anyone can draw impressive diagrams with a little practice, provided the right tools are used (cf. Chapters 8 and 9).

A well-prepared report can be a source of great personal satisfaction, a result that in the long run is probably just as important as the high grade or the praise it can bring to the writer.

# 1.5    Types of Reports

## 1.5.1    The College-Level Report

We conclude with a closer look at some of the *forms* a report may take, beginning with the *laboratory reports* typically required in conjunction with an assigned experiment in a college or university laboratory course.

Such a report would begin with a dated heading like that described near the end of Sec. 1.2.2, in which the name of the writer may need to be augmented by an indication of class, student number, laboratory section, or bench number.

Next, a brief *introduction* should explain what the experiment was designed to accomplish. The introduction should first place the exercise within a context and then dwell rather extensively on relevant background considerations, especially the underlying theoretical principles and historical precedents. At some point the *source(s)* of the procedure(s) utilized should be explicitly noted (in the form prescribed in Sec. 11.4 and Appendix J).

The writer would then describe what *steps* were carried out, what was actually *observed*, and what the *results* were, concluding with a thoughtful *interpretation* of the results and their significance. The writing style should be that which has come to characterize scientific writing generally, a subject elaborated briefly in Appendix B and treated at length (and in a beautifully lucid and readable way) by Schoenfeld (1986).

Various details regarding the exact form of the report might be more or less prescribed by the course instructor, but inclusion of all the foregoing components would almost certainly be expected, usually in the order given.

Clearly the factual and logical content of this or any other report is the reason for its existence. A disorganized, unintelligible, ambiguous, or misleading narrative is worthless, regardless of the value of the findings that led to it, because such characteristics obstruct the transfer of information. Moreover, as noted earlier, no report can have the desired effect if the reader finds its style offensive, or if it is difficult and wearisome to read. Attention to correct form and appropriate style can be interpreted as a sign of politeness as well: it is an indication of the writer's respect for the reader, and it shows a willingness to take some pains on the reader's behalf.

## 1.5.2    Research Reports

The major difference between a *research report* and a student laboratory report is that the former normally encompasses a whole series of interrelated experiments, many of which were never before *performed*, let alone reported. Experimental details may still constitute the bulk of the writing, but the discussion establishing their context and analyzing their significance is much more important, and its formulation requires a corresponding amount of thought and planning.

The recipient of a research report is usually a *research supervisor*, someone intensely interested in the work but not yet in possession of all the data. It must be borne in mind that the supervisor cannot possibly have been thinking about the project as single-mindedly as the researcher carrying it out. As a result, the report will be important not only as a record of newly-discovered facts, but also as a fresh source of background information, and because it contains at least a preliminary explanation and interpretation of the reported findings. The aim is to produce a report that satisfies all these needs and also conveys a positive impression of the researcher's competence and scientific potential.

Research reports are not expected to be "historical accounts". Research often proceeds in a disjointed, seemingly unscientific way. Results are obtained today explaining mysteries that impeded last month's work; meanwhile, other new findings create new confusion. The haphazard journey constituting the true course of events can be both fascinating and instructive, but only rarely is it reported, much less published. Instead, standard practice is to rearrange events in such a way that a coherent exposition results. Often the implied chronology is far more ideal than the one dictated by events. Organizational freedom of this kind applies both to the discursive parts of a report, in which progress toward a set of goals is outlined, and to the experimental section. Particularly the latter should always be organized *logically*, not chronologically.

A good research report need not be exhaustive (cf. Sec. 1.4.1). Results truly worthy of attention should be highlighted, while some omissions are quite permissible. If, for example, a given procedure has been repeated numerous times during a particular reporting period, one would normally describe it only once, choosing a typical example or perhaps the case that produced the best results.* An experiment now known to have been badly designed, and which produced no useful information, might well be omitted altogether from a research report. Falsification and deliberate suppression of embarrassing or disappointing findings are, of course, taboo, but a certain amount of selectivity is necessary. At the same time, one secondary function of a report is to account for the expenditure of a researcher's time, so time-consuming failures do need to be acknowledged and occasionally rationalized. In case of doubt, questions concerning how much to include and what to omit are best discussed with one's research supervisor.

## 1.5.3    Grant Reports and Proposals

### Types of Grant-related Documents

There are two distinct types of reports having to do with research grants and other forms of project funding. Some grant documents, like the research reports discussed in the previous section, relate primarily to work already *completed*. Others describe *planned* activities. In both cases the writing should reflect the underlying motive: concern for the financial support of one's research.

Outside funding is normally solicited by submitting an *initial proposal* to an appropriate agency. Most granting agencies receive many more proposals than they can possibly fund.† Final decisions about what will be supported rest with

---

* One should not, however, report a "best" experiment that is actually an inexplicable anomaly. Correct practice would be to describe a typical procedure, but to note in the discussion that one particular trial was extraordinarily successful for as yet unknown reasons. Seeking clues to the cause of the anomaly might well be an important future assignment.

† This ratio varies, however, and it can be a factor in one's choice of a target agency.

officials within the agency, but outside consultants are often asked to read, evaluate, and make recommendations concerning submitted proposals. The decision-making process tends to be complex and frequently requires many months.

Once a *grant application* (*grant proposal*) is approved, the first funds usually become available immediately, and the work they are designed to support is also expected to commence promptly. Some months later, the funding agency or grant authority may request an *interim report* on the work—which is assumed still to be incomplete—as a way of ensuring satisfactory progress. In virtually all cases, a *final report* is required at the end of each grant period (or fiscal year), a report that could pave the way for future grant applications.

Even if one's work is internally supported (as in industry, for example), proposals and reports analogous to those described above may well be required from time to time. Indeed, it is sobering to discover what an important role reports related to funding actually play in scientific research, especially as it is practiced in the western world. At universities, in government, and in private industry alike, the quality of written reports related to experimental work largely determines whether or not that work will continue, and, if so, at what level of intensity.

As with other reports, it is crucial that a grant report reflect what the *reader* is seeking. Especially in the case of an initial proposal, where a significant element of competition is involved, one must build the strongest possible case for the proposed project. The aim is to create a convincing impression that

- the proposed project is legitimate and worthwhile;
- the scope of the project has been carefully considered and is realistic;
- the goal stands a reasonable chance of being met if the work is conducted in the manner described in the proposal;*
- an adequate literature search has been performed;
- the applicant is reliable and has the background and the facilities needed to tackle the problem;
- the requested funds are truly essential to the achievement of the desired ends.

Many granting agencies issue fairly detailed guidelines for grant applications. Some are more prescriptive than others; government sources in particular are notorious for designing complicated application forms accompanied by 20-

---

* As a rule—however regrettable it may be—improbable ideas are unlikely to be funded no matter how intriguing they are.

page "explanations" and qualifications. All available information should be studied with the utmost care before anything at all is written. Once a proposal has actually been drafted, the guidelines should be re-read meticulously to make sure nothing has inadvertently been omitted. Particular attention should be directed to ensuring that every aspect has been given what the recipient will regard as its proper weight.

### Grant-writing Strategy

It is not our intent to present a comprehensive set of rules or strategies for writing successful grant applications; the subject is a complex and controversial one, and entire books have been devoted to it (e.g., Meador, 1985, Stewart and Stewart, 1984; cf. also the Chapter "The Awkward Proposal" in Rathbone, 1985). Nevertheless, the following suggestions should be of help when one is trying to decide what to include and what points to emphasize in order to maximize the probability of a favorable response.

We begin by considering *initial* applications. The first and most obvious piece of information required is the *identity* of the applicant (name, professional title or rank, employer, address, etc.).*

A *descriptive title* of the proposed project follows, preferably restricted to a maximum of two typewritten lines (cf. also sections 2.2.2, 3.3.1, and 4.5.2). It is wise to supplement the title with a specification of the *field* and the particular *sub-discipline* under which the project might best be categorized (e.g., inorganic chemistry: fluorine chemistry). This seemingly self-evident information may not be so obvious to a reviewer, but it could strongly influence the outcome of the proposal as a result of an agency's particular funding *priorities*. The planned *duration* of the overall project and the specific *time period* for which funding is requested should also be made clear at the beginning of the proposal.

The subsequent *elaboration* generally incorporates the following elements:

● A *summary*, consisting of a concise but intelligible overview of the intended work and its precise scientific objective.
● A *literature outline*, giving the reader a general picture of the current status of the problem.

---

* This simple matter of identification assumes particular importance in a grant application since, unlike most other reports, the document is intended for someone who probably does not know its author. For this reason the proposal should certainly contain at least some information about the writer's background and qualifications, including a *list of publications*.

A proposal is not the place for a comprehensive literature survey. Instead what is required is a brief, critical synopsis of those hypotheses and results which constitute the present state of knowledge in the area under consideration. One would be expected to include a balanced treatment of the work of all major investigators, not just one's own views and results. Nevertheless, the reader should here acquire an accurate knowledge of the applicant's previous contributions to the field.

● A *discussion* of the proposed investigation and the approach that is envisioned.

The main function of this portion of the report is to establish the importance of the scientific questions the applicant wishes to address and the likelihood that they would be answered if the requested funding were approved.

Other important components of the discussion include:

● an outline of any relevant *preliminary studies* the applicant has already carried out, both published and unpublished;
● a projected *timetable* for the envisioned work, listing specific key experiments and their planned sequence;
● explicit treatment of the *long-range goal* of the project.

The latter may well go beyond the scientific objective described at the beginning. If it seems appropriate, for example, one should call attention to (realistic) potential applications of the work, including any likely benefits to society.

Considerable attention should be paid to describing any *conditions* affecting the project. This is particularly true in the case of a first application to a given granting agency. *Facilities, equipment,* and *materials* should be described, as well as the terms under which these can be used, and any *restrictions* imposed on the project (by one's institution, for example) should be explicitly noted. The applicant should include information regarding the current and proposed make-up of his or her *research group,* and any plans or opportunities for *collaboration* with others. All presently available (or highly probable) forms of *support* apart from that requested in the proposal should also be listed in detail.

No proposal is complete without a firm *budget,* broken down into at least the following major categories:

> Personnel
> Scientific apparatus
> Supplies

Travel
Publication costs
Overhead

All items in the budget should be justified. In some cases this may require providing considerable detail, such as explaining why the expensive instrument manufactured by X has been specified rather than the less costly model available from Y.

Finally, many granting agencies require that applicants or their institutions submit with their proposals various signed *affirmations*. These may include formal statements attesting to the absence of alternative sources of funding for the proposed work, for example, or assurances that one's laboratory has a standing policy of non-discriminatory hiring.

In contrast to grant applications, *interim* and *final* grant reports are rather similar to ordinary research reports. A report of this kind should begin with a brief *restatement* of the scientific problem as it stood when current funding commenced, followed by a *discussion* of knowledge subsequently acquired. The latter should by all means include appropriate acknowledgment of any progress reported by others active in the field. All relevant work carried out under the grant is then described, but briefly and without experimental details. Unsuccessful experiments, too, class as relevant, although the emphasis should be placed on important new results and their significance. Any exceptional *experience* gained during the work should also be noted, as should, for example, unusual success attending the use of a particular method, or an important new application found for some piece of instrumentation. Above all, it is important to make clear the importance of the role played by the funding source to which the report is addressed.

As in other research reports, it is quite appropriate to conclude a final grant report with a brief statement of *future plans*, especially since these are likely to constitute the starting point for the next round in the seemingly endless cycle of reports related to financial support.

# 2 Theses

## 2.1 The Nature of a Thesis

The preparation, submission, and successful oral defense of a *thesis* is an important milestone in a scientist's career. Typically a thesis marks not only the last step in obtaining an advanced degree, but also the completion of a first major piece of "independent" research. More extensive than a report, a thesis provides the author with more freedom to select and structure the contents—a freedom that, like most, can have its frightening aspects.

The *purpose* of a thesis is to signal the completion of a major piece of original intellectual activity and to prove a degree candidate's ability to organize and present the work in writing, and on a professional level. Most graduate students in the sciences actually carry out the bulk of their research under fairly strict guidance and supervision, and even the thesis *topic* usually originates with an "advisor".* Nevertheless, theses are treated as if they were products solely of their *authors*, who are held responsible for every thought—and word —incorporated.

It is important to recognize that the document we are concerned with is supposed to *contain* a "thesis": an original *proposition* rendered plausible by the argumentation and purported facts surrounding it (cf. the closely related word *hypothesis*). The "thesis" in this latter sense ought to be viewed as the single most important reason for the document's existence, and it is summarized as succinctly as possible in an *Abstract* or *Summary* (discussed in Sec. 2.2.4).

---

* Graduate students in other disciplines are often surprised—and puzzled—to learn that this is the case, since they are commonly expected to work quite independently, and on projects *they* devise and persuade a faculty member to *sponsor*! In fact, one source (Davis and Parker, 1979) estimates that, on average, Ph.D. candidates in non-science fields spend about 1/3 of their thesis/research time selecting and narrowing a topic, 1/3 carrying out the chosen investigation, and 1/3 writing. For the chemist, a more typical time distribution might be 5%, 85%, and 10%, respectively.

A term commonly applied to *doctoral* theses\* is *dissertation*—derived from the Latin *dissertare*, "to discuss". The complex origin of the Latin word (*dis*, apart, and *serere*, to join) suggests that a dissertation might be visualized as "bringing together, in proper order" a set of disparate ideas or facts—by an author who is presumed to be an expert.

Why have we taken the trouble to explore what is surely only a matter of semantics? This excursion into word origins reveals precisely those characteristics required to set the *tone* for the lengthy document whose writing we propose to discuss. In a thesis, the author seeks to advance a major new proposition, marshalling all the evidence needed to establish its plausibility. Success means that he or she will subsequently be regarded as a full-fledged member of the fraternity of practicing scientists ("experts"). Thus, the aim is no longer merely to inform one's *superiors* of what has been achieved (as in a research report), nor is one quite ready to address a set of *peers* in the scientific world with a new discovery (the purpose of a journal article, cf. Chapter 3). Both these functions are to some extent served by a thesis, but they are secondary to the "proof of expertise" function.

How, then, should a thesis be structured? It is the writer's job to make the strongest possible case for the *validity*, the *novelty*, and the *significance* of a specific proposition. Thus, the proposition (thesis), stated clearly and unambiguously, should be the focal point of the writing. It should be set in a proper *context*, however, so a reader will have ample background for understanding its meaning and importance. Each newly discovered *fact* supporting the thesis will need to be invoked at the proper point, and the *evidence* for each must be presented—in a way that would permit independent verification by someone else.

The *argument* connecting all the facts, by definition an original one, is crucial, and it should be developed as fully and persuasively as possible. A solid and unbroken chain of reasoning is needed, so a major effort must be made to eliminate weak links that might evoke embarrassing questions. In the process, any information taken from *others* must be acknowledged in the form of references meticulously prepared according to a standard citation system (cf. Sec. 11.4).

---

\* Theses are sometimes required at the baccalaureate and masters levels as well. While the present chapter is mainly concerned with the doctoral thesis, most of its content also applies to theses for other degrees—whose primary distinction is often only that of length.

Finally, the whole must be assembled so that it has an attractive and professional *appearance*. The latter point is of special concern to institutions receiving theses—so much so that it becomes the subject of numerous rules and regulations (cf. Sec. 2.2.1 and Sec. 2.3.2, "What Problems Might Arise?").

A typical doctoral thesis in the sciences consists of a *typewritten* document* whose overall *length* is between 120 and 300 (double-spaced) pages, usually augmented by a number of *figures* and *tables*. Several *copies* of the original are prepared (generally by photocopying techniques), at least some of which are *hardbound*. One or more copies become permanent parts of the degree-granting institution's library, accessible there to anyone who might choose to examine them. A *microform version* may also be put on file at University Microfilms International, a clearinghouse whose mission is to make theses readily available to the public.†

It will be clear from the foregoing that the production of a thesis is no trivial matter. After having spent‡ months (more likely *years*) worrying about the details of individual experiments, calculations, or theories, the degree candidate suddenly finds that he or she is expected to write an entire "book" on the work and then to *produce* that book. All this is supposed to be done in a highly professional manner and without any advice or editorial assistance from a publisher. It is small wonder that theses vary tremendously in their quality, and that first attempts are sometimes rejected. Nearly all submitted theses are subjected to some demands for correction or modification. The goal of this chapter is to help the thesis writer begin the job and see it through to a finished product, one that will be accepted and of which its author can be justly proud.

---

* Or a document prepared on a word processor (cf. sections 4.3.5 and 5.2.2). The practice of requiring that theses be *typeset*, formerly common in Europe, is now all but extinct.

† University Microfilms is also the publisher of *Dissertation Abstracts International* (cf. Sec. 2.3.3).

‡ Note the past tense; the research is now complete! Only when this point is reached should serious writing begin. *Writing* a thesis is a full-time job, and the task can be properly approached only *retrospectively*.

## 2.2    *The Components of a Thesis*

### 2.2.1    *Overview*

Before we begin to consider the way a thesis should be *written*, it will be useful to examine how it should be *organized*: what *sections* it must contain and, at least roughly, in what order they should appear. Not surprisingly, it is impossible to specify a single, universally applicable format, though certain recommendations have received a kind of official sanction (e.g. Standard *BS 4821: 1972*). The optimal structure for a manuscript of any kind, including a thesis, depends in part on the nature of the material it contains. A largely theoretical investigation, or one that has as its principal result a series of novel computer programs, would be presented quite differently from an *experimental* study. We shall consider mainly the latter type of research because in all the sciences it is encountered most frequently.

The *components* that make up most theses, in a typical order of presentation, are

> Title Page
> *Preface*
> Abstract
> Table of Contents
> Introduction
> Results
> Discussion
> Conclusions
> Experimental
> *Appendix*
> *Notes*
> References
> *Vita*

The foregoing list is somewhat arbitrary: it could be extended to include additional entries like "Acknowledgments", "Abbreviations", "Explanation of Symbols", "Bibliography", "Glossary", or "Index". Alternatively, it could be condensed by deleting at least some of the italicized items, or by combining, for example, the Results and Discussion sections.

It is now necessary to introduce an important word of caution. The *order* and, to a lesser extent, the *form* and *terminology* of the sections comprising a thesis can be subject to rigid specifications laid down by the school or the department to which it will be submitted. This is a problem we shall mention frequently. Virtually all institutions supply, on request, explicit sets of guidelines for thesis preparation.* These must be studied carefully and repeatedly as the manuscript develops. The guidelines vary to a surprising degree from school to school, and any that apply must be taken as absolutely definitive regardless of the recommendations of this or any other book. An author's failure to adhere to university policy is sufficient reason for rejection of a thesis in spite of the value of its contents or the apparent triviality of the violation. Campbell, Ballou and Slade (1986) present a valuable set of recommendations and examples, although the focus of their attention lies outside the natural sciences.

Each of the above-listed sections of a thesis will be examined briefly in the following pages. Some are also discussed elsewhere in this book, in other contexts and in greater detail, and an attempt has been made to include cross-references to all relevant passages.

## 2.2.2     Title and Title Page

### Selecting a Title

The *title* of any document should be a very brief *summary* of the document itself (cf. also Sections 3.3.1 and 4.5.2); therefore it is best to postpone the choice of a definitive title until the thesis has reached an advanced stage of development. The focus of this particular document could well turn out to be rather different from what the author or even the advisor might have anticipated, particularly given that so much of the previous effort has centered around solving day-to-day laboratory problems related to specific experiments, leaving little time for reflection on what the results signify.† Consultation with one's advisor is recommended before making a final commitment to a title.

---

* The instructions can be remarkably extensive; we are aware of one instance in which they constitute a 50-page brochure!

† *Working titles* are often assigned to projects at the time they are first undertaken, but these rarely bear much resemblance to the final thesis titles. Virtually every research project undergoes considerable evolution in scope and direction as work on it proceeds. Working titles are useful, however. They provide an often much-needed restraint on seductive but ill-conceived (and time-consuming) diversions from the researcher's principal research goals.

The title should accurately describe the contents, but at the same time it should be kept as short as possible: usually no more than about ten words. If this seems impossible one should attempt to divide the title into parts, providing a short *main title* followed by a *subtitle*. Thus, rather than

> A Study of the Influence of Temperature, Solvent and Metallic
> Salts on the Degradation of Phenols by Yeast

a better title would be

> The Degradation of Phenols by Yeast:
> Influence of Temperature, Solvent and Metallic Salts.

In other words, the title as a whole should be inclusive, but qualifying details can be relegated to a subtitle.

A good title will contain as many *keywords* as possible. This rule was originally developed for journal articles, titles of which have long been subject to *indexing* in various ways by current alerting services (cf. "Abstracting Journals" in Sec. 3.1.2 and Appendix D.2). In the meantime, however, similar services have emerged for theses (based on *American Doctoral Dissertations* and *Dissertation Abstracts International*, discussed in Sec. 2.3.3), hence the concern for keywords here as well. As an example, consider the title

> A New Method for the Analysis of Solutions Containing Fluoride.

This may strike its author as a good way of expressing the *significance* of the work, or of arousing curiosity, but it would certainly not be as instructive to the reader encountering it in a random list of titles, as

> Automatic Photometric Fluoride Titration:
> Selective Indication by Thorium Nitrate and Alizarin S.

The likelihood that others interested in photometric titrations would discover this particular piece of work, were it to bear the first title, is slim indeed. Similar considerations dictate a corollary: titles should be purged of unnecessarily specific or obscure terminology, as well as of all but the most common abbreviations (e.g., IR, HPLC, LCAO; cf. Appendix H).

### Designing the Title Page

The title page introduces a thesis. Its appearance is very formal, and it has a certain amount of "display" character. No page of the thesis is so rigidly specified by individual universities as the title page. Most instructions for thesis preparation contain sample title pages and lists of the required elements. Although there is much diversity in recommended styles, a few generalizations are possible. Obviously the title page bears the full title of the thesis, as well as its author's name. In addition, the author's university and department are usually given, along with the date (or at least the year) of conferral of the degree for which the thesis is submitted. Other information on the title page might include a copyright statement* or a formal declaration of the university's right to make whatever use of the thesis it chooses.

The exact order and placement of the requisite items on the title page are usually subject to detailed specification, so we can only repeat the cardinal principle: check the official information available from the receiving university!†

## 2.2.3    Preface

Near the beginning of a thesis, usually immediately following the title page, the author may insert a brief personal statement. This *Preface* is used mainly for acknowledging technical, financial, and other assistance received during the course of the work. The page is sometimes titled *Acknowledgments* rather than Preface, or it may have no heading at all.

A preface can also contain a brief comment on the context and purpose of a work, but this is less common in theses than in commercial books. Some universities, especially in Europe, require that the preface include an explicit statement that the research reported is new and original, and that the work was actually performed by the thesis writer without undue help from others. Elsewhere, this assertion is usually taken for granted. A typical preface might read as follows:‡

> The work here described was carried out between
> June, 1980, and August, 1983, in the laboratories

---

* If no statement of copyright appears on the title page, it may be required on a subsequent page preceding the Table of Contents. The significance of copyrights is discussed in Appendix C.

† Even reference to the style employed in previously submitted theses found on the shelves of the university library is a dangerous substitute—the rules may have changed in the meantime.

‡ Not uncommonly the first paragraph is omitted as redundant, its content having largely been incorporated (by virtue of university guidelines) on the title page.

```
of the Department of Chemistry, Famous University,
Anywhere, Science Land, in partial fulfillment of
the requirements for the Ph.D. degree.
     I wish to express my deep gratitude to my
advisor, Prof. John J. Johnson, for his
encouragement and support, and for his willingness
to entrust so much of the development of the
project to my judgment. In addition, I am greatly
indebted to Sam Smith for his help in constructing
several pieces of special apparatus essential for
the investigation.

August, 1983                                    BBB
```

Appending the date and one's name—or initials—personalizes the page, but it adds nothing that has not already appeared on the title page.

### 2.2.4    Abstract

The *Abstract*, sometimes referred to as the *Summary*, is meant to provide a brief, intelligible encapsulation of the purpose, methodology and results of the work presented in the thesis. An abstract should be so written that it conveys a clear and complete picture of the associated manuscript.

The text of an abstract must be crafted with special care. Primary emphasis should be on the *results* that are reported, but the *goals* of the work and the extent to which they have been achieved cannot be neglected. It is particularly important that key *conclusions* be spelled out clearly and succinctly. The intent is to give the reader an accurate sense of the validity, extent, significance, and implications of one's findings. The writer should keep in mind that most readers use abstracts to decide in what ways, if any, a given document relates to their own interests, and whether or not it is worth the trouble to read further.

The abstract should be short (preferably no more than one page),* and should contain only text (i.e., no figures or diagrams). References to the document itself are strictly avoided, an injunction that includes references to

---

\* Many universities specify a limit of 350 words—or 2450 characters—since this is the requirement for abstracts to be published in *Dissertation Abstracts International* (cf. Sec. 2.3.3).

tables or figures. Often, an abstract page is *typed* more densely (one-and-a-half line spacing) than the double-spaced format usually mandated for the thesis itself. Again, however, university guidelines should be consulted.

An abstract generally *precedes* the text from which it is drawn, although occasionally placement at the end is specified. Regardless of its location, however, no abstract should ever be *written* until the entire manuscript is complete. This sequence ensures that the writer will actually be "abstracting" something, not anticipating (or dictating) the content of a document yet to be created.

## 2.2.5  Table of Contents

### Purpose

The *Table of Contents*, often referred to simply as the *Contents*, is a list in outline form of the *headings* (or *titles*) of all chapters, sections, and subsections, with the page number on which each begins. This is usually the last of the *preliminary pages* preceding the main text.

A table of contents at the beginning of a lengthy manuscript provides the reader with a sense of the logical structure of the work, and it simplifies the task of finding specific information. Therefore, the headings it contains must be selected with care, reflecting the same considerations stipulated for the title of the work as a whole (Sec. 2.2.2).

In effect, the table of contents is a modified form of the outline used in writing the manuscript (cf. Sec. 2.3.1). *Chapter titles* are likely to do little more than identify the large organizational blocks listed in Sec. 2.2.1, while the titles of the various chapter subdivisions (*section titles*) will reveal something of the thesis content.

### Structure and Form

It is generally best to restrict oneself to no more than two or three *levels* of subdivision within a chapter (see below: "Creating Subdivisions"). The level of a particular subsection should be clearly apparent from its heading. Most would now agree* that the preferred section designation system is the one used in this book. It is a system that has long been common in Europe, and is rapidly gaining universal acceptance because of its convenience.

---

* It must again be stressed, however, that university guidelines should be consulted before any system is adopted.

The system is based entirely on Arabic numerals. *Numbers* are assigned to divisions and subdivisions in such a way that their degree of complexity signals the relative level of the corresponding section with respect to the overall text. Each subdivision number is structured with periods (see below), and this gives rise to the name of the system: *decimal classification*. Unfortunately, the term is not completely satisfactory; it might imply that a given category would contain a maximum of *ten* subdivisions, while no such restriction exists.

Application of the system is simple. First, all *chapters* are numbered sequentially (1, 2, 3, ...). Then the major *sections* are numbered, chapter by chapter, each time beginning with "1", and these numbers, preceded by a period, are *added* to the chapter number (1.1, 1.2, 1.3, ..., 2.1, 2.2, ...). The same procedure is followed for further division into *subsections* (1.1.1, 1.1.2, ..., 1.2.1, 1.2.2, 1.2.3, ...). The numbers thus formed are arranged in *hierarchical order*. No period is used after the last number in a set, but the complete number is separated from the corresponding heading *text* by blank spaces (usually two).*

A portion of a Table of Contents constructed according to the system just described might look as follows:

```
3       Reactions in Nonpolar Media . . . . . 100
3.1     Condensed Phase Reactions . . . . . . 100
3.1.1   Photolysis Experiments Using
        Cyclohexane as Solvent  . . . . . . . 100
3.1.2   Photolysis in a Frozen Matrix . . . . 107
3.2     Vapor Phase Reactions . . . . . . . . 115
4       The Effects of Polar Media  . . . . . 123
4.1     Water . . . . . . . . . . . . . . . . 123
                  (etc.)
```

Note that the first words of all titles in the example are aligned so they can be scanned easily. The position of alignment is given by the longest section number. This would not be true of titles *within* the manuscript, however. Even though most chapters would begin with a *cluster* of headings, these would all be set against the left margin, and numbers would be separated from text by only two blank spaces; e.g.:

---

* There is one unfortunate aspect of this system. Particularly for manuscripts in chemistry, it is not unusual to find that a subsection heading begins with a number (e.g., "1,4-Additions"). In such a case the text fails to stand out as clearly as one might wish from the subsection identifying number preceding it.

```
3  Reactions in Nonpolar Media
3.1  Condensed Phase Reactions
3.1.1  Photolysis Experiments Using
       Cyclohexane as Solvent

(Beginning of the text of Chapter 3)
```

Occasionally, systems other than the one described above are employed for designating subdivisions. One traditional outline form in English, often used in typewritten manuscripts, contains a mixture of Roman and Arabic numerals combined with upper and lower case letters; e.g.,

```
III. Reactions in Nonpolar Media  . . . . . . . 100
      A. Condensed Phase Reactions . . . . . . . 100
         1. Photolysis Experiments Using
            Cyclohexane as Solvent . . . . . . . 100
         2. Photolysis in a Frozen Matrix. . . . 107
      B. Vapor Phase Reactions . . . . . . . . . 115
IV.   The Effects of Polar Media . . . . . . . . 123
      A. Water . . . . . . . . . . . . . . . . . 123
                 (etc.)
```

In this system, sometimes called the *number-letter system* (or sequence), chapters are labeled with upper case Roman numerals, while major sections of the chapters are identified and ordered using capital letters. The latter are further divided into subsections designated by Arabic numerals. If additional subdivision is required, lower case letters are first employed, then lower case Roman numerals. These are followed, if necessary, by identifiers set in parentheses: first lower case letters, then Arabic numerals. Notice the illustrated alignment conventions for an outline prepared using this somewhat more cumbersome and less straightforward, less self-explanatory system.

## Creating Subdivisions

Whatever rules for subdivision designation are adopted, they should be followed consistently throughout the manuscript. Doing so implies, for example, that it is inappropriate to introduce a chapter heading, follow it by several paragraphs of text, and only then begin with the first of a set of subdivisions. Constructions of this kind can most easily be avoided by devising a separate heading for any comments needed to introduce a chapter (e.g., "Basic Concepts", "Preliminary Remarks", "Introduction" or simply "General"). Each *level* of subdivision in a given chapter may contain as many *members* as necessary, but there should always be a minimum of two; it is pointless to introduce a subsection 2.1.1 (or II.A.1) unless there is at least a following subsection 2.1.2 (or II.A.2).

It is good practice to subdivide one's manuscript in such a way that no subsection contains more than about eight typed pages.* Insisting that large sequences of text be divided into smaller sections is a service to the reader, who expects headings to provide some guidance. Moreover, the requirement forces the author to organize the text carefully. Organizational shortcomings often first reveal themselves when one attempts to create subsections from lengthy unstructured text: sensible titles simply cannot be found! It will prove possible to introduce meaningful subdivision headings only if the text contains the right things in the right places, a quality the reader will appreciate. Text broken into subsections also has the advantage of having a more varied and structured appearance, its headings providing natural points of focus. As a result, the manuscript becomes easier to read and understand.

## 2.2.6    Introduction

The *Introduction* marks the beginning of the actual text of a thesis. In this first narrative section (sometimes more descriptively called the *Background*), the writer introduces the reader to the thesis topic and, in the process, describes the context of the problem as he or she saw it when the study commenced.

A well-written introduction is complete and lucid, and it bears some resemblance to a review article (cf. Sec. 3.1.2, "Review Journals"). The Introduction can be drafted only after one has meticulously read (perhaps re-read) as much  background material as possible. The author is almost sure to spend large amounts of time feverishly tracing at least one distressingly obscure original source of what earlier could be dismissed as "common knowledge" but now must be documented. Such searches have their frustrating moments, but the experience gained is valuable. One usually unearths some intriguing and enlightening reading—and, not uncommonly, a few cases of surprisingly "shaky" evidence that could serve as the starting point for future research.

---

* This rule applies to *text* pages; the presence of figures or tables may justify longer subsections.

A long introduction should be subdivided into several parts. For example, one might treat separately the historical context of the problem, the specific cases chosen for investigation, the reason for that choice, and the methodology employed. Whatever organizational scheme is adopted, it should be borne in mind that the intent is to assemble a kind of comprehensive panorama of the initial landscape. With the help of information and results drawn from other sources the reader should come to see the project from the same perspective that originally confronted the author, and thus understand and sympathize with the need for the study constituting the thesis.

Naturally, any relevant results reported by *others* during the course of the work should also be revealed. This injunction implies that the author has been diligent in keeping up with the current literature (cf. Appendix D). It should go without saying that one's own discoveries and thoughts must always be carefully and recognizably separated here, as in the entire manuscript, from anything already known or publicly surmised by others.

The last sentence of the introductory section should lead directly into the text that follows. For example, one might conclude with

> ...and it is for these reasons that a further investigation of the XXX reaction of compound YYY appeared to be warranted.

or*

> Thus, I considered it important to establish beyond question the amount of Pb(II) present in specimens from different sources.

## 2.2.7    Results

The next section, the *Results*, contains all the major experimental or other findings the author has accumulated during the course of the reported project, arranged in a systematic way (see Sec. 2.2.8). The only technical details that should appear here are those essential to a clear understanding of the information and its application. Descriptions of routine apparatus or analytical methods should be reserved for the Experimental section (cf. Sec. 2.2.10).

Whenever practical, sets of *related* results should be organized in tables, or interpreted through figures or diagrams (see Chapters 9 and 10 for guidance). If extensive data have been collected (e.g., measurements or analytical values),

---

* See the second footnote in Sec. 1.4.1 with respect to the use of the personal pronoun "I".

it is often best simply to *summarize* the results in the body of the text, perhaps augmenting the summary with representative examples. Details can then be provided in an *Appendix* (cf. Sec. 2.2.11).

## 2.2.8    Discussion

In the chapter entitled *Discussion* the results presented in the previous chapter are analyzed and compared with information appearing elsewhere in the literature. Conclusions are then drawn with special emphasis on those relevant to the "thesis". Occasionally it is practical to combine the results with the discussion in a single section bearing a suitably inclusive title. If this is done, however, it is particularly important (though sometimes difficult) to differentiate clearly between one's own findings and those reported in the literature. On the other hand, if the scope of the research project is unusually broad or diverse, the "Results" and/or "Discussion" can be further subdivided into a series of topics, just as was proposed for the Introduction (cf. Sec. 2.2.6). In this case, the actual words "Results" and "Discussion" might not even appear as headings.

One may wish to reserve *conclusions* derived from the project as a whole for a separate section (cf. Sec. 2.2.9). Even if this is the case, the discussion section cannot ignore conclusions altogether, and organizing such matters in an optimal way can be quite difficult.

The Discussion and the Results sections form the heart of the manuscript. The success of the thesis as a whole will be determined in large measure by the way this material is organized and written. The key to success lies in the discovery of an ideal *sequence* for the presentation of the experiments and their implications (cf. Sec. 2.3.1), a sequence that is *logical* from the standpoint of the intent and outcome of the study. In the interest of improving the logical argumentation and avoiding confusion one should never hesitate to deviate considerably from the order in which experiments were really conducted (cf. Sec. 1.5.2).

After presenting one's findings and elaborating on their significance, a scientist is usually anxious to conclude by engaging in a certain amount of creative extrapolation, including suggestions for future studies. This is appropriate if kept within bounds, but it must be remembered that a thesis is primarily an accurate report of a discrete set of scientific discoveries. This is not the place to carry out a long exercise in fanciful speculation. Furthermore, if any excursions into the future are launched they must be developed with care, and without omitting any steps. Readers must be led gently, because for them, unlike the author, the ground to be traversed will be new and unfamiliar.

## 2.2.9    Conclusions

Particularly when a thesis is long it can be useful to collect and summarize the most important results and their implications in a *Conclusions* section. Here the status of the problem at the time the work was initiated should be briefly reviewed before the new findings are enumerated, and some indication should be given of the extent to which the researcher's original goals and expectations were ultimately met. If a "Conclusions" section is to appear, the "Abstract" (Sec. 2.2.4) might take on a somewhat more descriptive form, stressing what has actually been done and achieved, because it can be assumed that even casual readers (or "browsers") will at least glance at a "Conclusions" section after reading an abstract. The conclusions would then serve to put the findings in their proper context. An "Abstract" is always limited in its ability to do this since it has so much ground to cover: it must also address objectives, methods, and results (cf. Sec. 3.3.3). The Conclusions section might therefore be viewed as an extension of the Abstract, summarizing contextual matter in a somewhat extended and easily readable form.

## 2.2.10    Experimental Section

Complete details of all one's experiments, including the methods, apparatus, and materials used, are presented in a separate *Experimental* section. The appropriate style is illustrated in Sec. 1.4.1. Every description provided should be so complete that other scientists will be able to repeat the work and verify the reported results solely on the basis of what is written. The accurate reporting of one's experimental results in a way that permits their verification can be considered even more important than deriving and explaining the conclusions (which others can do as well). Even though the Experimental section of a thesis may seldom be read by others, and will inevitably seem dull, it deserves a major share of the writer's attention.

Certain kinds of information are best summarized at the beginning, perhaps in tabular form. This includes generalizations regarding chemicals employed (e.g., their purity and the names of suppliers), important apparatus (type, brand, and model), techniques used for routine measurements (including standardization methods, solvent, temperature, etc.), and, in the case of biochemical or pharmacological studies, precise identification (including species and strain) of any experimental animals, plants, or microorganisms.* If possible, suppliers' addresses should be given, particularly where uncommon sources are involved. Any trade names used for chemicals should be supplemented by precise structural names (or, at least, *generic* names, cf. the first footnote in Sec. 6.3.2).

Experimental descriptions in a thesis are expected to contain a considerable amount of detail. In effect, the author (as well as the research advisor) should be able to reconstruct the entire course of the reported work without resorting to any laboratory notebooks. It is appropriate, for example, to include with an analytical result such information as the volume of titrant consumed and what conversion factors, if any, were employed, information that would surely be omitted from a journal article.

In theory, experimental descriptions at a comparable level of detail would also be desirable in journal articles. Unfortunately, however, due to space limitations imposed by editors, published descriptions tend to be significantly more brief. As a result, special "tricks" used in achieving optimal results are often omitted, and thus are lost to the wider scientific community (cf., however, Sec. 1.4.1). Similarly, published reports normally disregard minor modifications that may have been applied only to one experiment in a series, the assumption being that the differences were irrelevant. Such omissions are never permissible in a thesis. Here, space is not really a problem; the thesis writer is also the editor and publisher, so descriptions can be as exhaustive as necessary. It should never be forgotten that someday a thesis might serve as the final resort of a research scientist eager to find more details after having read a published article.

---

* It is also wise to mention other potentially relevant factors, such as diet or extent of exposure or conditioning to light, even if they are presumed to have had no effect on the actual experiments.

## 2.2.11    Appendixes

Material that a reader might find useful, but which is not crucial for understanding the thrust of one's arguments, or which might constitute a distraction, can be incorporated at the end of the thesis in one or several *Appendixes*.* Examples of material in this category are mathematical derivations, computer program statement listings or flow charts, spectra, circuit diagrams, and exhaustive tables of data derived from sets of experiments. An appendix can also be used for the elaboration of interesting side issues only alluded to in the text. Care is necessary in this regard, however. Information important for the development of the "thesis" should never be relegated to an appendix where it is likely to be overlooked.

## 2.2.12    Notes and Footnotes

Brief asides, or concise observations germane to specific points in the text, rarely warrant treatment as appendixes. Instead, they can be introduced as *notes*.[†] Such notes are either collected together in a separate section carrying that heading, or, more commonly, they are typed individually as *footnotes* at the bottom of the page, usually separated from the text by a horizontal line. As usual, university guidelines should be carefully followed; these may even specify the nature and length of the allowed dividing line! The "notes" device is a handy one, but it should be used sparingly since a profusion of notes can adversely affect the readability and overall appearance of a manuscript. Technical suggestions regarding the preparation and use of notes appear in Sec. 5.3.3.

---

* The traditional plural form of "appendix" is "appendices", but the spelling used above is now more common.—Appendixes, glossaries, endnotes, bibliographies and indexes are sometimes referred to collectively as *back matter* or *end matter*.
† Occasionally in theses the notes are expanded to incorporate literature citations (references) as well (cf. Sec. 3.4.5 and the fourth footnote in Sec. 11.3.1), although this practice is generally discouraged in the sciences.

## 2.2.13    *References*

The *References* section, sometimes—but inappropriately—called the *Bibliography*,* lists the precise sources within the public domain from which specific results and information have been drawn. Each use of such material is also appropriately signalled in the body of the text (for additional information regarding citation conventions cf. Sec. 11.3). Citing all previous work is essential to give credit where it is due and to lead the reader to additional sources of information. It also provides a clear distinction between that which is truly new and "one's own", and that which was previously known. A scientist is always expected to acknowledge the work and the ideas of others, and failure to do so constitutes plagiarism as well as a lack of professional courtesy. Unpublished information (cf. Sec. 11.4.5) or personal communications cannot be regarded as part of the open literature. Consequently, such sources should not be included in a reference list, but instead should be acknowledged directly in the text or through notes.

References are sometimes placed in *footnotes* rather than in a separate section of a document. Publishers discourage the practice, and *endnotes* are in any case easier to arrange. Nevertheless, there is one powerful argument in favor of using footnote references in theses. Readers frequently must content themselves with access only to a *microfilm* version of a thesis (cf. Sec. 2.3.3), and searching for endnotes with a *microfilm reader* is extremely awkward.

## 2.2.14    *Vita*

A *Vita* (or *Curriculum Vitae*) is a brief summary of the salient features of one's educational and professional background. A vita suitable for inclusion in a thesis should closely resemble what one might send as an introduction to a prospective employer.

The preferred format is usually an outline rather than a narrative. A typical vita begins with the date and place of one's birth, perhaps even identifying one's parents, and then proceeds to list the schools one has attended, including any degrees and academic honors awarded and the titles of any accepted theses. This is followed by an employment history, in chronological order and accompanied by job titles. Professional or educational affiliations and any

---

* The latter term should be restricted to lists of *general* references as opposed to specific citations.

noteworthy distinctions (fellowships, for example) should also be noted. The professional vita should conclude with an up-to-date list of one's publications.

All the above information should be given as concisely as possible and in the form of bare facts; one should suppress any urge either to elaborate or qualify the "data". Normally, students find that one page is more than sufficient, since their academic and professional careers have only begun.

The reason for including a vita is to give the reader more information about the author, presumably facilitating his or her identification should that ever be necessary. Not every American thesis is expected to contain a vita, though some universities insist on it; its inclusion is more common in Europe. While the placement is specified here at the very end of the thesis, a vita is sometimes regarded as part of the "preliminary matter" preceding the Table of Contents. Again, university guidelines should be consulted.


# 2.3    *Preparing the Thesis*


## 2.3.1    *From Outline to Final Draft*

### *The Need for an Outline*

In the preceding chapter (cf. Sec. 1.4.2) it was suggested that preparation of an outline is a helpful first step in organizing even a brief report. Here we can be more categorical: it would be foolhardy to try to write a thesis without first outlining it. The job is too big, and the need for rapid and successful conclusion is too great to permit one the luxury of simply writing "off the top of the head".

The starting point for a thesis outline is usually straightforward enough. It can be the list of chapters (given in Sec. 2.2.1) expanded to include the most obvious topical divisions. But this is only the beginning; the real problem is *organizing* one's results *within* the skeleton. Once that has been accomplished, and the general structure of the entire thesis is clear, the actual writing is not likely to be inordinately difficult.

Perhaps the best way to begin is to sit down with a stack of blank file cards, and to summarize briefly each independent conclusion derived from the research (minor ones as well as major ones). A separate card should be used for each conclusion. The resulting pack of "idea cards" can then be shuffled and reshuffled to produce various organizational patterns until one is discovered that incorporates all the results in the most logical and harmonious way.

A related way of organizing one's thoughts has been described by G.L. Rico in her book *Writing the Natural Way* (1983). Here it is suggested that the author take a blank piece of paper and write, with little thought or planning, a "cluster" of relevant words or phrases, all radiating from a central term. Patterns should be allowed to evolve freely, relying on the more associative, unconscious parts of the brain. Once a start has been made, the resulting cluster can be examined more critically and analytically. Terms or phrases are circled and then gradually connected to others, in the course of which the cluster gradually acquires more structure. Finally, the lines are transformed into arrows indicating logical relationships as perceived from the central entity—which may be the theme of an entire thesis or that of a single chapter. The results are then converted into a "Table of Contents".

Both strategies, the one based on "idea cards" and the "cluster" method, allow the writer to approach the task of organization in a relaxed way, and they permit ample scope for imagination and fancy—even daydreaming and random chance.* Rational thought can be saved for the later stages of writing. The importance of chance and creative fantasy in science is well documented, and there is no reason why they should not be a part of scientific writing as well. Indeed, if this notion were to take root many deep-seated fears about writing might be alleviated.

Once an optimum order for one's ideas emerges, outlining the proposed Results section can begin, followed (surprisingly quickly, in most cases) by the outline for the closely related Discussion. It will then be relatively easy to decide what background material should be sought for the Introduction, so it can also be outlined. The same method can be applied to each chapter in turn.

The most difficult kind of thesis to write is one comprised of several loosely related (or worse, *un*related) projects. Only when a satisfactory connecting thread can be found or devised to link the various parts will such a thesis gel into a satisfying whole.

---

* The same principle has been incorporated into various computer programs, e.g., ThinkTank™ and MORE™, distributed by Living Videotext, Inc., 2432 Charleston Road, Mountain View CA 94043, USA.

From the perfected outline, a first draft of the thesis is prepared, exactly as it would be for a report. In the process, the text will need to be divided into appropriate subsections as suggested in Sec. 2.2.5. Then follows the time-consuming and tedious job of "polishing" (Sec. 1.4.2). As noted earlier, the last parts of the thesis to be created should always be the Table of Contents, the Abstract, and the Title Page, probably in that order.

### A Curse Turned Blessing

Thesis-writing time proves an interesting eye-opener for the student who has habitually criticized a research advisor for insisting on *interim reports*. Suddenly that stack of old reports, most written months or years earlier, turns out to be a veritable gold mine of information. It yields clear descriptions of dimly remembered experiments and, more important, it provides contemporaneous interpretations of them. Results half-forgotten and assumed to be worthless suddenly take on new meaning, partly because re-reading a report written when a particular piece of work was fresh rekindles old enthusiasms, but also because the text is now being examined by a "new person", someone who has, in the meantime, gained considerable insight, perspective—even maturity.

One of the most valuable aspects of old reports is their point of view. Their writer was attempting to put the best possible face on what probably seemed, at least occasionally, a disappointing dearth of results. Every available result was cast in the most favorable light possible, and every conceivable implication was explored. That intensive exercise in rationalization would be difficult to perform on the same data now, long after the event. But the fact that it was done once—in writing!—will almost certainly mean that material is available for several interesting thesis subsections that would otherwise probably have been overlooked.

Old reports are also invaluable in another way: their Experimental sections can usually be salvaged almost without change. Stitching these together in a reasonable sequence will result in an approximation of the entire experimental part of the thesis. Once again, massive amounts of time will be saved, especially since even the best-kept notebook, opened months or years after the ink dries, reveals disconcerting shortcomings. The experience of first finding and then interpreting the notebook descriptions of six repetitions of a long-forgotten experiment in search of a "best" run (which also contains a complete data summary!) can be maddeningly frustrating.

## The Writing Itself

The writing of a first draft, once begun, should be pursued relentlessly and with as few interruptions as possible. Certainly no major projects should be allowed to intervene. Undivided concentration plays an important part in one's enthusiasm for the writing process, and it can be decisive as to whether or not the resulting text is continuous in a logical and stylistic sense.*

Some preliminary thought should be devoted to *how* and *where* the writing will occur. A modest investment in convenience and comfort is certainly warranted since the task is such a large one. It would be difficult to overemphasize the advantages of electing to use a word processor for one's writing (cf. sections 4.3.5 and 5.2.2), preferably one to which easy access is assured. Particularly with a document as long as a thesis it is an enormous help to be able to correct and revise one's writing repeatedly and easily, each time obtaining, within minutes, a pristine copy containing all the changes.

Like any major piece of writing, a thesis can be expected to emerge in stages. At any given time, some parts—one can hardly predict which—will be "ahead" of others in terms of their perfection. As improvements are made in one section, useful changes or additions to other sections will suggest themselves. Eventually, however, the whole will become fairly stable, and only then is it appropriate to show it to someone else—but probably not one's advisor; that would still be premature. Another student might be a better first choice, preferably a colleague with a reputation for being perceptive and critical. If the right person is chosen, the result is almost certain to be a blow to the writer's ego, because the perceptive critic *will* uncover faults. That is as it should be! After all, what is the point of asking for a critique unless one is willing to accept it, consider its merits, and learn from it? Every writer has blind spots. Rarely are raw manuscripts devoid of obscurities their authors have overlooked because *they* knew all too well what they were trying to say.

Finally the day will arrive when a legible draft is ready to be presented to one's research advisor for comments. Advisors vary tremendously in their handling of this assignment. Some read their students' draft theses fairly superficially, checking only to see that there are no major blunders. Others are more conscientious, freely (but more or less tactfully) offering general

---

* Trying to write or even complete a thesis is particularly difficult when one is in a new location and under the stress of becoming established in a new position, and it may well lead to disaster. Many have been tempted to accept a premature "once-in-a-lifetime" job offer while writing a thesis, but few would do it again given the chance.

suggestions of additions or deletions they think might strengthen the case to be presented. Still others take pride in their reputations as "nit-pickers", expecting their students' manuscripts to meet their own standards of writing in every way. The student writer has little choice but to deal with the particular traits of the research advisor he or she has chosen. Whatever happens, it can be a learning experience (often a very good one!) and, in any case, the thesis will have to be reasonably satisfactory in the advisor's eyes before there is any point in submitting it to the university. (Exceptions to this generalization are *very* rare.)

Once this final hurdle in the writing process is accomplished, the next problem is that of presenting the now acceptable text in the precise *format* required by the university.

## 2.3.2    The Finished Product

### Who Receives the Assignment?

The first thing to decide is whether or not a *typist* will be employed to produce the final version—or, for that matter, even an early draft. There are valid arguments on both sides of the question. Certainly a professional typist is likely to be able to work faster than the author, and hiring one can save the author from having to worry about typing altogether. This is no small consideration: technicalities such as proper typing of footnotes can cause the novice endless grief. Furthermore, a typist who has already prepared theses, especially for one's own department, will know what pitfalls to avoid with respect to form rules. Finally, a professional can be expected to do a "professional" job, and that is essential.

But the other side of the coin also needs to be examined. Obviously the use of a professional typist requires an investment, and a considerable one when the object is a multi-hundred-page technical thesis. Then, too, the author's own hand-written draft will have to be prepared with special care if someone else is to type it. No typist can be expected to be a mind reader or an expert at deciphering half-legible hieroglyphics.* This consideration is especially

---

* If a typist is employed only after a preliminary typed version of the thesis has been submitted to the thesis advisor, few "hieroglyphics" are likely to be present. However, the degree candidate may wish to delegate the typing job at an earlier stage, when the manuscript itself will present more problems. This raises another question: can one really afford to pay for having the thesis typed at least *twice*?

serious in the case of a scientific thesis, which will contain terminology totally foreign to the average typist—and thus will be a veritable magnet for typographical errors. *All* of the latter will have to be eliminated in the end, and that may well mean the re-typing of a great many pages, particularly if the university requires the submission of a "clean" original.* Then there is the problem of *structural formulas* and *equations*: who will insert them? Finally, entrusting the typing to someone else means the work is no longer directly under the author's control, a situation that can have scheduling implications.

In our opinion, the best solution today is to prepare one's own thesis using a *word processor*. Most universities now permit their use for theses, though with varying degrees of restrictions concerning the minimum acceptable quality of the print (usually requiring, in essence, that one have access to a printer of at least "near-letter quality"). One major advantage of using a word processor is that turning a final draft into a document normally entails little more than loading the printer with a different type of paper (that specified by the university), slowing down the output to improve the appearance of the print (or using a better printer), and spending a few hours creating the correct formats and introducing by hand a few structures and equations (cf. Chapters 8 and 9). Clearly this approach is the way of the future.[†]

### What Problems Might Arise?

Fortunately, most of the production problems that could lead to the rejection of a thesis on technical grounds can be avoided by a combination of common sense and *repeated, careful* reading of the university's thesis guidelines. These may seem formidable at first glance, but they were, after all, written for beginners, and hundreds of other students in the same position have already managed to comply with them.

Most of the rules are strictly technical. *Margins*, for example, will be rigidly specified, and that specification is taken so seriously by some universities that they hire people who, among other things, apply rulers to every suspect page of every submitted thesis, fully prepared to tell the author to try again. Similar

---

* If only photocopies are required it is often possible, for minor errors, to avoid extensive retyping by correcting the original in such a way that the correction is invisible in the copies (cf. Sec. 5.4).

[†] In fact, the "future" probably holds even more in store. The availability of so-called "laser printers" as output media for word processors may mean the disappearance altogether of the familiar "typescript", since these printers produce copy barely distinguishable from that created by typesetting.

considerations apply with respect to proper placement of chapter and section headings, tables and figures with their titles or captions, and footnotes. Moreover, specifications must be meticulously followed regarding type style and size, line length, and line spacings (e.g., for chapter and section headings).

The kind of *paper* to be used will certainly be specified. Usually, to be acceptable, it must have a high rag content and be acid-neutral, characteristics that assure it a long life. This is an important consideration. There is no way of knowing when someone might need the reported information, and a decaying thesis is difficult to replace. The popular "erasable" papers, sometimes (but rarely) permitted for reports, are invariably forbidden for theses since they produce virtually the opposite of "permanent" text. Strict rules also apply to the nature of any *corrections* in the final copy, especially ones made with *correction paper* or *correction fluid.* Such corrections are impermanent and unsightly, and are only permissible if photocopies of the original are to be submitted (cf. Sec. 5.4).

A high-quality hardcover custom *binding* is required for at least one copy of the thesis, that for permanent archiving. Many (but not all) universities attend to this matter themselves, usually through their libraries. A specified number of *extra copies* (usually two or three) is also required, not only by the university but often by one's department or advisor as well.

Finally, certain *deadlines* are normally imposed, and any that are not met can cause one's degree to be withheld for six months or even a year.

### One Last Hurdle

The last step in the process leading up to formal submission of a completed thesis is almost always an *oral defense* of the work before a panel of faculty members, occasionally augmented by one or more outsiders.* Some or all of the panel members will have been chosen by the candidate. In most cases the "defense" itself is almost a formality; by the time one reaches this stage, the eventual conferral of a degree is almost a foregone conclusion. Nevertheless, it is not unusual for the defense to reveal the need for at least a few last-minute changes in the thesis, and the occasion certainly deserves to be taken seriously, even if it does not warrant developing an ulcer.

---

* In some countries and universities the *defense* is accompanied by a true *examination.* This may be public, and its content may range well beyond the topics covered in the thesis itself.

The best way to prepare for one's oral defense is first to review the entire thesis and then to draw up a mental summary of its contents, one that could be delivered in an apparently impromptu fashion, calmly and fluently, in about 15 minutes (although such an opportunity may not arise). There is a high probability that at least one of the examiners present will raise questions like "Why didn't you instead ..." or "Suppose the result had turned out ...; what would you have done then?" Most of these can be anticipated with a bit of imagination, and doing so can save much stammering and embarrassing delay. Another obvious line of questioning might involve logical extensions of the reported work. Usually, at least one of the examiners will take pleasure in finding an intellectual corner to back the candidate into, and the others will watch with a mixture of trepidation and amusement to see what happens.

In a way, the thesis defense is a kind of ritual, a "rite of passage". Rituals have their place, and this one should be prepared for and accorded its proper degree of respect. Some surprises are bound to occur, but it is these that will make the experience memorable. Calm and self-confidence, balanced by a measure of tolerance and even humility, are prerequisites to a satisfying oral. The prospect should not appear unduly threatening, because while the panel members may know more facts within the discipline than the candidate, no one knows more about the actual work to be discussed than the person who carried it out.

## 2.3.3    Thesis Documentation Services

We conclude our discussion of theses by looking briefly at the way and the extent to which they can be classed as part of "the literature" (cf. sections 3.1.1 and 11.4.5). In the first place, theses could certainly be called at least "semi-public" documents. That is, they become a part of the open collection of the library of the university to which they are submitted.* The interested scientist who has established the existence of a given thesis could, in principle, travel to the university in question and read the thesis from cover to cover. Most university libraries make their thesis collections more open than the foregoing comment would suggest. Cooperative arrangements exist that permit the loaning of theses (either as bound volumes or microform copies) to other libraries requesting them on behalf of their own clients.†

---

\* Some universities reserve the right to restrict their *use*, however.

† One important restriction applies to this generalization: such cooperation at an *international* level is far more limited, often verging on the nonexistent!

Still, were there not other tools available it would be inappropriate to claim that theses really are part of the open literature. A major handicap to the potential reader would still be a lack of knowledge about *what* theses exist *where*. Two publication services in the United States attempt to deal with this problem by regularly compiling lists of recent doctoral dissertations. The first is *American Doctoral Dissertations*, a comprehensive registry of all theses (in all fields) successfully submitted to U.S. universities for doctoral degrees. It has appeared annually since 1934 and contains a list of titles of all new theses, arranged by discipline and cross-indexed by title keywords. No other information is supplied by this source, however, except the names of the author and school.

The second reference work, *Dissertation Abstracts International (DAI)*,* is considerably more ambitious in scope. It is both more and less comprehensive than *American Doctoral Dissertations*: it is *more* comprehensive in that, since 1966, it has included Canadian theses and, since 1975, European theses as well; unfortunately, it is *less* comprehensive, too, since not every thesis presented to every university is included. Some universities automatically submit thesis information to *DAI*, or at least make submission by degree candidates mandatory, but others only mention the possibility to their candidates, treating participation as strictly voluntary. Even worse, some ignore the possibility altogether.

There are major advantages to having one's thesis included in *DAI*, advantages which more than offset the associated nominal costs and additional "busywork". In the first place, the monthly issues of *DAI*, available in most research libraries, contain something far more useful than mere thesis *titles*: as the very name of the publication implies, every entry is accompanied by an *abstract* (limited to 350 words) of the contents of the thesis. In most cases this abstract is the same one that appears in the thesis itself. Even more important, the publisher of *DAI*, University Microfilms International (Ann Arbor, Michigan), requires submission of a *complete copy* of every listed thesis, permitting the firm to offer at a fixed price microfilm or printed reproductions of entire theses to anyone requesting them. Microfilm copies of the theses are kept in permanent storage (the originals are returned), so reproduction is possible at any time. Should the number of requests for a given thesis prove to

---

* *DAI* is the continuation of the publication formerly known simply as *Dissertation Abstracts*.

be large (more than six in a single year, admittedly a rare phenomenon), the author is even provided with royalty payments!

One obvious benefit of the *DAI* services is that thesis authors themselves can always obtain extra copies of their work, including hardbound ones, any time the need arises.* Authors receive a special discount when they order copies at the time of submission of the thesis. However, the real benefactor of *DAI*, is the scholarly public, because the existence of its services opens theses to literature search techniques comparable to those available for the regular "published" literature (cf. Appendix D.3). Thus, thesis abstract texts are extensively cross indexed and they can be computer searched for specific words or word combinations. With the advent of these possibilities there appears to be one less reason for maintaining that theses are distinctly inferior sources of information compared with journal articles.

---

* This statement should be qualified slightly by noting that the *quality* of a *DAI* copy rarely approaches that of an original thesis. Half-tone figures are particularly susceptible to poor reproduction.

# 3 Papers (Journal Articles)

## 3.1 The Nature of Scientific Journals

### 3.1.1 Publications and Publishing

*Journals*

"Scientific progress" as we understand it is contingent upon a competent, competitive, active, and curious *scientific community*. The key word here is community; more precisely, its cognates *communicate* and *communication* (cf. our opening remarks in Sec. 1.1). Meaningful scientific research demands practitioners who are kept mutually well informed. They must know what others who share their interests are doing and what specific research areas are current sources of dramatic results. It is also important that they keep abreast of the latest news from a wide range of fields. Effective communication and dissemination of scientific information is therefore crucial. The major responsibility for ensuring that such communication takes place has traditionally fallen not on newspapers or news magazines, but on *scientific journals*.

Journals are *public documents*: scientists all over the world have access to the journal literature simply by virtue of the fact that they are members of the "public". Thus, as soon as a journal article describing a piece of research appears, the work is said to have been "published", and its results become a part of "the literature". Indeed, it has become customary to use the related word *publication* to describe not only journals themselves but also, and more commonly, articles appearing in (or "contributions to") journals.

Publication in the traditional sense requires that a substantial number of copies of a document be *printed* and *distributed* (cf., however, "Media" in Sec. 3.1.2). Published reports thereby become widely accessible (by purchase or loan), both immediately and over a reasonable period of time. Implicit in the notion of publication is the further requirement that published material be *identifiable* and *retrievable*, criteria met in part by the use of International

Standard Serial Numbers and International Standard Book Numbers (ISSN and ISBN, see Appendix D).

Finally, *libraries* of various kinds are charged with maintaining *collections* of publications so published information will remain available to the public indefinitely. The organization and operation of a library is a fascinating subject in its own right, but one that goes beyond the scope of this book (apart from the aspects touched upon in Appendix D).

Most journals have traditionally limited their pages to contributions from one particular branch of science, although a few important journals pride themselves on spanning the often vague boundaries separating the sciences (e.g., *Nature, Science, Die Naturwissenschaften*). A "chemical" journal is obviously one devoted to chemistry, but even this category is not monolithic. For example, the *Journal of the American Chemical Society* is open to contributions from the discipline as a whole, whereas many chemical journals are devoted entirely to specific subdisciplines (e.g., the *Journal of Physical Chemistry*). Certain journals limit their coverage to exceedingly narrow subspecialties, though these are often interdisciplinary. Thus, a given journal might treat only a single type of substance, or a particular technique (e.g., the journals *Starch* and *Electrophoresis*).

The most widely read journals are those with a reputation for publishing results perceived as *exciting*, but at the same time likely to be *reliable*. It is difficult to say how such a reputation develops. As we shall see later in the chapter, the quality of a journal depends largely on the scientists who contribute to it and the nature of their contributions. The origin of a reputation thus becomes a "chicken and egg" question. Nevertheless, scholars quickly develop the ability to identify the "major" journals in their fields, the ones most likely to contain unexpected, ground breaking contributions. Such journals are avidly read (or at least skimmed) as soon as they arrive. On the other hand, those that tend to report mundane and largely self-evident or trivial extensions of well-established phenomena often remain untouched on a library's shelves for months, attracting attention only when a *literature search* shows them to contain some needed piece of information.

Readers expect a journal to appear on a *regular* and *continuing* basis. Individual journal *issues* are scheduled for release to the public at well-defined intervals or "periods", the reason why journals are classed as *periodicals*.\* The

---

\* Yearbooks or other volumes published only once a year, such as *Organic Syntheses* or *Annual Reports on the Progress of Chemistry*, are normally considered outside the category of periodical literature.

word "journal" (from the French *jour*, day) seems to suggest that scientific journals are published daily. Fortunately for the already overburdened readership this is not the case. Instead, the *monthly* journal is most common. Other accepted modes include *weekly*, *bi-monthly*, and *quarterly* publication, so that the German term *Zeitschrift* (from *Zeit*, time) is somewhat more apt than the English "journal".

## Authors

So far, our discussion suggests that journals exist primarily to serve the scientific community, but they also have another important function: providing individuals engaged in science with a vehicle for announcing and disseminating their research results (see, for example, Sindermann, 1982). For the practicing scientist this is probably the function of greatest interest. The familiar injunction "publish or perish" comes remarkably close to describing the facts of life as lived by a great many scientists.

That practicing scientists need to publish regularly—and thereby double as *authors*—is inherent in science: unpublished results are useless in a very real sense, and they certainly are not worth the investment required to obtain them (except in the special case of work whose purpose requires that it be secret or available only to a limited audience). The nature and content of a set of publications is what others examine for imformation about professional qualities that distinguish a particular scientist: unusual originality, for example, or perseverance, meticulous attention to detail, or thoroughness. No one will deny that the "publish or perish" syndrome has led to abuses; too many scientists are evaluated on the basis of the *quantity* rather than the quality of their publications. Nevertheless, publication of ongoing research *is* the primary means researchers have available for demonstrating the validity of their scientific credentials.

Finally, publications can turn out to be crucial for *dating* a researcher's work and establishing its *priority* (cf. Sec. 3.2.2). Timely and consistent publication is a scientist's best hope for receiving due credit for his or her efforts.

Few scientists can truthfully say they enjoy writing journal articles, but they really have little choice. This chapter, a combination of relevant background information and practical suggestions, should make the preparation of papers less difficult. We hope it will also help to enhance the quality of those papers that are eventually published, and simplify the lives of journal editors.

### Publishers

It goes almost without saying that if scientists are obliged to publish they should also understand something of the *publishing business*. No journal can exist unless it has a *publisher*. Already the subject becomes complex.* For example, a publisher could be an individual, but this is rare. In most cases it is a larger entity: a company (publishing house), a university press, an industrial corporation, a (scientific) society, or even an institution such as a museum. Some publishers are strongly influenced by a profit motive, while others are non-profit, service organizations.

We defer to Sec. 3.5.1 further consideration of the structure and management of publishing businesses in order to look more closely at journals themselves. Only the scientist who understands the aims and modes of operation inherent in journal publishing will find it easy to accept the associated idiosyncrasies.

## 3.1.2    Types of Journals

### Primary Journals

Of all the types of scientific journals the most obvious and indispensable are those called *primary journals* (the "primary literature"). These are charged with communicating new research findings, and in the case of chemistry they constitute the majority of roughly 15 000 (!) journals published worldwide. In view of their importance, primary journals and the articles they contain are the main concern of the present chapter (especially in sections 3.2 through 3.5).

An article in a primary journal *describes* new experiments or theoretical approaches, *reports* the results, and *discusses* their implications and the author's conclusions. Information is conveyed as *precisely* and *concisely* as possible, but presumably in a way that permits others knowledgeable in the field to follow the reasoning involved and understand the techniques employed. Any scientist with appropriate training should be capable of repeating a published experiment or derivation based solely on the contents of the article disclosing it. Variations among primary journals are explored in Sec. 3.2.5, "Types of Articles".

---

* The term "to publish" is ambiguous in itself. It may refer to the activities of a professional publisher or to an author's act of making a piece of work accessible to the public. Interestingly, the German language distinguishes between these two meanings by use of the two verbs "verlegen" and "veröffentlichen", respectively.

## Abstracting Journals

Although primary journals may be most important, there are other kinds of journals with significant roles. Indeed, a scientist today would be lost if the only source of published information were primary literature. So much is published that it has become literally impossible to keep abreast of the progress in even a relatively narrow field by attempting to read all the relevant primary journals. The development of *abstracting journals* represented one of the first attempts to deal with the problem of facilitating *current awareness*. These publications contain abridged versions (*abstracts*) of articles that have already appeared in the primary literature. The most important abstracting journal in chemistry today is without question *Chemical Abstracts*, and it is generally assumed to be absolutely comprehensive with respect to the world's chemical literature.*

Various other important aids to current awareness have also been devised, particularly in recent years. These include *title listing* services (e.g., *Chemical Titles* or *Current Contents*) and even "footnote" indexes (*Science Citation Index*). Together with abstracting journals, such "derivative" but more or less comprehensive publications comprise what is called the *secondary* literature.

Abstracting or other secondary journals have a far wider function than simply encouraging current awareness. They also provide an important means for accessing and searching the primary literature; i.e., they facilitate *literature retrieval*. For this reason, publishers of secondary journals prepare various specialized *indexes* to permit efficient location of desired information within the secondary sources and, ultimately, in primary literature. Searches can be based on a number of criteria, including subject (using *keywords*), authors' names, and patent numbers. Chemical substances are searched by name, composition, registry number, or structure. A brief introduction to efficient use of the secondary literature is provided in Appendix D.

---

* In fact, *Chemical Abstracts* is probably the largest and most elaborate abstracting journal available to any branch of knowledge. The journal itself is only one part of an entire *system* of information and documentation services provided by the Chemical Abstracts Service (CAS) on behalf of the American Chemical Society (for further details, see Schulz, 1985).

## Terminological Confusion

A chemist accustomed to considering protein structure at primary, secondary, and tertiary levels will, at this point, probably suspect the existence of a class of *tertiary literature* as well. The suspicion is well-founded, but there is far from universal agreement on the best and least ambiguous way of defining or characterizing what constitutes tertiary literature; in fact, opinions on the subject vary widely. Most literature specialists begin by assigning *review journals* (described below) to the tertiary literature.* It would seem logical to conclude that monographs, handbooks, etc., would fall in the same category. On the other hand, some authorities treat all of these, including reviews, as part of the secondary literature, placing abstracts in the same category as bibliographies and catalogs (Mücke, 1982).†

## Review Journals

Rather than delving further into the "science" of literature classification we conclude this brief overview of journal types by looking more closely at review journals. A *review* is an article providing a concise, structured overview of some relatively narrow topic. Scientists read reviews to obtain *background* in areas with which they are unfamiliar, or as sources of *references* to the original literature. Unlike articles in primary journals, reviews are usually written at the request of the editor by a specialist whose own original relevant papers testify to his or her interest, knowledge, and competence in the field to be treated. Ambitious authors should aspire to write a review article at least once in their lives, and it is something of an honor to be asked to do so.

Not all review journals are alike. Some have a tradition of publishing what might be termed *comprehensive* reviews (e.g., *Chemical Reviews*), while others deliberately solicit reviews that are *critical* and *selective* (e.g., *Accounts of Chemical Research*). Reviews in journals intended for a purely academic or professional readership are very concise and specific, often documented by

---

* Review abstract journals such as the *Index of Biochemical Reviews* or even the *CA Reviews Index* might, as a further extension of this logic, qualify as "quaternary", but we certainly do not advocate introducing the term "quaternary literature"!

† The confusion surrounding literature classification terminology does not end here. Consider one additional illustration: Herner (1982) includes monographs within the primary literature. He then defines the secondary literature as comprised of abstracting and indexing publications, review journals, reference works, and even textbooks. This leaves something he calls "descriptive compendia of primary and secondary sources of information" as the tertiary literature.

hundreds of literature references. A journal that is more educational or "general academic" in nature, on the other hand, will solicit articles appealing to readers with a scientific background but little previous knowledge of the subject and thus no interest in minute details. The journals *Science* and *Nature* contain reviews of this kind. Finally, a scientific *magazine* (such as *Scientific American*) is intended for an even broader audience, primarily laymen. Its success depends on authors who are willing to put most of their effort into carefully explaining very basic ideas and their interrelations, while at the same time devoting special attention to historical background and future prospects related to the subject.

Review articles have much in common with books, the subject of Chapter 4. Thus, most technical reviews are rather like small monographs, while a well-written review for a largely "educational" journal (e.g., for publication in the *Journal of Chemical Education*) is analogous to a textbook chapter. A "general academic" review, or a magazine article, is roughly comparable to a popular science book.

## The Limits of Classification

It will surely come as no surprise that some journals defy strict classification, often because they serve several purposes simultaneously. Different sections in a given issue of such a journal could be regarded as part of the primary, secondary, or tertiary literature. A case in point is *Angewandte Chemie*, the English-language "International Edition" of which regularly publishes both review articles and original communications, or *Science*, which goes even further in that it also contains news articles.

Classification schemes are at best somewhat artificial and oversimplified, and it is no wonder that scientific periodicals are difficult to compartmentalize. "The Literature" is a very complex entity. In a sense, the vast network defined by the 500 000 or so research papers, reviews, and books added annually to the chemical literature could be said to correspond to an enzyme, a catalytic protein with a structure characterized by a perplexing multitude of bond distances and angles (cf. our analogy under the subheading "Terminological Confusion" earlier in this section). In both cases, the whole is remarkably successful in achieving a desired end, and the fact that it continues to operate smoothly is more important than knowing exactly *how* it functions.

*Media*

Journals can be considered representative of a particular *medium* of communication which, at least theoretically, could take a variety of different material forms. Usually, a journal consists of a collection of *printed pages*. These are assembled into single *issues*, and a certain number of issues (e.g., 12 from a given year) comprise one *volume*. A journal's print may be produced by standard *typesetting*, or it may be the result of direct *photomechanical reproduction* of authors' typescripts (as in "letter journals", cf. sections 3.2.5 and 3.4.6).

In some cases, however, the "medium carrying the message" from author to reader is not paper at all. Perhaps it is microfilm, or even a set of magnetic records in the files of a database.* Nonetheless, the term "journal" can remain applicable provided the *function* is unchanged (cf. the description of the *Journal of Chemical Research* in Sec. 3.2.5). Differences in medium can have profound implications, especially for librarians, but they are of less concern to a prospective author anxious to learn how to *write* for a journal.

## 3.2    Basic Decisions for the Author

### 3.2.1    Why Publish?

Before one tries to publish a piece of work in a journal there are decisions to be made which will influence the likelihood that the planned publication will be well received. These decisions, approached in the form of questions, constitute the framework of this and the following sections.

The question "Why publish?" is easily dispatched. Publication makes one's results available for others to build upon, and it allows one to become more widely known (cf. Sec. 3.1.1).[†] The latter function may provide the less

---

* Traditional paper editions of many journals (including those published by the American Chemical Society and others) are today backed for retrieval purposes by full-text magnetic tape records.
† Publication can also be seen as fulfilling a responsibility toward the institution or organization that has sponsored or supported a piece of research, since sponsors normally assume the wider scientific community will benefit from their efforts.

altruistic motive, but it is usually foremost in the young scientist-author's mind, partly because employers are particularly sensitive to the number and "visibility" of a candidate's publications when hiring or advancement (promotion) decisions are being made. Publishing in a scientific journal is akin to self-advertising: it attracts attention to the quality (and quantity) of a scientist's work. The U.S. Postal Service may have taken this function too literally several years ago: it was used as an argument to defend an increase in postal rates for scientific journals!

The average scientist invests a substantial amount of time in writing articles, even though the effort rarely results in any direct reimbursement. There are a few review journals prepared to pay honoraria for invited articles, but they are exceptional. More typical (especially in the United States) is the reverse situation: authors are forced to pay *page charges* for the privilege of having articles printed and distributed.* The aspiring author should keep this odd circumstance in mind; with a little planning it may be possible to arrange a research grant so that it will cover subsequent publication costs (cf. "Grant-writing Strategy" in Sec. 1.5.3).

## *3.2.2    When to Publish?*

As soon as one has significant and clearly reliable research results in hand, when all essential data have been collected and interpreted and sound conclusions have been reached, it is time to consider publishing. Prompt publication is especially important for the researcher whose reputation is for some reason at stake, or who senses that others are on the verge of obtaining and announcing similar results. *Priority*, after all, accrues to the first research group submitting a given piece of information to a publisher.†

A scientist's "maiden article" is particularly significant since it generally represents his or her first meaningful exposure to the scientific public. This first publication is likely to be based on a thesis project, although the

---

* The following note appears regularly in such journals as *Proc. Natl. Acad. Sci. USA*: "The publication costs of this article were defrayed in part by page charge payment. This article must therefore be hereby marked '*advertisement*' in accordance with 18 U.S.C. §1734 solely to indicate this fact." The relationship to the remark regarding "self-advertising" is obvious.

† Most journals for original research make a point of noting in each published article the date the manuscript was received by the publisher (cf. Sec. 3.5.2). Even a small, seemingly insignificant article submitted earlier by another research group can undermine the importance of one's publications or one's claim of originality.

description must be severely curtailed to make it suitable for a journal. If a thesis can be regarded as a scientist's "journeyman piece" (professional qualification), then a first publication is analogous to a "masterpiece" in the original sense of the word: qualification as a "master of one's trade".

The same high standards that universities impose on theses also apply to journal articles; indeed, publisher's requirements can be far more stringent. Only those papers that meet the editor's particular set of standards will be accepted for inclusion in a journal. Science as a whole clearly benefits from such "quality control", but so do individual scientists. A strict editor might well be responsible for sparing someone the agony of trying to live down a bad reputation acquired through careless or premature publication, or consistently poor writing. A proud announcement of unscreened results later discovered by others to be a duplication of information already published might have even worse consequences. Selectivity on the part of journals minimizes the likelihood of such situations. The risk entailed in submitting a manuscript for the first time to an editor's scrutiny and judgment may be a frightening prospect, but the newcomer can take comfort in the fact that more experienced colleagues have all been subjected to the same ordeal, and survived it. They can also usually be counted on for help and moral support.

We have said that publication should be high on a researcher's priority list, but occasionally there are good reasons for withholding a manuscript. For example, one might prefer not to tempt others quite yet to enter a particular interesting avenue of investigation. Or, there may be good cause for making sure that every conceivable objection to a particularly startling set of results can be countered before it is exposed to the criticism of others. Charles Darwin provides a classic example: he elected not to publish his *Origin of Species* until more than 20 years after the voyage on the research ship *Beagle* which provided him with the key observations underlying his theory of evolution. The exercise of optimal "timing" is truly an art, and some of the factors involved can be very personal.

## 3.2.3    What to Publish?

This question is closely related to the preceding one. Only results that can be described with assurance and a clear conscience as *reliable, significant,* and *new* should be submitted for publication in a primary journal.

The *reliability* criterion means that results are "publishable" only after they have been subjected to extraordinarily careful scrutiny. If derived from experiments they must be verifiable, and the published description must permit such verification by anyone trained in the discipline (cf. the remarks under "Primary Journals" in Sec. 3.1.2). Ideally, "true" and "false" would never be a subject of debate for published experimental results in the exact sciences. Controversy, if any, should be limited to the *conclusions* an author has drawn or the way data have been *interpreted* (i.e., to the "Discussion" section of an article). Consequently, important results, especially surprising ones, should always be reconfirmed before they are announced to the public.

The second test of publishability, *significance*, is more difficult to formulate in a clear-cut way. Its purpose is certainly not to stifle publication. Rather, it is meant to help control the virtual flood of publications issuing from the scientific community. Space in journals (and libraries) is precious. It is generally agreed that priority should be assigned to those results most likely to prove valuable. The more "prestigious" the journal, the more strict will be its definition of "significant".

Finally, the third test warrants comment, even though it may appear self-evident: papers should be submitted for publication only if they report results that are truly *new*. Unfortunately, some scientists, and even some journals, appear to have trouble exercising self-control with respect to this criterion. It is not at all uncommon to find the same results presented first as a so-called *note* or *short communication* (see Sec. 3.2.5) and later to discover them again described in detail in a "full paper". Most would concur that this practice should be discouraged except in unusual cases of important discoveries, if for no other reason than because it contributes unnecessarily to the expansion of the already overgrown literature. Some scientists are tempted to use double publication as a way of "padding" their personal publication lists. It is not a particularly effective device, however, because both the practice and its motivation are all too transparent.

## *3.2.4   With Whom to Publish?*

Research today, more than ever before, tends to be a team effort. Anyone who has made a significant contribution to a set of results should be included as a *co-author* of any derived papers. For example, a publication based on a thesis project should normally have the thesis advisor as a co-author. After all, this is the person who probably thought of the project in the first place and who guided much of its subsequent development. It is true that the hierarchical relationship of graduate student to professor has been relaxed somewhat, but it still represents a partnership that warrants expression through joint authorship.

When several people have contributed to a research result it is necessary to establish the *order* in which their names will be cited. In the past, the author with the highest *rank* usually insisted on being listed first, regardless of who did the most work, but this practice is changing. Today, the order generally reflects a mutual assessment of the relative importance of the authors' respective roles, although alphabetical arrangement is also common.* The first-named author enjoys a particular advantage, in that the article is likely always to be associated with his or her name ("Johnson et al.").† Agreeing on an order can be difficult at times, and both generosity and vanity are occasionally involved. The only piece of advice we can offer is that the problem be worked out in advance.

## 3.2.5    In What Form to Publish?

### Types of Articles

Earlier (in Sec. 3.1) it was implied that there are various kinds of primary journals publishing various kinds of articles. For results that are limited in scope, or ones important enough to deserve especially rapid dissemination, or, on the other hand, for work that still needs to be regarded as somewhat tentative, the best form of publication is what is known as a *short communication* or a *note* (sometimes called a "preliminary note").‡

Articles of this kind are limited to a few paragraphs and are rarely subdivided into formal "Results", "Discussion", and "Experimental" sections, as is done with "full papers" (see below). Furthermore, the "Introduction" (if any) is often so condensed that only an expert in the field would be likely to grasp why the work is significant. In many cases experimental details are omitted,

---

* In some fields it is customary for the "senior author" to be named last. The first position is then assigned to the team member most actively involved in the reported work.
† The registry *Who is Publishing in Science* (ISI, Institute for Scientific Information) completely ignores all but first authors.
‡ Many journals distinguish between "short communications" and "notes". The former category is reserved for brief but nonetheless *complete* papers, usually describing secondary results of more extensive studies discussed elsewhere. "Notes" are then brief *announcements* of work. These contain little detail because it is assumed they will be amplified later in more formal publications.

the implicit assumption being that a more complete description of the work will appear later (an obligation some researchers fail to fulfill—to the aggravation of their peers).

Some journals publish *only* short communications (e.g., *Journal of the Chemical Society: Chemical Communications*). Often the papers they contain are direct copies of submitted typescripts that have merely been subjected to a *photoreproduction* process. *Tetrahedron Letters* is a case in point. The major advantages of the technique are that it minimizes the time between manuscript submission and journal publication, and involves significantly lower production costs. Such journals (*letter journals*) are designed to have wide circulation among active researchers anxious to keep abreast of the very latest results. If one chooses this vehicle for one's publication it is especially important to pay close attention to all editorial guidelines. No changes whatsoever will be made in a manuscript once it has been accepted by the editors (see Sec. 3.4.6 for more details).

Comprehensive or lengthy descriptions of significant results or discoveries deserve to be written more formally and with a certain amount of "polish". Such manuscripts should be submitted to a standard primary journal for publication as a traditional "paper" (sometimes called a *full paper*). Sections 3.3 through 3.5 are devoted mainly to articles of this type.

A variety of factors has recently led to a rash of novel experiments within the publishing industry. Some may drastically alter publication patterns in the future. The *Journal of Chemical Research* is an example of a periodical of a totally new type, and it presents the author with an interesting alternative. The standard edition of this journal contains only one- or two-page summaries ("synopses") of full research papers submitted in the usual way to the publisher. The detailed reports themselves are not circulated to all subscribers, but they are available for separate purchase from the publisher as *microform* editions (either *miniprint* or *microfiche*).* It is possible to obtain a regular subscription to the microform supplements, but most readers order only papers that appear to be of special interest. Following the same trend, several American Chemical Society journals have recently begun encouraging their authors to consider relegating peripheral information to microform supplements available from the publisher on request.

---

* Full-sized copies of individual papers can also be obtained from the publisher upon special request, but at a higher cost.

## Choice of Language

In recent years the *language* of publication has also become a matter requiring serious consideration. No longer can the choice of language be made solely on the basis of an author's own nationality or linguistic fluency. Instead, one must take into account the role language plays in determining a publication's effectiveness. For many, *English* has become the language of first choice, primarily because it is understood by more scientists worldwide than any other. Indeed, its dominance appears destined to become even more pronounced.* Today many journals originating in countries where languages other than English predominate (e.g., Germany, Japan, or Czechoslovakia) are accepting (or even insisting upon) papers submitted in English.

It seems safe to say that if one's object is to reach the largest possible audience, or if one is anxious to establish an international reputation, then at least occasional publication in English is advisable. Some view the trend toward universal use of English with alarm, for reasons varying from linguistic chauvinism to concerns about the accuracy and clarity of descriptions produced in a language with which the author is uncomfortable. But the die appears to be cast.

## 3.2.6    Where to Publish?

In selecting a target journal, characteristics such as scope, the kinds of articles included, and language should be considered first; they alone should suffice to narrow the choice considerably. Nevertheless, there are other important considerations.

Most authors want to publish in highly reputable journals. All practicing scientists know what are the most important journals in their field: they use them almost daily. Beyond this rough yardstick, however, interesting objective criteria also exist for judging the relative importance of a given journal. The starting point is *Science Citation Index* (*SCI*), a secondary periodical developed (in 1955) by the Institute for Scientific Information (ISI) in Philadelphia. *SCI* is intended to be used for locating articles (cf. Appendix D), and it provides comprehensive access, in the words of the editors, to work

---

* In 1984, 69.1% of the papers covered by *Chemical Abstracts* were written in English; in 1961 the corresponding share was "only" 43.3% (based on data reported in the brochure *Chemical Abstracts Service Statistical Summary 1907-1984*).

cited in "all of the most important technical journals in the world". The index is distinguished by its unusual breadth: it covers physics, chemistry, the life sciences, medicine, agriculture, engineering, technology (applied science), and environmental science. More than half a million publications are processed annually, drawn from (in 1985) roughly 3250 journals. Although these represent less than 10% of the science journals published worldwide, it is estimated that they contain over 90% of the important new scientific findings.

Journals are chosen for inclusion each year in *SCI* (and also in *Current Contents*, produced by the same institute) on the basis of a very subtle system that utilizes an annually revised statistical survey of the previous volume of *SCI*, called *Journal Citation Reports*. From this compilation it is possible to determine an *impact factor* for each journal, "a measure of the frequency with which the 'average cited article' in a journal has been cited in a particular year" (definition taken from *SCI* ). If the assumption is granted that *cited* papers are *important* papers, then those *journals* whose papers are cited frequently can be regarded as "important" journals.

Another *SCI* criterion by which journals can be evaluted is the *immediacy factor*. This is "a measure of how quickly the 'average cited article' in a journal is cited". Statistical citation information is being used in an increasing number of ways despite certain obvious inherent dangers. Since such data are now available for a relatively long time span it has also become possible to use them for analyzing publication *trends* over the years.

Probably the quickest way to verify a given chemistry journal's success based on citation criteria is simply to scan recent weekly issues of *Index Chemicus* (also published by ISI) to see if the journal in question is considered worthy of inclusion in the ISI information and indexing system. The total number of such journals is roughly 100.

Publishing in prestigious journals is obviously tempting, but it has drawbacks. First of all, such journals tend to be overwhelmed with submitted manuscripts. As a result, one has to anticipate long delays between submission and publication, as can readily be seen by looking at any issue of a prestigious journal and comparing the cover date with the submission dates of articles. In other words, if immediate publication is critical, one should consider lowering one's sights. Furthermore, highly regarded journals tend to have high *rejection quotas*; more than half the articles submitted may be returned to their authors. Rejection does not necessarily mean that a manuscript is "bad", but it does mean that it will need to be resubmitted to another journal, causing further delays. The risk of receiving a rejection notice from a major journal is often worth taking, but alternatives should also be considered, especially if one is more interested in establishing priority than in attracting attention.

Finally, one other criterion for journal choice cannot realistically be ignored. There are authors who feel such a strong sense of attachment to particular editors or publishers that they choose their journals on strictly personal grounds. The factors entering into this kind of decision are particularly difficult to categorize or assess, though some may become more understandable as we examine the roles played by various members of an editorial and publishing staff (Sec. 3.5).

## 3.3    The Parts of a Paper

### 3.3.1    Title

A journal article, like any other scientific document, needs a clear, descriptive *title* to serve as a kind of "mini-abstract" of the contents (cf. Sec. 2.2.2). From a title a reader should be able to ascertain the scientific subdiscipline to which the work relates and at the same time gain insight into the goals, methods, and results of the reported investigation.

It is rare for anyone, even an individual *subscriber*, to read every article published in a journal. Readers more commonly rely on article titles to guide them in refining their reading lists. This is one of the clearest indications of a title's importance, but there are others as well. For example, various *bibliographic services* rely heavily on titles for classifying and indexing published articles. *References* in journals and books frequently include the titles of cited papers in order to help readers decide whether or not a given source is likely to be worth further investigation.

Perhaps most important from the author's standpoint, however, is the fact that a number of major alerting services (such as *Current Contents*, *Science Citation Index*, and *Chemical Titles*), as well as selective abstract distribution services (e.g., the *CA Selects* assembled by the Chemical Abstracts Service), depend for their effectiveness on the appropriateness and accuracy of titles of newly released papers. The "Permuterm Subject Index" of *Science Citation Index* is a typical example of a source in which the wording of incorporated titles plays a crucial role. This index is built using all possible permutations, listed in alphabetical order, of *pairs* of significant words from recent publication titles. The underlying premise is that many scientists will scan the lists in search of words central to their research interests, presumably because they will thereby locate the new papers relevant to their work. (Word pairs are regarded as more definitive in this respect than single words.)

Another example relates to the increasingly popular computerized literature search systems, such as Chemical Abstracts Service's *CA Biblio File,* or *MEDLARS* (Medical Literature Analysis and Retrieval System), maintained by the U.S. National Library of Medicine in Bethesda, Maryland. These also were developed on the assumption that relevant articles can nearly always be located on the basis of "key words" in their titles. The increasing dependence of scientists on tools such as these makes the careful choice of wording in a title more critical than ever to ensure that those who should be aware of a paper will in fact encounter it.

The title is one element of what is known as the *bibliographic portion* of an article (cf. Sec. 11.4.1). Others are the list of authors and the relevant *citation data* (in part available only after publication has occurred), consisting of the name of the journal, volume number, page, and year.*

A typical journal article title should consist of at least four, but not more than twelve, words. If more than eight seem necessary it is usually better to say less in the title, but append a *subtitle* (cf. Sec. 2.2.2, where the subject of titles is addressed in somewhat more detail). It is perhaps worth reiterating that formulas and special symbols must be avoided in titles because of the problems they cause in indexing. Moreover, some journals insist that certain words that tend to be overused (e.g., "new", "improved", etc.) be excluded from titles.

No title is likely to be capable of highlighting every interesting aspect of an article. Consequently, many journals request authors of accepted papers to supply lists of up to 20 additional relevant words (*keywords*) to supplement words in the title for indexing and bibliographic purposes. Any journal that maintains this practice usually mentions the fact in the introduction to its annual index. Some journals also request authors to do just the opposite: provide an abridged title, which then appears at the top of each page of the printed text as a so-called *running head* (or *column title*, cf. Sec. 4.4.2).

---

* Some also consider the abstract (Sec. 3.3.3) part of the bibliographic information.

### 3.3.2    Authorship

A complete list of authors' names (cf. Sec. 3.2.4) follows the title of an article. Authors should always publish under precisely the same name; variation in the use of initials should be avoided. Thus, David L. Smith, if that is how he prefers to be known, should be careful never to identify himself in an article as D.L. Smith, D. Lamar Smith, or David Smith, Jr. Consistency prevents much potential confusion for other scientists as they search the literature, and it also ensures that an author's publications will all be grouped together in indexes and other bibliographic sources.

Scientific journals ordinarily refrain from including academic and professional *titles* with authors' names, although full names with titles and even *addresses* may be printed separately at the bottom of the first page of an article, or at the end. If a paper has several authors, one name is often marked with an asterisk identifying this person as the *corresponding author* with whom an interested reader should make contact to obtain further information about the work.

Sometimes the "address" given for an author consists only of the name and location (city) of the institution or firm where the work was carried out (the author's *affiliation*). Address information is also subject to indexing; it is used, for example, by *Index Chemicus* to construct a "Corporate Index" that contains entries like

UNIV MARBURG
Inst Pharmazeut Chem
D-3550 Marburg/Lahn, W. Germany

Each such heading is followed by a list of all current publications originating from that particular institution. Its use as a source of mailing addresses is one of the most valuable applications of this index. Another source of addresses is *Who is Publishing in Science*, a publication (previously mentioned in a footnote to Sec. 3.2.4) issued annually by ISI, the Institute for Scientific Information. However, this source is limited to *first* authors of papers.

Fortunately, it is becoming increasingly common for journals themselves to include authors' complete mailing addresses rather than simply institutional affiliations. The practice is most welcome, since it clearly facilitates correspondence with authors, either for the purpose of exchanging information, or simply to request reprints.

### 3.3.3    Abstract

Just like a thesis, a journal article is usually preceded by a short summary, or *abstract*. The problems entailed in creating an abstract have been elaborated upon in Sec. 2.2.4. Briefly, an ideal abstract ought to be

• complete,
• accurate,
• objective,
• brief, and
• intelligible.

Writing a good abstract is far from easy, an assertion supported by the fact that the major abstracting services employ professional "abstractors" rather than relying solely on authors' efforts. There is general agreement among editors that the published abstract of a journal article should not exceed 250 words.* Its content should focus on

• principal objectives and scope of the investigation,
• methods employed,
• major results, and
• principal conclusions.

As in the case of the title, one should avoid formulas and abbreviations in an abstract. Furthermore, no references to tables, figures, or formulas within the paper proper should be given, because an abstract is expected to be completely self-contained.

Most scientists find that browsing through abstracts is one of the best ways to keep their "current awareness" at a reasonable level; in doing so, however, they are tacitly assuming the adequacy of the abstracts. The preparation guidelines above are far from arbitrary. Experience has shown that abstracts prepared according to these standards have the maximum potential for conveying information, but at the same time catching the interest of at least some readers who, judging on title alone, might overlook a useful or interesting article.

---

* Standard *ANSI Z39.14-1979* admonishes: "Keep abstracts of most papers and portions of monographs to fewer than 250 words, abstracts of reports and theses to fewer than 500 words (preferably on one page), and abstracts of short communications to fewer than 100 words." In Sec. 2.2.4 we quoted (from another source) 350 words as the upper limit of a thesis abstract.

As noted previously, some bibliographic sources draw their information from abstracts as well as from titles. This is another reason why abstracts must be formulated with care.* Certain publishers (the American Chemical Society, for example) have very close working relationships with major secondary publications and alerting services, supplying them with edited abstracts of all papers about to be published. These abstracts very quickly become available for wider circulation and processing, occasionally leading to the anomaly that the abstract of a recent piece of work will be in the hands of readers before the article itself.

Finally, there is one additional reason for devoting considerable attention to the preparation of the abstract for a journal article, and that a very pragmatic one: a good abstract can make a distinctly positive impression on referees to whom an article is sent for evaluation (cf. Sec. 3.5.3). This is a consideration of no small importance in an age when a great many manuscripts are rejected!

Many internationally prominent journals require that all submitted articles be accompanied by an abstract in *English*, regardless of the language of the article itself or the principal language of the journal (cf. "Choice of Language" in Sec. 3.2.5). Often, foreign language journals print two complete abstracts, one in the language of the article and the other in English. Sometimes English abstracts are required even when they will not be printed. Non-English-speaking authors may not appreciate this extra writing assignment, but in practice they usually find it less difficult than expected. Fortunately, it can be accomplished adequately with a minimal knowledge of technical English, since abstracts are not intended to be showcases for the display of elegant style.

The term "abstract" is actually a rather broad one, and it is often applied in contexts very different from the one so far considered. For example, a scientist who plans to present a paper orally at a conference may well be asked by the organizers to submit in advance an "abstract" of the proposed talk.† An abstract of this kind obviously serves a purpose different from that of a journal abstract;

---

* The subject index of each issue of *Chemical Abstracts* is prepared entirely on the basis of key words taken from the titles and abstracts of that issue's articles. The annual indexes, however, are drawn from full texts.

† Such a request usually means that a volume containing all the abstracts from the conference will be made available to conference participants. The abstracts may even be published later in a book so people other than those who happen to be present at the lecture will have access to its highlights. The unabridged lecture will still be available only to attendees, however, since it is the *abstract* that will be printed.

in fact it more nearly resembles a *short communication* (cf. "Types of Articles" in Sec. 3.2.5). For this reason it may need to be prepared in ways that violate the usual rules for article abstracts. A lecture abstract might well contain formulas (structures), figures, tables, or literature references. The same holds true for *poster abstracts*, brief texts that describe research results presented in the form of poster displays at scientific conferences (cf. Appendix A.5).*

### 3.3.4    Body of the Article

The basic considerations applicable to the *body* of a journal article are not unlike those outlined earlier for reports and theses (especially as discussed in the relevant parts of Sec. 2.2). Nevertheless, since articles vary a great deal in length, format, and style they permit ample opportunity for both author and journal to display personal idiosyncrasies.

An author who has selected a "target" journal should make a point of becoming well acquainted with it, including examining its literary and stylistic characteristics. This makes it possible to adapt one's writing style accordingly. Editors often provide help (on request) in the form of a pamphlet of "Instructions to Authors", and writing suggestions may also be found printed in a recent issue of the journal itself (e.g., on the back cover).

Editors particularly appreciate receiving manuscripts carefully patterned after current articles in their own journals because this saves them much time-consuming editorial revision, and it hastens the publication process. Referees are also sensitive to matters of style. Even the author's task is made simpler if a submitted manuscript is in proper form; otherwise, copy that later arrives for proofreading will be cluttered with countless editorial markings.

The principal section of a journal article is usually subdivided (formally or implicitly) into three major parts: an *introduction*, a *body*, and a *conclusion*.[†] One observer (Bennett, 1970) has suggested an analogy between these three parts of a narrative and the three phases of a trip by air: the take-off, the flight,

---

* Poster abstracts are also sometimes incorporated into printed "conference proceedings" volumes.

† The "conclusion" should not be referred to as a "summary"; the latter term is ambiguous since it is often interpreted as a synonym for "abstract".

and the landing. Just as most airline passengers are curious about the kind of plane they are boarding, where it has come from, over what route and at what altitude it will fly, readers about to start working their way through technical articles react positively to a certain amount of preparation designed to help them relax during their "journey". The introduction to an article serves this purpose. It sets the stage by briefly summarizing the state of knowledge before the author began work on the project and showing the relationship between this starting point and the goal. One or two sentences may well suffice, but they need to be very carefully formulated.

Coverage of the intervening terrain is a job reserved for the "flight". Here, the challenge is to provide a description which leads "passengers" to their destination without "losing them in the clouds". It is important for authors to remember that their "passengers" are less likely to become confused or impatient if the route chosen is a direct one, avoiding detours and side excursions (or at least making them optional by clearly labeling them as such).

As a part of the landing routine, a commercial pilot usually makes a brief farewell announcement that includes a few words about whether the flight is on time, what weather conditions are expected at the destination, and the availability of connecting flights. In the same way, it is appropriate to end a journal article by discussing briefly in a "conclusion" the extent to which the questions posed earlier have been answered and the significance of the results, perhaps also forecasting likely future developments. Under no circumstances, however, should the conclusion be used to introduce important ideas. That should have occurred in the body of the text, although it is wise to *reiterate* major findings in the conclusion, since readers often consult this section prior to undertaking a careful study of the entire text.

For a possible distinction between sections headed "Results" and "Discussion" we refer the reader to the relevant portions of Chapters 1 and 2 (esp. sections 2.2.7 through 2.2.9). The remainder of this chapter is devoted to matters of manuscript preparation and subsequent steps in the publication process.

# 3.4 Preparing and Assembling the Manuscript

## 3.4.1 Text

In the following discussion it will be convenient to distinguish (as publishers do) between a "text manuscript" of an article and the "manuscript" as a whole. The *manuscript* in the larger sense is the entire document, including figures and tables. The *text manuscript*, on the other hand, consists only of that portion which has been *typewritten*, and will later be *typeset*. It is frequently called the "typescript", a term more appropriate than "manuscript" (cf. Latin *manus*, hand, and *scribere*, write).

Editors expect text manuscripts to be *double-spaced* (cf. Sec. 5.3.1) with wide *margins* and typed on plain (i.e., without letterhead) U.S. letter-format (8 1/2 in. x 11 in.) or international standard A4 (210 mm x 297 mm) paper. All pages must be consecutively *numbered*, and typing should be on one side only. It is usually recommended that each page contain about 30 lines of text, with no more than 50 characters to a line (blank spaces included) and most of the resulting extra space on the *right*. The reason for insisting on this particular format is that it provides ample space throughout the text for later insertion of editorial comments and corrections. The entire text should be prepared in this form, including the experimental section, figure captions, and reference list.

Careful *pagination* (page numbering), preferably in the upper right-hand corner, is important to prevent any confusion about the completeness of the manuscript. Every page that contains text to be printed must be numbered, because unnumbered pages could easily be misplaced. If any pages are *eliminated* after numbering—e.g., as a result of deletions or reorganization—this fact should be clearly noted, perhaps by "numbering" the last surviving page as a *range* of pages covering the gap up to the next existing page number. *Inserted* pages may carry the number of the preceding page, followed by an a, b, c, ..., although this device does not guarantee that the existence of such pages will be apparent.* If the manuscript is relatively short it is probably better simply to renumber the pages whenever changes are made.

---

\* One can ensure clarity by indicating at the *bottom* of each page the number of the *following* page.

Certain portions of the text may need to be set in small type, in which case the typesetter must be so informed, either by the author or the editor. The message is usually conveyed by drawing a vertical line in pencil in the left margin beside the appropriate text and annotating it with "((pt))".* Other indications of *type style* may also be required; e.g., for words or symbols to be set in italics or boldface (cf. Sec. 4.3.7, "Choice of Type"), but many editors prefer that authors leave this task to the editorial staff (a point normally clarified in the "Instructions to Authors").

### 3.4.2    Formulas

#### Mathematical Equations

*Formulas* present special problems in scientific texts. In chemistry, two distinct kinds of formulas can occur: mathematical and structural. *Mathematical formulas* (*equations*) are statements that connect (physical) quantities, units, mathematical operators, and numbers. Typesetting machines can handle these more or less as they would regular text, although more care is needed on the part of the typesetter than for standard prose. More sophisticated equipment may also be required. Still, formulas of this kind can be treated as part of the body of the text and can therefore be incorporated into the text manuscript. Special characters not available on the author's typewriter should be entered by hand (cf. Sec. 8.1.3). The author who prepares written copy with a computerized text editor (*word processor*), especially a primitive one, may find it necessary to insert complex mathematical expressions with a typewriter, although more advanced systems often make this unnecessary (cf. sections 4.3.5 and 5.2.2).

It is wise to *number* any equations present in a journal paper. One advantage in doing so is that it permits very concise text references; e.g., a simple mention of "eq. 10" (equation number 10). Numbered equations are preferable from a technical point of view as well. In the absence of numbering it is mandatory that each equation be positioned exactly where it belongs in the text. This can create difficulties, particularly near the end of a page, because

---

* The suggested use here of double parentheses is explained in Sec. 3.5.6 and illustrated in Appendix I; "pt" stands for the French *petit*, small. Even parts of an article that are to be typeset in small type must be double-spaced in the manuscript.

insufficient space may be available. Editors consider it important for aesthetic and economic reasons to avoid unnecessary blank space on a page, so providing for a single awkwardly placed formula may require redesigning as much as several pages of text. For this and other reasons, considerable time and expense can be saved in the preparation of a printed article by avoiding equations within sentences. The best solution is to place equations at the ends of paragraphs, and then refer to them by number.

In highly mathematical texts, however, it is not always practical to follow the latter suggestion; rather, sentences or fragments of sentences such as "of the form ... where", "since ... ", "gives ... ", "if ... then", or "from which ... " are often intermingled with equations in a continuous flow of reasoning which cannot be interrupted by rigid typesetting rules. It is also common in text of this kind to number only the more important equations, those to which reference must be made.

### Chemical Formulas

More typical (and troublesome) in chemistry journal articles are *structural formulas* and related symbolic notations (e.g., reaction schemes), here grouped under the general heading *chemical formulas*. With the exception of empirical formulas of the type "$H_2SO_4$", all such formulas and any equations containing them must be kept *strictly separate* from the text manuscript and submitted as a group in a *formula supplement*. There is good reason for this injunction: chemical formulas must be *processed* separately, because they cannot, as a rule, be typeset.* This generalization does not hold, of course, for "camera-ready manuscripts", since these are never typeset or subjected to processing of any kind (cf. Sec. 3.4.6). Chapter 8 contains advice on how authors should prepare in-text formulas and equations for camera-ready copy.

We will later and in various contexts explore how printed copy originates, but for the moment two important generalizations will suffice:

---

* Only formulas that occupy straight text lines and are comprised exclusively of letters, numerals, and standard type symbols can normally be typeset. Even the presence of a diagonal or curved arrow rules out routine typesetting. In fairness it should be pointed out that very recent technical innovations have opened the possibility of producing a complete range of chemical formulas by machine. The process is an application of what is known as *computer-aided design* or CAD.

● All manuscript components that require special technical processing need to be isolated both from the body of the text and from other supplementary material; and

● editors prefer that the *number of items* within each discrete category of "special parts" be kept to a minimum.

Manuscripts for submission to conventional journals require some thought on the part of the author with respect to chemical formulas. The effort is worth investing, however, since it may save much time later. Suppose, for example, that inspection of the chemical formulas in recent issues of a target journal shows that the style is consistent throughout, and all formulas contain exactly the same kinds of letters and other symbols. One could then safely assume that final drawings were all prepared by an expert engaged by the journal's editor. In other words, all formulas will be redrawn, so it would be pointless for an author to expend extra time being artistic. Even free-hand pencil sketches would serve the purpose, provided they were unambiguous.

The author who chooses this kind of efficiency must keep in mind that the draftsman who will later work from the sketches is not trained as a chemist. All letters, numerals, and other symbols must be made perfectly legible to someone who is accustomed to working swiftly. Extra attention should be devoted to subscripts and superscripts (*indices*), since these could easily be overlooked or misinterpreted. Moreover, all bonds should neatly connect one atom (or ring position) with another. None should be allowed to end "somewhere" in space—near a hastily drawn $CH_3$-group, for example. Also, those bonds that should stand out for stereochemical reasons must be clearly emphasized.

If, on the other hand, one's target journal accepts formulas for reproduction as submitted by the authors, then exceptionally careful work is called for (cf. Sec. 8.2). The extra effort is time-consuming, but may result in a bonus: some of the same drawings may prove useful in preparing illustrations for a lecture (cf. Appendix A.3).

All structures in a manuscript should be numbered sequentially using *Arabic numerals*. These numbers are then used exclusively in subsequent discussion; even if a structure is *drawn* a second time it should continue to be identified by its original number. To the extent possible, structures should be grouped together into *blocks* or *reaction schemes*. New structures within a given block or scheme are numbered from left to right and from top to bottom, regardless of logical arguments to the contrary. Most journals have adopted the practice of printing structure numbers (the *formula numbers*) in *boldfaced* characters. These are indicated in a typescript by either a normal or a wavy underline, depending on which marking convention the editor follows (cf. Sec. 3.5.4 and "Choice of Type" in Sec. 4.3.7).

The proper placement of a given formula block (preferably between paragraphs) is indicated on a separate line within the text by means of a note in *double parentheses* (cf. Sec. 3.5.6 and Appendix I), e.g.:

((formulas 17-20))

((reaction scheme 4))

The corresponding copy (*block material*) must, of course, be readily identifiable as an individual item in the formula supplement. The inclusion of structural formulas in tables should be avoided; formula numbers should be used instead.

Every structure shown must be *referred to* by number in the text. Such numbers are placed in parentheses whenever they *follow* the name of a compound, e.g.:

4,4-dimethyl-5-hexen-2-one (**88**).

If one's chosen journal employs boldface type for formula numbers, parentheses are omitted from a number which *replaces* the name of a compound or class of compounds, e.g.:

A solution of **14** in ...

Even a number of the latter kind would need to be enclosed in parentheses if boldface is not available, however:

Compound (5e) was treated ...

Lower case letters a, b, c... may be added to the numerals (as above) to differentiate various derivatives of a parent compound, such as ones bearing the substituents $R^1$, $R^2$, $R^3$, ... or having varying numbers of ring members $n = 1, 2, 3, ....$ Use of this technique has the advantage that a basic molecular framework need be drawn only once and assigned one Arabic numeral, perhaps augmented by lower-case letters for differentiation (e.g., **14a**, **14b**), thereby saving production costs. At the same time, it can be a subtle way of emphasizing important structural relationships among compounds.

A special problem is posed by the need to designate specific atoms appearing in a formula in order to make reference to them in the text. Standard subscripts (e.g., $C_3$) should be avoided since these normally imply *multiples*. Instead, the following forms are recommended:

$$C_{(3)} \quad \text{or} \quad \text{C-3}$$

### 3.4.3    Figures

*The Choice Between Figures and Tables*

Often, quantitative data must be displayed in the form of *data sets*. In principle this could be done either with a *table* (where the data would be presented in "digital"—*numerical*—form) or by showing data points in a *figure* (i.e., in "analog"—*graphical*—form). A choice is always necessary: redundancy is inappropriate both from the reader's point of view and because of the ensuing costs. It is up to the author in each case to decide which of the two approaches is better from the standpoint of lucid presentation. Figures, the subject of Chapter 9, have the general advantage that they attract a reader's attention, and often they are more effective for emphasizing *relationships* (cf. Sec. 9.2.1). Tables, on the other hand, are more economical to produce. Their use is discussed in the following section and in more detail in Chapter 10.

Like structural formulas, tables and figure artwork must be kept separate from the main typescript. In the case of figures the reasons are obvious: all figures, whether *line drawings* (cf. Sec. 9.2) or *photographs* (cf. Sec. 9.3), require photomechanical processing prior to printing, and drawings often need to be reworked by a draftsman before reproduction. Tables also present special problems, primarily relative to format and placement.

*Preparation and Submission of Figures*

Most journal editors prefer that the *number* of figures (*illustrations*) accompanying an article be kept to a minimum, and that the *lettering* within each also be restricted, mainly for reasons of cost. The printing of a figure involves more extensive processing than that of the text. Use of figures should therefore be restricted to situations in which, for example, a functional relationship or a piece of apparatus is better "illustrated" because the alternative would be a wordy or tortuous explanation.

Authors are normally expected to submit one *original* and as many *copies* of each figure as there are copies of the text manuscript. The original should be a black India ink drawing on high-quality paper (preferably vellum), or a photograph with good contrast. Figure copy may vary in size, but any drawing smaller than standard page format should be glued onto its own sheet of standard paper.

Editors of some journals, particularly prestigious ones, routinely have all submitted artwork redrawn by professional draftsmen. Authors should inquire whether this is the policy of their target journals, because they may be spared the time and effort otherwise required for producing perfectly drawn and lettered copy.* The situation is analogous to that of chemical formulas (cf. "Chemical Formulas" in Sec. 3.4.2). Most journals, however, simply arrange for photomechanical reproduction of the original drawings submitted by authors, even though this results in a certain amount of non-uniformity of style and quality. In this case the authors themselves hold the responsibility for preparing the best figure copy possible (for detailed advice see Chapter 9).

Reproduction of *spectra* is rarely permitted in journal articles, although exceptions are sometimes made if important new substances are involved or when a newly discovered relationship between structure and spectroscopic properties requires clarification. In the event that spectra are to be incorporated one should consider augmenting them with structures of the corresponding compounds, neatly drawn near the curves using templates (cf. Sec. 8.2.1).

Reproduced spectra are always considerably smaller than original plots obtained in the laboratory. Appropriate journal guidelines should be consulted and recent issues examined in order to ascertain what scaling is appropriate and how axes should be lettered (cf. Sec. 9.2.3). Many modern spectrometers are equipped with microcomputers and plotters that allow for the production of graphic plots to any desired scale, although the quality may vary with size. Some journals prefer to undertake necessary size reductions themselves, using film negatives of original-size spectra.

Because of printing constraints, no published figure can ever be larger than a standard journal page. Most are considerably smaller. Nevertheless, it is recommended that drawings submitted by authors be large, often twice the size desired in print. This permits a linear reduction by 50% during processing. An excerpt from a typical set of "Instructions to Authors" reads as follows: "Figures should be planned so that they reduce to 7.5 cm column width; the

---

* The advantage may have its price, however: authors are sometimes held responsible for the expenses incurred in redrawing.

preferred width of submitted line drawings is 12 to 15 cm, with capital lettering 4 mm high, for reduction by one-half." In any case, one should try to prepare all figures to a single scale.

Publishers often advise their authors to refrain from incorporating any hand-lettering in reproducible illustrations. Instead, text should be supplied on a duplicate copy or on a transparent overlay so a skilled draftsman at the publishing house can prepare the final lettering (cf. Sec. 9.1.3). The author will have an opportunity to approve the results (and any other changes that may have been made) as a part of the process of *proof correction* prior to final printing (cf. Sec. 3.5.6).

### Figure Numbers and Captions

Figures should be numbered sequentially throughout an article. All submitted figure copy must be annotated with the appropriate *figure numbers* and with the name(s) of the author(s), preferably as blue pencil notes in the margin. The respective figure number also becomes a part of the *figure legend* or *caption* (sometimes also called the *underline*). Since legends are always typeset, one might assume they could be treated as part of the text manuscript (typescript), perhaps placed between appropriate paragraphs. Most journals prefer, however, that they be typed sequentially on separate sheets of paper.* In no case should legends be made a part of the figures themselves.

Figure legends require considerable care in formulation. The object is to make them fully self-explanatory, but at the same time brief. A caption should be *self-explanatory* because the impatient "reader" is likely to concentrate on the figures in an article before reading the text. Authors may wish it were otherwise, but the facts demand that every effort be made to ensure that figures and their legends are intelligible even in the absence of a host article. At the same time, their content should encourage "browsers" to read the complete text. The insistence on *brevity* is also a subtle way of reducing the likelihood that figure legends will contain information more properly placed elsewhere. Important explanations, observations, and conclusions belong in the body of the text, not in figure captions.

Every numbered figure must be referred to (in a *figure quotation*) at least once in the body of the article, e.g.:

---

* The reason is that this makes it easier to compare legends and the corresponding figures during the first proofreading stage (Sec. 3.5.5).

... is shown in Fig. 3.
... (cf. Fig. 8).

The word "figure" is usually abbreviated "Fig." as in the examples above. References of this kind should never contain phrases such as "... illustrated in the *following* figure ..."; in printed form the "following" figure may appear on a subsequent page, or it may actually have *preceded* the figure quotation.

Additional guidance for proper and effective design and preparation of figures is reserved for Chapter 9.

### 3.4.4    Tables

Tables also require a certain amount of special consideration. In most cases, every table should be typed on a separate sheet of paper, and all tables should be collected as a group, separate from the main text.* If a table fails to fit normally on standard format paper it may be typed sideways, but no tables wider than the larger dimension of U.S. standard (8.5 in. x 11 in.) or international size A4 (210 mm x 297 mm) letter paper are permissible. However, a long table may be continued over two or even more typewritten pages.

Tables have much in common with figures. Like the latter they must be capable of standing as independent entities, and each should be provided with a *table number* and a brief explanatory text (the *table title*) subject to the same considerations as a figure caption. As with other numbered elements, the author must ensure that each table is *referred to* at least once in the body of an article. Table design and organization are the subject of extensive comment in Chapter 10.

### 3.4.5    Footnotes and Endnotes

*Notes* (more properly: *substantive notes*, as distinct from the *citation notes* related to literature references) are used for supplementary material that falls outside the main flow of ideas constituting an argument. They have aptly been

---

* Separation was long justified by the argument that processing of tables required special skills and equipment. This is actually no longer true in most cases, thanks to computerized typesetting. Consequently, small tables can be incorporated into the body of the text.

described as "the carpet on which a story walks". Properly used, notes are a great help to the reader. They permit the introduction of a certain amount of peripheral information, but in a way that minimizes the disruption of a train of thought. Thus, the reader who is not interested in every detail may, at least on a first reading, ignore altogether anything relegated to a note.

The most convenient notes are *footnotes*, those printed at the bottom of the page. Notes appearing at the end of a document are technically *endnotes*, though the word "footnote" is often used loosely to refer to both types. Unfortunately, in today's rational and cost-conscious publishing world, more and more editors find it necessary or expedient to discourage the use of true footnotes, despite their advantages to readers. The author who has personally suffered the inconvenience of trying to type a thesis containing a large number of footnotes (cf. Sec. 2.2.12) will understand why this is so. Footnotes can be most awkward to deal with, and the problems that arise in their typesetting are, in some ways, even more troublesome than those accompanying typing. To appreciate why, one need only consider briefly what is involved in the final stages of page layout prior to printing (cf. Sec. 4.4.2). Each page must be filled completely, but page breaks must not occur at inopportune places—in the middle of a table, for example, or preceding the last line of a paragraph. Inevitably it becomes necessary that various bits of text be shuffled from one page to another in order to achieve acceptable placement of some uncooperative figure or table. If footnotes are present as well the problems become far more complex, especially since changes are likely to entail alteration of (non-numerical) footnote marks—leading to more work, frayed nerves, and increased costs. Consequently, most journals restrict all notes and literature references to the ends of papers, so that the word "footnote" in its original sense is no longer applicable. Some publishers even refuse to make a distinction between substantive notes and citation notes; notes of all kinds are simply numbered consecutively and printed at the end as uniform blocks (cf. Sec. 11.3.1).

### 3.4.6    A Special Case: The Camera-ready Manuscript

In Sec. 3.2.5 it was noted that many journals publish direct *photo-reproductions* of the textual material submitted by their authors. Especially stringent requirements apply to such *camera-ready* typescripts. Absolutely nothing is done to a piece of camera-ready text by the editor once it is accepted for publication. It is simply photocopied and printed, and the author alone bears full responsibility for the quality and appearance of the finished product. There is no copy editing (cf. Sec. 3.5.4) and no proofreading (cf. Sec. 3.5.6), so any peculiarities in the author's writing style and any errors in grammar, spelling, or punctuation in the original typescript will appear in print. The same obviously holds true for flaws and structural weaknesses in the argumentation.

Ordinary typescripts are double-spaced, but this is not true for camera-ready copy. No extra space is required in which to mark corrections, and it would be foolish for a publisher to reproduce so much blank space. The cost would be exorbitant, and vast amounts of paper would be wasted. Instead, it is customary for such manuscripts to be typed with *one and one-half* spaces between lines. One should look in recent issues of a target journal to find examples of the preferred format of text blocks, and for guidance on the optimal numbers of lines per page and characters per line. It will quickly become apparent, however, that individual articles vary significantly, partly because different authors use different typewriters. Consequently, only average values can be determined by direct inspection. The best policy is to have the journal's "Instructions to Authors" close at hand during manuscript preparation. These instructions may be extremely detailed,* but the recommended guidelines should be followed meticulously.

Typescripts for photoreproduction are normally *reduced* during the production process to 70% of their original linear dimensions. The main reason for doing so is to save space, but it also has the effect of producing a reasonable imitation of printed text. Thus, a *typed* capital letter (typically 2.5 to 3.0 mm in height) after reduction becomes about the size of a standard *printed* capital letter (ca. 2 mm). An additional benefit of reduction is that the finished copy acquires a "clearer" appearance.

---

* The first four paragraphs in the "Instructions to Contributors for the Preparation of Manuscripts for Direct Reproduction" of *Inorganic and Nuclear Chemistry Letters* read, for example: "(1) Type manuscript (one and a half line spacing) on good quality, white bond paper that measures at least 27 x 36 cm. (2) It is imperative that a black typewriter ribbon be used—blue does not reproduce. The typist should ensure a clean, clear impression of the letters. Avoid erasure marks, smudges, pencil or ink corrections and creases. (3) Typing area of page 1 must be 20.5 x 24.5 cm (see instruction 4), typing area for all other pages must be 20.5 x 27 cm. (4) The title should be all in CAPITAL LETTERS, centered on the width of page 1 with the top of the title 2.5 cm from the top of the typing area. After one line spacing, type the authors' names and addresses on separate lines, Capitalizing the First Letter of All Main Words."—Note that "line spacing" is used here in the sense defined in the third footnote to Sec. 5.3.1.

Aesthetic flaws that might be overlooked in a typed document tend to stand out when the same copy is reproduced in a journal or book. For this reason, camera-ready typescripts need to be exceptionally well prepared if they are to reproduce clearly and with a pleasing appearance. Use of a typewriter equipped with a black single-use ribbon and having clean type faces is an absolute prerequisite. The most uniform copy is produced by an electric typewriter (cf. Sec. 5.2.1), but a good mechanical typewriter, properly handled, can also give acceptable results. No erasures are permissible, nor are changes made with the help of commercial correction fluids.* The only way an error can be corrected is by typing replacement text on clean paper and pasting it over the error. However, even this procedure must be very carefully executed (cf. Sec. 5.4).

*Line drawings* (cf. Sec. 9.2) are normally the only figures permitted in camera-ready manuscripts.† Ideally one should paste photographs of one's original drawings directly into the typescript in appropriate spaces left between text blocks. The author's name and the figure number should be marked on the back of each such photo in case it should become detached from the paper. Figures, just like the text, undergo size reduction, so thought must be given to proper scaling of the originals. Lettering, for example, must be sized so that standard capital letters (or numerals) will have a height of 2 mm or a bit less after reduction.

# 3.5    From Manuscript to Print

## 3.5.1    Journal Management and the Editorial Office

We have already seen (in Sec. 3.1.1) that journals are indispensable to science as we know it. Nevertheless, all journals—at least those published in non-socialist countries—must be run on a commercial basis regardless of the fact

---

* Type fails to make a proper impression on paper treated with correction fluid, and the correction may even flake off. A *correction ribbon*, available on most modern typewriters, actually removes letters from the pages, so this technique is allowed (cf. Sec. 5.2.1).

† Figures of the half-tone type are excluded since these require "screening" (cf. Sec. 9.3) prior to reproduction, a complex operation inconsistent with the desire for simplification implicit in the camera-ready journal process.

that they clearly serve a public interest. This generalization includes journals sponsored by scientific societies; societies also operate under budgetary constraints, and they are forced to keep economic losses incurred by their journals to a minimum (or, in fortunate cases, to maintain revenues at a maximum). As a result, some *editorial* decisions ultimately have *economics* as their basis, sufficient reason for our directing some attention to business matters.

We noted in Sec. 3.1.1 that every journal has an *owner*: the *publisher*. Journal ownership is no small matter; the *title* of a journal alone may be worth hundreds of thousands of dollars, and proper management of the assets is essential.

The publisher delegates day-to-day responsibility for a particular scientific journal to an *editor*, who is usually a scientist. Some editors do their journal work on a part-time ("free-lance") basis while maintaining a regular appointment elsewhere: at a university, a hospital, or some other scientific institution. Editors of major journals, however, are usually full-time employees of a private publishing house or of the publishing office of a society.

The editor's name and address* can be found in the *masthead* of each issue of a journal. Unfortunately, the location of this key imprint is not standardized; it may appear on one of the very first or last pages, or it may be on the cover. Apart from the editor's name and address the masthead contains other valuable information, sometimes in very small print, and it warrants careful inspection. For example, from the masthead one might learn that a particular journal maintains its editorial offices in a city entirely different from that of the publishing house.

The editorial office of a major journal is typically run by a *managing editor* in charge of organizational matters, but it is the *journal editor* (the "editor-in-chief") who makes all decisions of a scientific nature.[†] It is his or her responsibility to define the subject area and scope of the journal. These are usually outlined in the "Instructions to Authors", although they may occasionally be elaborated upon in an editorial column in the journal itself.

However, it is misleading to assume that an editor has a completely free rein. Editors are obliged to follow the general publishing and business policies of

---

* Some journals have regional (e.g., *Journal of Chemical Research*) or sectional (e.g., *Spectrochimica Acta*) editors. In such cases, authors should direct any correspondence to the most appropriate editor.

[†] *Staff editors* and *assistant editors* may complete the editorial staff of a large journal.

their journals' proprietors, and most editors also receive guidance from a group called an *editorial board*. The latter is comprised of a select list of scientists, usually with varied backgrounds and interests. They are appointed by the publisher, often for fixed terms, to help supervise the activities of a given journal. Their primary role is to ensure the availability to the editor of a wide range of expertise. It is also expected that their presence will lend a certain amount of prestige to the publication, since the scientific reputation of a board member reflects on the stature of a journal. In addition, board members may be called upon to help in the refereeing of papers (see Sec. 3.5.3), or to perform in other supportive roles. The "Instructions to Authors" of one journal contains the statement, for example, that "Authors may, at their option, submit manuscripts to any member of the editorial board for advice or consultation prior to submission to an editor."

Reputable journals invariably receive many more manuscripts than they can possibly publish. Economic considerations dictate constraints on both manpower available for processing articles and the number of pages that can be produced in a given year. Only manuscripts which meet certain minimum *quality standards* are accepted; the remainder are rejected (cf. Sections 3.2.2 and 3.2.6). It is the editor who must ultimately decide what material is "acceptable", and the editor's position must therefore be regarded as a powerful one.

The fact that a great many manuscripts must be rejected actually carries certain advantages. On one hand, a high degree of *selectivity* can help maintain the quality of a given publication. Beyond this, however, there is the further consequence that quality in research itself is encouraged by imposing restrictions on the flow of dubious, uninteresting, or poorly presented results. In a sense, editors can be seen as "valves" helping to prevent a flood of unreliable or poorly digested information from drowning the scientific community. Most researchers outside the publishing business acknowledge the need for such measures, though they may be reluctant to accept the consequences when their own manuscripts are affected.

The role of the editor as judge has been lucidly described by Day (1983), largely from the perspective of U.S. and British journals, whose editorial practices differ somewhat from those in other countries (cf. also Michaelson, 1982; Gastel, 1983). The editor of a typical U.S. journal normally takes an "all or nothing" attitude toward submitted manuscripts. If an article is accepted, it will move into production very rapidly. Only a minimal amount of formal *copy editing* takes place, and this is usually done by someone without scientific training.* If a submitted manuscript is judged to contain too many flaws in organization or consistency, or if it lacks clarity, it will simply be returned to its author as unacceptable. It is then up to the author to try to make the necessary improvements in the manuscript—or else submit the work to another, perhaps less demanding journal. The details of the process leading to acceptance or rejection of a paper are described more fully in Sec. 3.5.3.

## 3.5.2   Submitting the Manuscript

As soon as an author has selected a target journal and has a manuscript that is *complete* and *clean* (cf. Sec. 3.5.4), the manuscript should be mailed in *duplicate* or *triplicate* (i.e., an original and one or two[†] copies) with an appropriate *cover letter* to the editor (cf. the first footnote in Sec. 3.5.1). One additional complete copy of the manuscript (including text, figures, and all other supplementary material) should be kept by the author as a safeguard against possible loss.

An author's cover letter need not be lengthy, but certain basic pieces of information must be present. Obviously it should request, in a straightforward manner, the publication of the enclosed contribution (identified by title— perhaps an abridged version—and authors) in a particular journal. The letter should also affirm that the submitted manuscript is new and is not being considered for publication elsewhere. A short explanation might follow, introduced by "The reported findings are an extension of earlier work

---

* The nature of copy editing is described in more detail in Sec. 3.5.4. In the English-speaking world it is considered a profession in its own right, and the argument is maintained that it is unnecessary, and perhaps even a disadvantage, for copy editors to understand the texts with which they work. In other countries different traditions prevail. A typical editorial office of a journal or book publisher in Germany, for example, usually employs a number of copy editors and staff persons with scientific training. They not only have the assignment of attending to grammatical and other formal matters but also of improving the technical clarity of submitted texts. It is probably debatable which system is best. The latter approach offers the opportunity for more substantial editorial improvements in submitted manuscripts, but perhaps at the risk of introducing editors' biases, and—inadvertently—of "encouraging" authors to put forth less than their best efforts. The Anglo-American system is simpler from the publisher's standpoint, but authors suffer from the fact that it leads to higher *rejection quotas*—according to one source (Day, 1983), the fraction of manuscripts accepted as originally submitted may be as low as 5%. It probably also results in the printing of more misstatements and even outright errors (i.e., all those that go undetected in the reviewing process, cf. Sec. 3.5.3).

† Some journals request two copies for reviewing purposes (cf. Sec. 3.5.3).

published in ..." or "The essence of this article has been presented as a Discussion Paper at the Z Conference." In the case that rapid publication is desired (e.g., in a letter journal), an explanation of the urgency would be appropriate, perhaps even essential.

For protection, the entire manuscript should be placed between sheets of cardboard. It should be packaged in a strong envelope or a padded *mailing bag* and firmly sealed with reinforced tape. Since a publisher may produce dozens of different journals, the mailing address for the manuscript (as well as for any correspondence) should include the name of the journal. This will help ensure prompt arrival at the proper office. The package is valuable, so it deserves to be sent by first-class mail—airmail if the destination is overseas. (Surface mail often suffers incomprehensibly long delays.)

As implied in the above suggestions for the cover letter, no manuscript should ever be submitted to several journals at the same time. In submitting a manuscript to a journal the author is understood to be giving that journal a *first option* to publish it. Only in the event that it is rejected in the form submitted is the author free to send the same manuscript to another journal.* Hopefully, this step will seldom prove necessary.

Most journals send a dated *acknowledgment of receipt* card to the author confirming the manuscript's arrival and offering a formal assurance that it will at least be considered for publication. The recorded date of receipt at the editorial office can be of some importance, particularly with respect to *priority* claims, and it is usually made a part of each published article. If one is unsure of a journal's acknowledgment practices it may be wise to enclose a self-addressed confirmation postcard with the manuscript cover letter. Should four to six weeks elapse without any word from the publisher an inquiry is justified concerning the fate of both manuscript and acknowledgment.

At this point the scene shifts to the offices of the journal, leaving the author to wait (impatiently) while a series of formal events runs its course. First, the editor or a member of the editorial staff checks to see if the manuscript has been properly prepared from a purely technical point of view. An author's failure to provide the required number of double-spaced copies, for example, or obviously careless preparation of formulas, figures, or tables, may be

---

* The masthead of a journal often includes a statement such as the following, taken from *Inorganic and Nuclear Chemistry Letters*: "It is a condition of publication that manuscripts submitted to this journal have not been published and will not be simultaneously submitted or published elsewhere. By submitting a manuscript, the authors agree that the copyright for their article is transferred to the publisher if and when the article is accepted for publication."

grounds for immediate *rejection*. A more scientific basis for early rejection would be an editor's judgment that a manuscript fails to fall within the subject area and scope of the journal as currently defined. Should this occur the author will be notified promptly. It would then be logical to submit the manuscript to a different journal—this time exercising greater care in its choice (cf. Sec. 3.2.6). If, on the other hand, the manuscript passes the preliminary screening it will be subjected to a systematic *reviewing process*, described in the following section.

## 3.5.3   Referees

Copies of the entire manuscript are normally sent by the editor to one or, more often, two *referees* (or *reviewers*) for their independent evaluations of the quality of the contribution. This "reviewing"* process is highly formalized, especially in the United States and the United Kingdom. Referees are chosen on the basis of their presumed expertise in the area of the research to be reviewed. Often they are members of the editorial board. They remain anonymous, although the author's identity will be known to them. Reviewers' comments are solicited on special forms designed to aid in the evaluation of manuscripts. Comment is requested on such matters as[†]

- significance of the research question or subject studied;
- originality of work;
- appropriateness of approach or experimental design;
- adequacy of experimental techniques;
- soundness of conclusions and interpretation;
- relevance of discussion;
- efficiency of organization;
- adherence to style as set forth in the current "Instructions to Authors";
- adequacy of title and abstract;
- appropriateness of figures and tables;

---

* The term "review" is used here in a sense different from that in Sec. 3.1.2 ("Review Journals"), being closer to its use in the context of a "book review" (mentioned briefly in Sec. 4.5.3). The analogy is not perfect, however, for the latter is prepared *a posteriori*.
[†] These particular (admirable!) criteria were formulated by a committee of the Council of Biology Editors (CBE). Many journals rely on less detailed questionnaires.

- length of article;*
- adherence to correct nomenclature;
- appropriate literature citations.

Referees are requested to submit their recommendations and comments within a few days (typically no more than three or four weeks). If they detect minor problems in a manuscript they are also encouraged to suggest appropriate changes. Every report is expected to contain a clear-cut recommendation concerning whether the paper should be accepted, accepted only after revision, or rejected. The final decision rests with the editor, however (cf. Sec. 3.5.1), and it can be a delicate one, especially if (as frequently happens) two reviewers arrive at different conclusions.

If the referees propose improvements or amendments, the editor will usually write to the author suggesting *modification* of the article along the lines proposed. The author is given to understand that compliance prior to an established deadline will result in a reconsideration of the manuscript. The author receiving such a response has three options available: (1) to submit to the editor an argument questioning the validity of the criticisms, (2) to make the suggested changes—often the wisest choice—or, (3) to withdraw the manuscript and submit it to another journal. If the second alternative is chosen it is important that *all* the requested changes actually be made, and that the specified time limit be met. A new cover letter should be written to accompany the resubmitted manuscript. It should contain explicit reference to all previous correspondence and call attention to the various modifications made since the original version was submitted. In all likelihood the editor will accept the revised manuscript and initiate the process of transforming it into a printed article.

## 3.5.4    The Edited Manuscript

Once a manuscript is *accepted* for publication it must be *typeset.*† Figures sometimes need to be redrawn, and reproduction proofs, films and offset

---

* One reviewing form ("refereeing report") subdivides its inquiry on length into the following two questions: "Is the manuscript sufficiently concise? Is the experimental part nevertheless detailed enough to render the experiments easily reproducible?"
† The term "typeset" is somewhat of a misnomer since publications today rarely rely on "setting (movable metal) type" in the traditional sense. This technique has largely been replaced by highly automated photo-offset technology.

plates must be prepared before printing can begin. The first step in the process is called *editing* or *copy editing*.* This consists of making technical corrections in the manuscript and adding various editorial markings. The art of editing and copy editing has been expertly described in many books (e.g., *The Chicago Manual of Style*, 1982; Bennett, 1970; Butcher, 1981; Judd, 1982; Freedman and Freedman, 1985) and we will not dwell on it at length since authors of journal articles have little direct contact with it. Book authors, on the other hand, can expect to do a certain amount of their own editing, hence the techniques involved are reviewed briefly in Sec. 4.3.7.

Publishers and editors expect the journal article manuscripts they receive for copy editing to be in "final" form. That is to say, manuscripts should be as clean and as nearly perfect as possible. Every submitted manuscript should have been carefully checked for errors and be utterly free of hand-written corrections. The demand is not an unreasonable one, particularly since a manuscript for an article rarely exceeds a dozen pages.[†] Only special symbols not available on a typewriter should be hand-written.

Good copy editing demands very careful *reading*, not only for the purpose of spotting errors but also to ensure *consistency* in the text. Copy editors are free to make whatever changes they think necessary with respect to *grammar*, *spelling*, and *punctuation*. They are also expected to verify the correct and consistent use of terms and symbols, and to see that stylistic idiosyncrasies characteristic of the journal have been observed throughout. A case in point is use or omission of periods at the ends of figure captions, or punctuation

---

* The verb "to edit" has several meanings, as has the noun "editor". "Editing" can be taken to include any of the activities and responsibilities described in Sec. 3.5.1 and Sec. 3.5.3 for journals and in Sec. 4.1.3 for books. Work performed on an actual manuscript at the editor's office, or under the editor's supervision, is usually called "copy editing", and is carried out either by specially trained and experienced staff members or by free-lance agents (cf. the last footnote in Sec. 3.5.1). Moreover, a scientist who invites other scientists to engage in a joint effort to write a book under his or her guidance is also an "editor".

† Preparing "perfect" copy is particularly easy for the increasing numbers of scientist-authors with *word processors* at their disposal (cf. sections 4.3.5 and 5.2.2). Book manuscripts are not subjected to the same rigorous expectations as journal manuscripts, primarily because of their greater length. Editors realize that it is unrealistic to insist that a book manuscript, running to hundreds of pages, be "spotless". Nevertheless, if corrections appear they are expected to be made according to a standard set of conventions discussed in Sec. 4.3.7.

accompanying *quotation marks*.* The work of a copy editor could be described as *production-oriented*. That is to say, it is done mainly for the benefit of the production staff, even though the copy editor is likely to be working in the editorial office.

Once a manuscript has been edited it proceeds to the printshop, where certain of the *markups* introduced by the copy editor are translated into instructions for the printer. These include directives regarding *type size* (e.g., in section headings) and *type style* (e.g., for highlighting keywords), and guidance on such aspects of the general *layout* of the text as spacings and indentations. (For a given journal such details will be handled as a matter of routine and always in the same way, at least over a prolonged period of time.)

Mathematical equations in text are a potential source of major problems for both editors and printers (cf. Sec. 8.1). International and national agreements and standards (IUPAP, 1978; IUPAC, 1979a; Standard *ISO 31/0-1981*; Standard *DIN 1338, 1977*; Standard *DIN 1338, Beiblatt 1, 1980*; Standard *DIN 1338, Beiblatt 3, 1980*) specify that symbols representing mathematical and physical *quantities* should always be printed using distinctive slanted letters (*italics*), as discussed in Sec. 7.1.3. A typesetter has no means of knowing which characters in an equation or formula stand for quantities and which do not. Careful and intelligent marking at an earlier stage is thus essential. Usually this entails annotating each *quantity symbol* with a short underline in colored pencil (green is often specified), one character at a time. It is unlikely that a paper on synthetic chemistry would contain many such symbols, and an experienced copy editor would probably recognize those few that occur. The situation may be very different with a theoretical paper, however. Here—as in papers dealing with mathematics, physics, and the engineering sciences—at least some understanding of the subject matter comprising the text is necessary to mark the copy correctly. Typically the journal editor must personally find the time to do the job.

Editing in the broadest sense may involve altering any of those aspects of an article addressed by the reviewers (see Sec. 3.5.3), although major

---

* With respect to the latter point there are two schools of thought: the *conventional* (or *traditional*) and the *logical* (Fowler, 1984). The conventional approach requires inclusion *within* quotation marks (single or double) of all commas and periods, regardless of origin (cf. *The Chicago Manual of Style*, 1982), on the grounds that this confers the most pleasing appearance on a piece of text. The logical approach, as its name implies, punctuates according to sense. That is, commas and periods are set *outside* quotation marks except when they actually form part of the corresponding quote. The American Chemical Society follows the logical system (cf. Dodd, 1986), and we have elected to do likewise.

manipulations are more appropriately termed *editorial changes* (or, occasionally, *revisions*). An editor who sees the need for substantial changes will almost certainly want to have the author's reactions prior to making the alterations and subjecting the manuscript to further steps in the production process.* After inspecting the changes the author may want to negotiate controversial points, either in writing or by telephone; if not, the amended manuscript can simply be returned to the editor with an indication of approval.

## 3.5.5 The Typeset Contribution

An edited manuscript containing all necessary markings is ready to be *typeset*. "Proofs" (sample printings) of all material for the emerging article are prepared, and copies are sent to the author for approval, usually along with the edited manuscript. These first proofs, the *galleys* (or *galley proofs*), consist of text blocks arranged in a single column and printed with wide margins and temporary pagination. Tables and figure captions are often printed separately from the text, and chemical formulas and figures always are. The author may not even receive galleys of the latter parts if the originally submitted materials will simply be reproduced photomechanically.

It is the author's responsibility to "proofread" (cf. Sec. 3.5.6) the entire set of materials, make corrections where needed, and return the corrected proofs to the address given in the accompanying cover letter. This may, incidentally, not be the address from which the proofs were sent: they may have been dispatched from the publisher's production department, or from an independent printing house, while the author will be requested to forward all corrected material to the editorial office. The editor then verifies that all needed corrections have been properly marked and no obvious errors have been overlooked, and that no unwarranted corrections (not to mention changes or additions!) have been ordered. Then the corrected galleys are passed back to the printer.

---

* A case in point is the redesigning or renumbering of chemical formulas. Particularly in a long review article, minor changes of this type can sometimes greatly reduce production costs. Editors themselves may well be prepared to undertake the necessary work, particularly in the case of an invited manuscript, where it would be unreasonable to make excessive demands on a "guest" (the invited author).

The editorial office keeps records of the dates of all the above movements of a manuscript and initiates appropriate action if it appears that *deadlines* are not being met. Since each issue of a journal must go to press on a specific date, delays cannot be tolerated. Authors are urged to return their corrections within a specified time period (e.g., ten days). If this deadline is not met the editorial staff may undertake corrections on their own without further consultation with the author, whose lack of response is interpreted as forfeiture of the right to pass judgment.

A second set of proofs, the *page proofs*, is next prepared. These are intended to represent the article in its final form. All text, formulas, figures, tables, and notes will have been put in their correct places, and final page layout decisions made. The editorial staff verifies that the arrangement of the pieces is reasonable and that all corrections on the galleys have been properly incorporated, that pagination is satisfactory, and that preceding and subsequent articles are properly joined to the text. As soon as all material for one issue of a journal is assembled in final form at the printer's, plates (usually *offset plates*) are prepared, and *printing* begins.

Authors are usually not involved at the page proof stage.* Science journals have firm publishing deadlines to meet, and their editors are unwilling to risk unpredictable postal delays connected with sending out page proofs. Experience has also shown that proofs have a way of arriving just as the author has set out on a round-the-world lecture tour—but publication cannot wait! In any case, most authors lack the experience needed to judge, for example, whether a manuscript has been broken into columns and paragraphs in the optimal way, or whether tables and figures have been set in the best possible places, and these are the principal issues to be addressed.†

Often the contact between editor and printing house is indirect and managed by the publisher's production department, which in turn assumes some of the responsibilities described above.

---

* On the other hand, some journals have a policy of sending *only* page proofs to their authors. In the past, a few that did so then insisted their authors return them within 48 (!) hours. We are unaware of any current instances of this practice among scientific journals.
† The situation is different in the case of a book, where the author assumes greater responsibility. Book authors are always asked to examine and approve the corrected page proofs.

## 3.5.6    Proofreading

### Learning to Be Critical

We return now to the subject of *proofreading*, touched upon in the previous section. Proofreading is the art of detecting, marking, and correcting errors in typeset text. The approach and methodology are quite distinct from those applicable to copy editing, and it is very exacting work. Each of the thousands of printed characters comprising an article must be regarded as suspect, and each must be examined with care. It is the job of a meticulous proofreader (the author!) to make sure—not just hope—that every letter, every symbol, even every space is set correctly.

*Reading* in the usual sense is inappropriate for error detection. Instead *close inspection* is required. Unfortunately, the odor of print seems to act hypnotically on many authors, reducing their level of alertness. In some ways, an author is the worst possible person to ask to proofread a manuscript. Someone less familiar with the text is less likely to be blinded by foreknowledge. An outsider cannot read the proofs superficially, seeing what "ought" to be there, but instead constantly needs to make comparisons between the proofs and an authoritative version of the text. The author, by contrast, knows all too well how each sentence should read, the way the various abstruse terms are to be spelled, and what factors and symbols belong in the equations. Authors with good visual or phonetic memory need to be especially cautious because they know large parts of their text literally by heart. They are almost certain to find it an excruciatingly boring and therefore difficult task to concentrate on the manuscript anew. Indeed, scientists in general tend to be rather careless proofreaders. Their professional training has prepared them to capture at a glance the meaning of an entire sentence, while proofreading demands that they read in a totally different way.

Some research groups counter all these obstacles by insisting that proofs of submitted articles be read by *two* members of the group simultaneously. One reads the typeset text aloud, word by word, enunciating all punctuation marks, while the other follows along in the manuscript. This method is commendable; if it were followed everywhere journals would contain fewer typographical errors.

Even if the luxury of two readers is not available, proofreading must be done with the (edited) manuscript close at hand. It is particularly important that all printed *symbols* and *numbers* be compared directly with those in the original text, because these are a common site for errors. Moreover, such errors may be especially serious, particularly errors in numbers.* Proofs should be read in conjunction with either the author's unedited copy of the manuscript, or, better still, with the edited version, if this is available. Edited manuscripts in the author's possession must always be returned intact to the editor, usually packaged together with the corresponding proofs. They should under no circumstances be further changed or "corrected" because they may be required as "evidence" in subsequent discussions with the publisher or printer (e.g., if an author receives an unexpected bill for costs supposedly incurred as a result of excessive changes made at the proofreading stage).

Errors detected in a set of proofs should always be indicated[†] with standard *proofreader's marks* applied to the erroneous word, letter, symbol, or space. The corrections themselves are written by hand in the margin. Proofreader's marks are shorthand symbols with meanings understood by all typesetters. Only if these symbols are used (and used properly!) can it be assumed that the changes made later at the composing machine will be those the proofreader had in mind. A proofreader should *never* write a correction directly *within* a text block, because it is likely that such a mark would be overlooked by the typesetter. This admonition is in sharp contrast to the advice given with respect to manuscript correction *prior* to typesetting, as discussed more fully under the subheading "How Clean is Clean?" in Sec. 4.3.7.

All handwritten proof markings must be perfectly legible, because a typesetter cannot be expected to be familiar with technical terms and usages, and is not paid to decipher hieroglyphics. Printed block letters are preferred over cursive script for corrections. Often the latter is explicitly reserved for instructions to the typesetter (i.e., for comments that are not to be typeset), allowing these to be more easily distinguished.

Corrections should always be kept to a minimum and must be limited to the removal of *errors*. Any changes in wording or style at this stage of production would be costly and might result in substantial charges levied against the

---

* An incorrect number (e.g., "5.0 g" instead of "0.5 g") could easily destroy the validity of a reported procedure or even make it hazardous! Furthermore, errors of this kind are far less easily recognized than typographical errors in words, and they can, if undetected, easily pass into the "canon" of science.

[†] A colored ballpoint or felt tip pen, preferably red, is useful for these markings. It should be pointed out, however, that some journal and book publishers—especially in the United States—reserve to editors and in-house proofreaders the right to use ink for corrections. If this policy applies, authors are cautioned to use only soft pencil. Proofs sent to the author may already contain correction marks introduced by a proofreader at the printing plant.

author. *Review articles* are an exceptional case; here an editor may permit minor additions of new information even after proofs have been prepared, although every effort is made to add such amendments only at the ends of paragraphs or in new paragraphs. Late additions to articles for primary journals are accepted only with great reluctance. Such material is usually printed in small type at the end of an article as a "Note added in proof".

## *An Excursion into the Realm of Cryptic Symbols*

Examples of typical proof errors with the corresponding proofreader's corrections have been collected in the form of a table in Appendix I. Proof-correction shorthand is unfortunately not consistent from country to country, a circumstance made the more regrettable by the increasingly international character of the modern publishing industry. Two different correction systems are shown in two separate columns in Appendix I. One is headed "American" and contains those symbols traditionally used in the United States. The other is designated "Alternative". Here we show proof correction techniques employed in continental Europe, especially Germany.

The examples given in Appendix I are limited. No attempt has been made to cover all the errors that may occur in a typeset text, nor is the full range of technical markings familiar to professional proofreaders and typesetters shown. Nevertheless, the table should provide the author with guidance sufficient to deal with the most common problems, and the range of examples will certainly demonstrate the kinds of strategies required for making intelligible corrections.

An example of the versatility of the symbols is provided by the *circle* ("American"), which is roughly equivalent to the *double parentheses* symbol found in the "Alternative" column.* A proofreader uses a circle (or double parentheses) to house *instructions* to the printer; e.g.[†]

- wrong font
- set in small capitals
- exponent
- subindex

---

* Double parentheses have one advantage over circles: they can be used to house longer comments. Copy editors use this symbol as well, employing it to convey messages to the author, for example.

[†] Recall that such instructions are usually written in normal cursive script rather than the block letters used for corrections.

- align type
- straighten line
- put in center of line
- reset
- less space between words
- equalize spacing
- insert comma
- insert omitted text
- see manuscript
- asterisk
- damaged

The "Alternative" system (described fully in Standard *DIN 16 511*) is somewhat easier to learn than the "American" system, and its use requires less explanation.* Cryptic symbols are largely avoided, the major exception being the "delete mark" or *deleatur* (from the Latin word meaning "to be deleted").† This particular symbol is also used by American and British publishers, though often in a form that bears little resemblance to its origin (a stylized letter "d"). All the other marks in the "Alternative" system are of a generalized nature, allowing them to be used quite freely in a variety of ways. For example, marks such as:

may all be employed at will to specify various types of corrections.

Once a given mark has been applied to the text (as a *textual mark*, or *mark in proof* ) it is repeated in the *margin* at a point as close as possible to the line containing the error. It is here—and only here!—that the correction proper is executed (assuming the mark is not self-explanatory).

---

* We are of the opinion that the "Alternative" system of proofreader's marks has sufficient advantages, including lack of ambiguity, to warrant its use whenever possible.

† Apart from its usage shown in the examples in Appendix I, this symbol, coupled with an "add" symbol at the end of the preceding (or beginning of the subsequent) line, can be used to correct *end-of-line hyphenation*.

## *A Few Details*

If a large number of corrections is necessary, it is important that markings be carefully chosen. They should include as much variety as possible to minimize the likelihood of confusion. If one particular correction must be made repeatedly within a given proof sheet it is often possible to use a single marginal note to specify the change. The symbol is then augmented by the appropriate number, and the whole is set off by double parentheses or a circle. If individual digits within a *number* require correction the whole number should always be replaced. Special care should also be exercised when correcting *mathematical expressions* and *equations* (formulas). A typesetter is more likely to understand what is intended, for example, if the correction is first shown in the conventional way, and the entire expression is then *rewritten* in the margin, wrapped in a curved line open to the edge of the page.

Correction marks must always be placed accurately within the text. Absolutely no doubt should be left concerning the character(s), punctuation mark(s), or space(s) to be altered or deleted. Many proofreaders adopt the policy of specifying the deletion of one more than the requisite number of characters and showing by the marginal correction the fact that this character is actually to be retained, followed perhaps by a space.

All components of the manuscript must be proofread, even the most minute footnote to a table. Any *figures* that have been redrawn under the editor's supervision need to be checked with special care to make sure that the artwork has been cleanly and accurately done and that all lettering is correct. The author is usually supplied only with a photocopy of the original figure, so corrections can be made without fear of damaging valuable figure copy. *Tables* should be carefully scrutinized to be sure that all entries are correctly spaced in their proper rows and columns, and that decimal points are aligned (cf. Sec. 10.2.2). It is especially important to ensure that subtle mistakes have not crept into quantitative data during typesetting. Finally, the author should make a complete *photocopy* of the edited proofs and store it in a secure place as a safeguard against loss of the original.

When the authors and their partners in the editorial office have finally completed the tedious, meticulous work of two stages of proofreading they can enjoy the prospect of soon seeing in print another first-class journal article.

# 4    Books

## 4.1    Preliminary Thoughts

### 4.1.1    What is a "Book"?

Everyone associates a set of images with the word *book*, but it may come as somewhat of a surprise that even professionals in the publishing industry would be hard pressed to supply for it a concise, comprehensive definition. Books are incredibly varied in their form, origin, content, purpose, and availability. For example, consider a few of the descriptive terms applicable to various kinds of scientific books: textbooks, laboratory manuals, problem supplements, monographs, dictionaries, compendia, programmed learning materials, reference works, handbooks, encyclopedias, continuing series, loose-leaf supplements, spectra collections. The list could be made much longer, and its breadth is astonishing.

One observer resorted to describing a book as "a complex repository of meaning", which may come close to all that its various manifestations can be said to have in common. Technological advances ("non-print media") have complicated the picture further: it would seem strange if a data collection were *not* a "book" simply because a computer had transferred the information to a set of microfiches now housed in a filing cabinet.

From an editor's point of view the most important distinguishing characteristic of a book is that it is an *independent publication* of its authors. In other words, an author of a book, or even the editor of a collective work, is, in the bibliographic sense, an "independent agent". Obviously this does not mean that a book author produces an entire work without any help or encouragement, simply placing a finished manuscript on the publisher's desk; nor should one infer that the same author, if he or she were to write a journal article, would thereby become more "dependent" on someone. What it does signify is that the author of a book enjoys greater freedom (coupled with greater responsibility!) than the author of a journal article, and at the same time establishes a different and more direct relationship with the publisher.

## 4.1.2    When is Publishing a Book Justified?

What *motivates* a scientist to write a book in the first place? Many factors can be involved, but the most important should be that the writer has ideas he or she wants to share, ideas that will be of significant value to a specific readership. Recognition of the need for a book on a hitherto untouched subject may be sufficient motivation. At the same time there is no point in denying that most book authors are interested, at least to some extent, in establishing or enhancing their professional reputations. Finally, unlike the author of a paper, the author of a book has an *economic* stake in the venture. The book will be "marketed", and its writer will receive a share of the income.

Financial considerations play a central role in the life of the publishing house that will produce the book. A publisher may see serving the public as its primary mission, but if the firm is to survive, the books it produces must generate earnings. Publishers thus place a high premium on careful and realistic *planning*. They attempt to forecast as accurately as possible, and on a project-by-project basis, both probable *costs* and potential *revenues*. The first step in making such a forecast is to assess the size and nature of the *target market* for a given book; i.e., the publisher tries to determine what people or institutions might be induced to buy it. This is important, because the market must be large enough and the customers sufficiently interested and affluent to more than cover all publishing and marketing costs (unless, of course, a benefactor is willing to underwrite losses on the project).

Once the market has been identified and characterized, various parameters related to the book itself must be established. One of these is a reasonable *printrun* (the number of copies to be produced)—"reasonable" based on the publisher's past experience with similar works. Author and publisher then need to reach a fairly firm understanding about the *size* of the book (in number of pages) and its *format*. This is not always easy; what may seem most desirable to one of the parties might not be regarded by the other as feasible.

At this point it becomes possible to extrapolate the book's minimum *selling price*, taking into account a modest profit margin. The price calculation is crucial, because a book project can be regarded as realistic only if the potential price falls within the range prospective buyers would be willing to pay, and is *competitive* with that of any comparable works.*

It is clear, therefore, that the decision to produce a book is not one made by an author alone. A *publisher* must agree that the idea is a good one. Whether or not agreement is reached often hinges on factors in addition to theoretical marketing and production considerations. For example, the publishing house will consider carefully whether or not it really has the ability to forecast and, later, to service the market in question. An experienced publisher will also try to assess a potential author's abilities as a *writer*, in addition to his or her level of *expertise* in the field to be treated and *record* of relevant past accomplishments. Even an author's *renown* ("star appeal") is a factor, as is the author's ability to be *persuasive* in presenting the case for a project.

The actual decision-making process and the roles and strategies of the participants will vary somewhat from publisher to publisher, but the principles outlined above are fairly universal. It should thus come as no surprise that many book proposals are ultimately rejected. Rejection can be for any number of reasons, including ones prospective authors may be unprepared to accept.

Not all book projects are proposed by authors. Some are conceived and eagerly pursued by publishing houses themselves, but even these sometimes fail to materialize. Often the scientist with the expertise needed to write the envisioned book is not prepared to accept the challenge.

The following discussion will proceed from the optimistic premise that the idea for a book has been accepted by a publisher and the book's production is assured. This will give the reader the opportunity to follow it from its inception as nothing more than a vague idea in a scientist's mind to its realization as a bound volume on a book dealer's shelves.

## 4.1.3    Authors and Publishers

As we have seen, a proposed book can become a reality only if the prospective author establishes a satisfactory partnership with a publisher. This step should *precede* the drafting of any significant amount of text.† But what publisher

---

* A publisher faces very similar considerations with respect to journals, but only in the case of a book is the author directly involved in the process.
† Interestingly, the situation is quite different in the humanities, for example, where it is rare for a publisher to show serious interest in a project before examining a completed manuscript.

should the author approach? The answer to this question will involve many of the same concerns raised with respect to journal articles (cf. Sec. 3.2.6). Probably the best advice is to seek a publisher with an established reputation in the field of the envisioned book. A publisher with appropriate expertise in both technical and marketing matters is invaluable. General knowledge or even hearsay may serve as a starting point in the search for a publisher, but it is usually more fruitful to browse through current *catalogs* of various publishing houses for recent book titles one recognizes and respects.

A publisher's reputation and position in the marketplace are not the only considerations, however. Final choice of a publisher should be withheld until one has actually met those people at the publishing house with whom it will be necessary to work. A good author-publisher relationship requires mutual respect and confidence. The parties will often need to consult with one another, and compromises will occasionally be necessary when disagreements arise. This means that personalities can be of considerable importance.

A *senior editor* (also called *publisher's editor*, *house editor*, or *editorial director*) is the author's chief contact at a publishing house, especially during the planning of a book. The senior editor recommends what books should be produced and what authors engaged. In fact, it is often a senior editor rather than an "outside" scientist ("outside" from the perspective of the publishing house) who provides the initial impetus for publishing a book. For this reason, the more descriptive term *acquisition editor* is occasionally used. The role of a senior editor is somewhat analogous to that of the editor (or editor-in-chief; cf. Sec. 3.5.1) of a journal. Compared with their journal counterparts, however, senior editors are much more involved in the overall publication process, and they have considerably more contact with their authors.

In addition to the senior editor the author will also have occasion to work with the *director of publications* (the "publisher" in a more restricted sense of the term), the *production director*, and the *marketing director* (sales manager), as well as their associates. One should take the opportunity at an early date to talk with as many of these team members as possible.

Not all the author's partners on the production side will necessarily work in the firm's own offices. For example, a publisher will sometimes find it useful to refer a given book project to an *out-of-house* editor, whose responsibilities may be more or less comprehensive depending on the circumstances. A typical case is a book designed to consist of a large number of separate contributions; preparation of such a work would normally be coordinated by a *volume editor* (or simply *editor*). This editor—usually an esteemed scientist—would assume responsibility for defining the topics to be treated and assigning their writing to specific authors, and would then serve as the link between authors and publisher. A book that will be part of a *series* usually falls under the jurisdiction of a *series editor*, partly to ensure that the new volume's style and approach is consistent with others in the same series. Some publishers also make use of one or more free-lance *editorial consultants* to initiate first contacts with prospective authors.

An author writing his or her first book will find there is much to learn about the mechanics of publication. One's first job after selecting a publisher should be to find out as much as possible about the organization and operation of the firm and develop a clear understanding of every step in the processing of a book prior to its printing and distribution. Most find the experience an interesting one, and it certainly reduces the likelihood that serious misunderstandings will arise.

## 4.2    The Planning Stage

### 4.2.1    Provisional Table of Contents and Tentative Preface

As soon as a firm contact has been established with a publisher and the general nature of the book has been agreed upon, the author should begin drafting a *plan* for the book, a sketch known as a *content summary* (also called a *chapter outline*, or *disposition*). What was initially conceived only in broad terms now begins to take concrete form.

First, the necessary *chapters* and their major subdivisions (the *sections*) should be identified and assigned tentative names and numbers. Detailed outlining of the individual sections can wait, although it is useful to jot down on the summary appropriate *topical keywords* to act as reminders during the writing process. No logical order is necessary for one's preliminary notes; they are only intended as "raw material" for later assembly. Much of what is planned now is likely to undergo radical change as the manuscript develops. Nevertheless, a general pattern for the book will begin to emerge, and the summary itself will much later form the basis for a *table of contents*.

The next step is to estimate roughly how many pages of manuscript (and, ultimately, of printed text) each chapter and section will contain. The *total length* of the book will already have been established in discussions with the publisher, but now the individual chapters need to be *weighted*. Those covering the most intellectual ground deserve the largest numbers of pages, and rough estimates of available space should be known when the time arrives to begin writing.

The publisher will want to see at least the author's preliminary content summary. Some publishers also pose certain formal questions to their authors at this point, using a document called a *Project Evaluation Form*. Typically, the questions center on a book's projected readership and probable contents. If the publisher makes a practice of having publication projects evaluated by outside consultants a different form may become important: a *Referee's Report*. With its help the publisher will attempt to obtain independent judgment as to a realistic printrun number, a reasonable list price, and any existing literature that must be regarded as seriously competitive.

More surprisingly, perhaps, the publisher may also request that a *tentative preface* be prepared and submitted. In the preface a book's author speaks directly (but briefly) to the readers, describing his or her intent in writing the book. Why and for whom was it conceived? What is it about? What makes it unusual and worth reading? Readers are interested in the answers to these questions, but so are publishers because of the effect the author's views may have on potential buyers.

After reading the chapter outline and the tentative preface the publisher may request that changes be made. Perhaps prior experience in the marketplace will suggest that the author's approach is not right for the intended target audience, or that the planned *level* of presentation—something to which the preface should at least allude—is inappropriate.

## 4.2.2     Sample Chapter

If the initial outline appears promising, the publisher is likely to ask for a *sample chapter* (*specimen chapter*). The sample text (usually about a dozen manuscript pages) should be very carefully prepared. The content should also be chosen with care so it is as representative as possible of the entire book. For example, if figures, tables, equations, and formulas will occur frequently in the book, these should also appear in the sample.

The author need not feel overly constrained by established "norms" in preparing a text sample: each book is unique and each is expected to have its own peculiarities. Nevertheless, it is important that *internal stylistic consistency* be maintained. Consistency is most easily achieved if, in the course of writing the sample chapter, the author begins to formulate a working list of *conventions* to be followed—rules for using hyphens, for example, or notes concerning which of two alternate spellings will be adopted for certain words. Such a list is handy to have available, and it ought to be shared with the senior editor for additional suggestions.

The importance of writing a sample chapter cannot be overstated. In the first place, it gives the author a true taste of the awesome task that lies ahead. Books are not as easy to write as one might suppose in a rash moment of inspiration. At a more mundane level, a text sample allows publisher and author to begin discussing and determining such points as *style*, *terminology*, and *page layout*.

The chief reason for preparing a sample chapter, however, is that it gives the publisher some substantive evidence of the probable character and content of the anticipated book manuscript. Now is the time to uncover any major problems; it is both awkward and time consuming to make drastic or fundamental changes in a large book manuscript once it is completed. As a case in point, consider the risk associated with a manuscript to be prepared by an author whose native language is different from that intended for the text. The senior editor would be most anxious to find out, from a few sample pages, to what extent provision might need to be made for linguistic editing (*language polishing*).

Finally, a sample chapter is useful to the publisher for a technical reason: it can reveal unusual problems that may accompany typesetting or artwork preparation, and this permits refinement of production cost estimates. For example, if a book contains an exceptionally large number of mathematical symbols or formulas, publication is certain to be complicated and costly.

Author and publisher will want to take every possible precaution to ensure that, when it is completed, the manuscript can be *adapted* for publication with the least possible effort and friction. Every last-minute *modification* or correction requires an investment of time by both parties, and time is money. The later in the production process changes are made, the higher the costs. Furthermore, extensive revisions in a manuscript can delay publication by weeks, or even months. If the timeliness of the work is impaired, sales may be reduced, and the author's royalties will be correspondingly curtailed or delayed.

### 4.2.3    Sample Printing

Publishers often arrange to have at least a portion of the sample chapter converted into *sample printed pages*. This involves having the senior editor first annotate the appropriate manuscript segment with various instructions, and then pass it on to the production director, who supervises setting the text in type.

Typesetting will be facilitated if the set of sample pages is accompanied by a clear list of required special *symbols* to help avoid errors. The list may well form the basis for a *List of Symbols* to appear in the book itself, usually placed conspicuously either at the front or the back. The *nature* of the list is also important, because it may influence the publisher's choice of typesetting procedure or typeface.*

From the sample printing the author will be able to gain a first impression of the appearance of the finished book.† The sample may clarify aspects that warrant special attention during the writing process, such as the size and placement of figures, the best treatment of equations, or the optimal length of paragraphs.

### 4.2.4    The Publishing Agreement

After all the preliminaries have been settled, and provided management is still convinced that the project is sound, the publisher will submit to the author a formal *publishing agreement* (*publication contract*). This is a legal document outlining the rights and responsibilities of the parties involved, and it is designed to be binding within the framework of *copyright* and *publication rights* legislation (see Appendix C). The nature and significance of publishing agreements is the subject of the following comment in a very readable brochure for authors by Sowan and Horwood (1983):

---

* In certain typefaces, for example, there is virtually no distinction between the numeral "1" and the letter "l" (typewriters frequently use the same character for both). The printer should be notified promptly if this kind of ambiguity presents a potential problem.

† One speaks in this context of the book's *layout*, meaning the overall optical impression conveyed by the typographical arrangement, the type face(s) and size(s), and the organization of headings, tables, figures and other special features.

"... (a publishing contract) is a document of mutual trust and undertaking, whereby the author and publisher make an honourable compact. It is generally standard throughout the industry, and represents the fruits of long-established experience of working together. Its wording may seem unnecessarily legal, implying mutual distrust, but this is not in fact so. Any clause will be explained or justified by the publisher if it is misunderstood."

Publishing agreements are so highly standardized (often consisting of computer-processed standard text fragments, or even printed forms) that some authors bristle when they receive them ("What has happened to my cherished individuality?"). The reaction is uncalled for. Use of a form is in no sense meant to demean a book's (or its author's) uniqueness. On the contrary, it should be regarded as evidence that both book and author will be accorded fair treatment, treatment equal to that given all others.

In the contract, the proposed book will be identified by name, but this is generally understood as only a *working title*. The final choice of a title is a delicate matter best deferred until later (cf. "The Subject Title" in Sec. 4.5.2). The contract also specifies an approximate *length* for the work, consistent with the verbal agreement already reached. Normally—and this also will be stated— deviations by as much as 10% from the specified length will be tolerated, but anything beyond that must be with the written approval of the publisher. The reason for such a clause is that manuscripts have a tendency to grow beyond the bounds originally envisioned by their authors. Such an enlargement can seriously upset a publisher's marketing calculations, because a longer book means higher production costs. This dictates a higher list price, and opens the possibility of reduced sales, particularly if the target market is highly cost-conscious.

Another important clause in the contract defines the book's *authorship*, noting in particular whether it is to be single or multiple. *Royalty* terms are also stated, and if more than one author is involved the contract stipulates how the royalties will be divided.

Finally, the publishing contract establishes a *deadline* for submission of an acceptable manuscript. It is not unusual for a publisher to allow two or more years for preparation of a complete manuscript. A deadline so far in the future may seem absurdly generous to the novice author, but in fact it is probably quite realistic. Serious writing should begin at once and should be pursued as rapidly as possible, because the job is almost certain to prove far more difficult and time consuming than the author realizes.

A typical publishing agreement contains many clauses in addition to those discussed above, but this overview will suffice for the present (see Appendix C for further details, however).

## 4.3    Preparing the Manuscript

### 4.3.1    Becoming Organized

We have already stressed that anyone undertaking to write a book (at least if it is to be a good one!) must be prepared for a vast amount of work. It may seem that we are belaboring the point, but the decision to sign a publishing agreement should not be taken lightly. Far more time and dedication are required to write a book than to write a journal article—or even several. A "full paper" in a journal averages perhaps ten to twelve typewritten pages, with possibly a few figures or tables, while a book manuscript usually consists of hundreds of pages. The amount of reference literature that must be processed can be equally enormous, and it may include some with which the author is quite unfamiliar. Successful completion of a book manuscript can easily extend over years—especially since the average scientist is of necessity a part-time author. One lesson to be learned from these deliberately sobering comments is that aspiring authors must *organize* their work well if they are ever to see it come to fruition.

Writing is an intensely intellectual process. Consequently it is highly individualized, and it would thus be pointless for us to attempt to devise a set of "writing rules" and claim that they would guarantee success. There are, however, a few *organizational* matters every author should attend to before beginning to write in earnest.

First, one should take the time to select a comfortable, private place to work, where all the necessary tools—pencils, paper, typewriter, key reference books, notepads, literature collection (cf. Sec. 4.3.2)—are readily at hand. Serious thought should be given to the possibility of acquiring a word processor for reasons elaborated upon in Sec. 4.3.5. It is also important that one establish a daily (or weekly) schedule in which regular *blocks* of time are available for undisturbed writing.

As the manuscript develops, each new section—whether written at home, in the office, or on vacation—should be carefully *stored*, preferably in a file folder or a loose-leaf notebook. Even one's first drafts should be neatly filed and preserved, not just the later, more polished versions. Not all "polishing" represents progress, and a time may come when it would be useful to consult earlier work in search of passages that may have been rejected prematurely.

In the beginning it is difficult to predict when particular chapters will actually be drafted, or even in what order, and different chapters will certainly be at different stages of development at any given time. For this reason it is best to treat chapters independently of one another, designating a separate *file* or binder for each, and numbering the emerging pages separately within each chapter. Every draft should be *dated*, and dividers should be placed in the files so that earlier versions of a chapter are clearly separated from later ones. Figures should also be stored by chapter, but separately from the text.

At the very outset it is important to project a *timetable* for the work, and to begin keeping a *diary* of one's progress, comparing it periodically with the goals outlined in the timetable in order to ensure that the contract deadline will be met. To be realistic, the timetable should contain ample provision for coping with unexpected intrusions, but only a real emergency should be allowed to interfere with projected target dates.

If the actual pace of one's work makes it clear that the original deadlines cannot be met it is best to discuss the situation promptly with the publisher. Conscientious observance of deadlines is particularly important for books with *multiple authors*, because the work of the entire team is adversely affected if one member falls behind. Deadlines related to *handbooks* and other volumes issued at regular intervals are the most critical in this regard. Such works have rigid production schedules, and the editor in charge might even be compelled to reject a section submitted late, either omitting it from the work entirely or arranging for a substitute.

## 4.3.2    Collecting the Literature

Once plans are laid and a "workshop" has been established, the author's next task is to ensure the availability of necessary background materials. An attempt to write a scientific book without convenient access to a well-organized *literature collection* would be almost like trying to walk without feet. Crutches make that kind of walking possible, at least for short distances, but it is an alternative one would certainly prefer to avoid. Similarly, some sort of "crutch" might supplant ready literature access, but the author's work would be seriously impaired, and the resulting book would probably suffer.

The text of virtually every scientific book begins with *facts* in order to develop *concepts*, and it usually leans heavily on previously published ideas or experimental data. The book in effect tells a gradually unfolding "story" made up of countless facts, seemingly unrelated until the author reveals their connections. To use another analogy, an author usually has the task of weaving a pattern out of isolated threads of information gathered from the *literature* (cf. Sec. 3.1 and Appendix D). Key published findings and the record of the supporting evidence must be constantly at the writer's disposal so facts can be checked and then *documented* in the approved *bibliographic* fashion (cf. Chapter 11 and Appendix J). A conscientious author will repeatedly feel the need to consult original documents, or at least copies of them. Lengthy abstracts or a reliable set of extensive personal notes can be very useful, but they cannot substitute for original publications.

Almost every scientist possesses at least a rudimentary literature collection.* It probably had its origin many years earlier when as a graduate student he or she first encountered the literature and discovered the value of a personal set of systematically arranged *reprints*.† Some private archival systems are extremely complex, with provisions for expansion and repeated reorganization as research interests change. Well-developed collections of this type are very handy whenever an extensive report or a journal article must be prepared, and they are nearly indispensable to book authors. Specific suggestions for organizing and maintaining a literature collection are reserved for Sec. 11.2.

---

* The term may be used here with a double meaning; it may refer to the books, journal articles and other documents on the author's shelves, or to a *catalog* of these and other documents.

† The word "reprints" in this context is actually a colloquialism, and refers to both *photocopies* and *offprints*. The latter are printed copies of individual journal articles prepared by the publisher at the time the corresponding journal was issued. In publishing terminology, however, a *reprint* is something quite different; it is a copy derived from the original printing plates, but prepared *later* in order to meet a continuing or renewed demand.

## 4.3.3    Structuring the Argument

### Sorting Out One's Ideas

The most difficult and time-consuming task of all is the composition of the actual text of the book. It is folly to sit down and simply begin writing with the expectation that a perfect book will quickly unfold. Even experienced authors freely admit their inability to formulate "finished copy" on the first try. Virtually no one has the skill to develop spontaneously an argument that is both logically sound and stylistically polished. One reason for this is that putting an idea into words for the first time forces one to examine the idea itself more critically than ever before. As one writes, one acquires new insights and understandings, and this often requires rethinking and restructuring previously written text. Seasoned writers acknowledge that the best manuscripts evolve gradually through a long series of revisions, and editors who have dealt with many authors know this to be true.

Some writers prefer to begin by jotting down, in random order, fragmentary notes identifying all the relevant ideas they can think of on each topic in the planned book. The notes may be nothing more than single words, though short phrases are more useful for capturing appropriate contexts for the thoughts. A separate sheet of notes should be prepared for each heading identified in the content summary (cf. Sec. 4.2.1), but at this point no time should be spent considering the detailed *arrangement* of ideas within each topic area. Alternatively, one might apply the "cluster method", described briefly in Sec. 2.3.1: scattering one's handwritten and encircled idea fragments across a page and connecting them with lines or arrows as structures suggest themselves. In any case, systematic organization of one's topics is a task better deferred until the scope of the whole work is more clearly established.

Once begun, the note-taking process should be continued, section by section, chapter by chapter, through the entire book. Gradually, a sketch or *topic outline* of the book emerges, a further development of the content summary. Arriving at the sketch in this way (rather than trying to proceed directly from start to finish) reduces the probability that important ideas will inadvertently be omitted, or that the author will belatedly recognize that some topic not yet covered should have been placed nearer the beginning of the manuscript.

## On the Length of Sections

One should also remain alert to possible *section* headings as the sketch is being prepared. A rule of thumb states that no more than about five printed pages should appear under a given heading (cf. "Creating Subdivisions" in Sec. 2.2.5), and preferably only two or three. It is particularly important to adhere to this rule if *section numbers* will be used to *cross-reference* the text. Overly long sections are a hindrance to the reader who wants to make use of the cross-references. An alternative way to cross-reference a book is to cite specific page numbers, but this creates technical problems. Among other things it means the author must provide the proper page numbers in all cross-references throughout the manuscript in the course of correcting page proofs (Sec. 4.4.2). Furthermore, production costs are higher when page references are used rather than section references because the numbers must be inserted at such a late stage in book production.

Preventing sections from becoming too lengthy can be difficult. One solution is to create more sections at the same hierarchical level. Another obvious remedy is further division into subsections, as discussed in Sec. 2.2.5. However, subdivision of this kind should normally not exceed three or four levels; beyond this point the associated numbering system becomes too cumbersome for the reader to scan.

Occasionally, however, more extensive breakdown of a topic is truly necessary. In this case there are two possible approaches. One—of which frequent use is made in this book—is to introduce extra *unnumbered* sub-headings, where the resulting segments are treated as parts of the numbered heading under which they fall. Alternatively, and at a somewhat lower level of systematization, one can employ *paragraph titles*, words or phrases placed directly in front of the first words of paragraphs. These serve the same descriptive purpose as section headings, but they become parts of the paragraphs themselves (though each is separated from the actual text of its paragraph by a period, a colon, or simply an extra space). The presence of unnumbered headings means that some cross-references will include words or phrases along with numbers (cf. the first cross-reference under the present subheading). For emphasis, unnumbered subheadings or paragraph titles are usually set in a distinctive typeface (e.g., boldface or italics).

Another way of reducing the complexity of a section numbering system is to bring large *parts* of the book together, labeling them with Roman numerals (to save "wasting" a digit within the main Arabic numeral classification scheme), e.g.:

Part I:    General Methods

Part II:   Applications

Chapters and subsections within these parts would be designated by the usual numbering system. That is, all chapters are numbered consecutively and continuously—according to the decimal system (Sec. 2.2.5, "Structure and Form")—throughout the entire book. Part II might therefore consist, for example, of chapters 4 through 6.

## 4.3.4    Writing the Text

### The First Sentences

Once tentative subdivisions have been designated it is time to begin preparing a *sentence outline*, a rough but nonetheless coherent sketch of the whole book. The topics to be treated will already have been selected, but now they should be placed in their proper order and expanded into sentences.

The sentence outline should not be elaborate: it is, after all, still an "outline". The purpose is to bring additional order to one's thoughts and to begin visualizing them in context. No thought whatsoever should be given at this stage to grammar, punctuation, or spelling, nor should any time be wasted in seeking "elegant" ways to express the ideas. Instead, the aim is to develop as much of the book as possible in the shortest possible time.

One should, however, try to recognize places where *figures* or *tables* might be helpful in clarifying concepts or shortening unwieldy descriptions. The appropriate locations should be marked and annotated with brief notes summarizing exactly what is to be inserted (as a concession to the shortcomings of the human memory). Also, assertions that seem to warrant documentation (through a *citation*) should be marked (cf. Sec. 11.3.4 for suggestions). At this stage it is always better to err on the side of too many citations; unnecessary ones can always be deleted later. As far as possible, these citation alerts should mention specific sources and even specific pages. Nothing is more irritating to a writer than being unable to locate a particular piece of information known to exist somewhere, or vaguely recalling an argument that would be perfect to incorporate (or criticize) if only it could be found again.

## The First Draft

Once completed, the sentence outline should be set aside for a few days, a time during which additional background material can be read. (The need for such reading will probably have become apparent during the outlining process.) Later, the overall structure of the outline should be examined critically. Any shortcomings (organizational weaknesses, for example) should be corrected at once. In the process, figure and table locations should be confirmed. Numbered lists of these should also be prepared so it will be easy to keep track of them. The lists also provide a record of the way each figure and table will be designated when referred to in the text. This is the time to devote thought to the writing conventions one intends to follow, and to problems of *nomenclature* (cf. Chapter 6) and *terminology* (Chapter 7), so usages will be consistent (and correct!) throughout the book.

Only when this point is reached is it sensible to begin work on a true *first draft*, a complete manuscript whose text is written (or perhaps dictated) in carefully formulated sentences. In other words, now that the major organizational issues have largely been resolved, it is time to devote attention to *style*.

While writing, one should keep in mind the reader one is addressing. Just as a good lecturer never delivers a private soliloquy, but addresses and responds to a particular audience, so a successful author attempts to visualize future readers, consciously employing a writing style appropriate for their needs, expectations, and backgrounds. If the work is intended as a textbook, for example, certain arguments would properly contain a measure of redundancy as an aid to learning. If, on the other hand, the average reader is expected to be a busy specialist, the information is better presented as concisely as possible.

This is probably the first occasion on which it is appropriate to use a *typewriter* (or word processor)*. The author who can type rapidly has a great advantage, because this ability permits the preparation of neat, legible drafts (ones that are easy to rework) without the need to involve anyone else. *Speed* in typing is what counts, not accuracy, especially if one has a word processor. Modest typing skills are sufficient to enable one to transfer spontaneous thoughts to paper more rapidly and more easily with a keyboard than with a

---

* An exception to this generalization is the author accustomed to using *outlining software* in conjunction with a word processor. Certain commercial programs of this type are quite sophisticated and offer numerous advantages to the author willing to experiment with unconventional methods (cf. the first footnote in Sec. 2.3.1).

pencil.* Even the author privileged to have a secretary or typist available is probably better advised to type his or her own drafts: serious errors will be less numerous and the copy will be ready when the author needs it, not when a typist finds it convenient.

All typed copy (*typescript*) should have extra wide *margins* and be double-spaced, with typing on one side of the paper only. These suggestions may appear wasteful, but their validity will soon become evident (Sec. 4.3.6). Care should be taken to incorporate all necessary *table* and *figure references*, consulting the previously prepared lists to ensure proper numbering. *Literature citations* should also be included, as well as *footnotes* if these are to be employed (cf. Sec. 3.4.5). The latter are most conveniently typed within the main text, but separated from it by dividing lines. Footnotes should be placed at the bottom of the page only if one has a word processing program equipped to do so automatically (cf. also Sec. 5.3.3).

### 4.3.5    The Benefits of Computers

In recent years the application of computer technology has made possible the adoption of radically new ways of organizing and writing lengthy manuscripts. So-called "personal computers" (or *microcomputers*) are now available to many authors, and the result has been a revolution in the entire writing process (cf., for example, *Chicago Guide to Preparing Electronic Manuscripts*, 1986; Zinsser, 1983; Roth, 1984; and Biedermann, 1984). Manuscript text is composed directly at the keyboard of a *word processor*[†] and stored on *floppy disks* (*diskettes*)—or, increasingly, on *hard disks*—often in multiple versions representing various stages of revision (cf. Sec. 5.2.2). File folders and ring binders thereby become almost obsolete, to be used only as repositories for computer outputs of early drafts or for saving duplicate copies. It seems safe to say that the time is not far off when most authors will use computers almost exclusively for their work. This is how the present book was developed, permitting the three of us involved to exchange text revisions across the

---

* One of the present authors formerly prepared all of his manuscripts in *shorthand* up to the stage of the final draft, and then dictated the text into a tape recorder. Shorthand is rapidly becoming a lost art, however, and it is for this reason that we mention the technique only in passing.
† A "word processor" can be defined as a computer-controlled *system* (hardware and software) for preparing, revising, storing and printing any type of manuscript.

Atlantic in the space- and cost-saving form of diskettes rather than printed copies (which of course meant that we needed to ensure the mutual compatibility of our equipment).

For those not yet acquainted with "computer-aided writing"* we provide here a brief review of its advantages;† the subject is examined again and in greater detail in Sec. 5.2.2. A word processor is best thought of as an "intelligent typewriter". Text is typed almost as with a standard typewriter. Unlike a typewriter, however, a word processor functions under the management of a sophisticated "program" (an "editor"), which provides a multitude of convenient features. Thus, the program makes it not only possible but even easy for a writer at any time and at any point in a manuscript to

- *correct* text (single characters, words, or longer passages)
- *insert* pieces of text (single characters, words, paragraphs—even whole chapters);
- *delete* text (characters, words, etc.);
- *move* text elements (including relocating text from one document to another);
- *find* instantly any desired sequence of characters (symbols, words, etc.)— for indexing purposes, perhaps, or to locate places where changes are to be made; or,
- *replace* any or all instances of a given sequence of characters by a different one (*correcting* a repeatedly misspelled word, for example).

One is also able in most cases to change without difficulty the *format* of the text (e.g., line length and spacing, or paragraph definition), and with some word processors it is even possible to change *typefaces* and/or *sizes* at will.

The flexibility and power of such a system is enormous. One interesting consequence is that word processing permits the development of long manuscripts not in discrete stages, but rather in a continuous, "floating" manner, with all parts evolving and undergoing revision concurrently. This reduces our description of the various "stages" of manuscript preparation to little more than a reminder of some of the intellectual processes involved. We have drastically modified our own ways of writing as we have increasingly taken advantage of the flexibility inherent in word processing technology.

---

* We propose the acronym "CAW" as a short designation, analogous to CAD (Computer-Aided Design) and CAM (Computer-Aided Manufacturing).
† Any author reluctant to make use of the new technology is strongly advised to read in Zinsser (1983) the delightful case history of one writer who was transformed from a skeptical and timid witness of word processing into a committed disciple.

Every serious author should take time to explore a variety of alternatives and then adopt the one that seems most natural and efficient. Every facility available should be tested and considered, although with due consideration to the time and place in which one can most comfortably work and to limitations imposed by the needs and desires of others—secretaries, for example, or co-authors.

The subject of word processors and computers in general is one to which we will return often and in various contexts (in particular see Chapter 5). Word processors are revolutionizing writing practices not only for authors, but for the publishing and printing business as well,* and not only with respect to books. The reason we devote part of this particular chapter to extensive comment on their use is because the benefits are exceptionally significant for large scale manuscripts.

## 4.3.6    Revisions

The first draft, like the sentence outline, should be set aside temporarily after it is written. A fairly long break is important; it establishes a certain distance between author and manuscript, and facilitates objective self-criticism. The passage of time causes even the most experienced writer to become (painfully!) aware of shortcomings in a new piece of text (see Sec. 1.4.2). However, one usually also finds that appropriate changes quickly come to mind. The importance of re-working a manuscript can scarcely be overstated. Reading the results of one's work with an open mind a few days after it has been put on paper is nearly always a humbling experience.

All *corrections* and *revisions* should be written directly in the manuscript, preferably between the lines, but using the margins as needed. It is for this reason that drafts are typed with generous spacing, a practice that carries over to final copy as well (cf. Sections 3.4.1 and 4.3.7). Even major insertions will probably fit in the wide margins, but if not they should be typed on properly numbered addendum pages (cf. Sec. 3.4.1).

Many pages of the manuscript—perhaps all of them—will undergo significant change, necessitating extensive re-typing or "repair work" on the word processor. The result will be a *second draft*, not uncommonly followed by a third draft and very likely a fourth! With each new version, previously

---

* A case in point is the concept of *desktop publishing* described, for example, by Bove, Rhodes, and Thomas (1987) and, rather differently, by Seybold and Dressler (1987).

overlooked problems (weak arguments, rambling or improperly located paragraphs, awkward sentences, meaningless "filler" words or phrases) will become apparent. When these are corrected, a new set will mysteriously surface, but only after a clean, easily read copy has been prepared and a certain amount of time has been allowed to pass.*

Eventually, however, a draft will be produced that is reasonably† satisfying upon a second reading, a few days after it is finished. At this point it is appropriate to invite one's colleagues to read and criticize the manuscript, especially those parts on the fringes of the author's expertise.

Almost anyone who has written a technical book will acknowledge that there is a great deal to be learned in the process—particularly from other people, including subsequent readers of the book. This kind of personal gain can even be a valid motive for writing a book. However, the *book* stands to benefit most if the greater part of the learning *precedes* publication. For this reason, experts should be sought and induced to read chapters that fall within their realms. At the very least their reactions will help one identify potential areas of controversy, and it is quite possible that unsuspected shortcomings will emerge.

After all critical comments have been collected and digested, one last revision should be undertaken. Once this has been completed, the manuscript is ready to put in "final form".

## 4.3.7    The Final Copy

### How Clean is Clean?

The last step in developing a manuscript is preparing it for the publisher, in the cleanest form possible. Publishers are less demanding with respect to the

---

* In principle, a text prepared with a word processor can be read and corrected entirely from the screen (and keyboard), eliminating the preparation of printed copies altogether. However, many authors prefer to have the chance at least occasionally to hold the evolving work in their hands, and to make changes and corrections in the way to which they have long been accustomed. More than nostalgia is involved; at the very least it can be said that a printed copy facilitates glancing back and forth between various sections of text, a tedious and unsatisfying operation with a word processor.

† The qualifying adjective is an acknowledgment of the fact that most authors are never *completely* satisfied with their work, just as a scientist is always aware of "one more experiment" that would make a set of results "absolutely air-tight" (cf. Sec. 3.2.2). Nonetheless, insistence on perfection must be kept within bounds if publication is ever to take place.

appearance of *final copy* (*fair copy*, *clean copy*) for book manuscripts than they are for journal articles (cf. Sec. 3.5.4). Nevertheless, it is expected to be carefully typed (double-spaced), and any hand-written or typed corrections must be neat and unambiguous.

Just as in the drafts, all corrections should be written between the typewritten lines. It is inappropriate to use marginal proofreader's notes like those described in Sec. 3.5.6 and Appendix I because these are strictly reserved for *typeset* copy, a rule which—unfortunately—is often ignored. Every correction should appear immediately above the text to which it pertains, and a hand-drawn line or inverted "v" should be used to show precisely where any added text belongs. Fig. 4-1 is a sample text page on which a copy editor has made a number of typical *typescript corrections* in the manner suggested. Notice that nothing is written in the margins except instructions to the typesetter.

There is good reason for giving careful thought to the way typescript corrections are shown, particularly with respect to their legibility and placement. Final copy must be further processed by a *typesetter* at the printing plant. Typesetters are specifically trained to work rapidly, capturing large blocks of text at a single glance. A manuscript that can be scanned continuously is the easiest kind to typeset, and convenience translates into lower production costs. Hours wasted because a typesetter cannot decipher an author's notes are extremely expensive, as is time consumed unnecessarily in the process of looking back and forth from text to margin. These are simple economic facts, but few authors seem to appreciate them.

In some cases final copy is published just as it is received from the author. A photomechanical (offset) reproduction process is employed, the same as that described for letter journals (cf. Sec. 3.2.5, "Types of Articles", and Sec. 3.4.6). The practice is an increasingly common one, particularly for highly specialized monographs or conference reports; i.e., works expected to have a circulation too limited to justify the extra expense of typesetting. Final copy of this kind must obviously meet rigorous specifications. Publishers usually supply their authors with specially formatted *manuscript typing paper* for the purpose, with all necessary instructions for its use. The paper typically contains pre-printed lines for the text, but these are of a color to which the photographic process is blind, so they will be absent in the printed version. The author preparing such copy should carefully review Sec. 3.4.6, where camera-ready typescripts for journals are discussed.

Many publishers offer special typing paper even for manuscripts that *are* to

- 58 -

4.3.6  The Final Copy

The last stage in manuscript development is to prepare
it in the cleanest form possible for the publisher.
Publishers are less demanding with respect to the
appearance of final copy (fair copy, clean copy) for
                                                  they
book manuscripts than are for journal articles (cf.
                                         or
Sec. 3.5.4). Some last minute hand-written typed
                will
corrections should usually be tolerated, provided they
are neat and unambiguous. Just as in the first and
second drafts, all corrections should be
              written legibly
between the (double-spaced) typewritten lines. One
          not
should use the marginal proofreader's notes described
in Sec. ☐ . These are strictly reserved for typeset
manuscripts. All corrections should be placed
immediately the text above to which they pertain, using
hand-drawn lines or inverted "v"'s          to
show precisely where any added text belongs. A copy
editor amends a text in exactly the same way. The
sample page shown as Fig. 4-1 illustrates of typical   a number
    typescript corrections. Final copy should
    facilitate continuous scanning by typesetters
at the printing plant. These specialists have been
trained to capture text at a single glance, because
              s
doing so minimizes the time required for their work.
Extra hours trying wasted in to decipher author's notes
are costly, and editors often lament the an fact that

[(insert from page 60)]

be typeset. This is paper that has both a pre-printed "frame" to enclose the text and numbered lines on which to type. The service is provided because editors hope it will ensure proper typing of all submitted copy: double-spaced, and with wide margins. Copy typed on "normed" manuscript paper also has the advantage that it allows the editor and the author to estimate the number of *typeset* pages that will result. Manuscript forms usually accommodate about 30 lines of 50 characters each. In the case of a "pocket edition", one printed page is equivalent to about 1 1/2 standard manuscript sheets, while it takes 2 or 2 1/2 sheets to produce a printed page for a typical (ca. 17 cm x 24 cm) single-column textbook or monograph. An even higher factor of 3.0 to 3.5 may be applicable to a two-column reference work set in small type. The exact conversion factor depends on the choice of type size and page layout, and it is also highly sensitive to special circumstances such as the need to isolate a large number of equations—which produces more blank space on a printed page than in a typescript. In some cases the conversion factor can be as low as 1.

## *New Demands and Promises*

Sec. 4.3.5 introduced some of the ways technology has begun to revolutionize manuscript preparation, but there is much more to the story. For example, as increasing numbers of writers and their secretaries acquire access to computerized word processing equipment, publishers become less tolerant of final copy strewn with corrections. Today's well-equipped authors are expected to read carefully what they prepare and to take full advantage of the ease with which perfect pages can be printed by "computer-driven typewriters".

Publishers are even more intrigued by another exciting development: typesetting equipment controlled directly by the diskettes on which a modern author has stored a "typed" text (cf. Sec. 5.2.2). Many printing houses are already in a position to exploit this technique, and the day is probably not far off when the majority of books will be printed without the need for a second "typing" at the printing plant.* In principle, all the necessary technology

---

* In its latest set of instructions to authors (cf. Dodd, 1986) the American Chemical Society provides special guidelines for authors wishing to submit journal articles on diskettes. Modern phototypesetting equipment has long been computer-driven; the novel aspect is using an author's own diskettes to bypass the expensive "re-typing" step. The major complication is that authors (or publishers) must modify the stored manuscripts somewhat to ensure proper line spacing and indentation, as well as to permit correct interpretation of features such as subscripts. Usually this entails introducing special *code symbols* throughout the text (cf. Sec. 5.2.2, "The Benefits of a 'Magnetic Manuscript' ").

already exists, but its application has been delayed by the lack of compatibility between individual parts of the system. Progress is likely to be rapid, however, and authors with access to word processors should certainly discuss the matter with their publishers.

### Choice of Type

One may want to include in final copy recommendations for the use of special *type*. For example, in many books—including this one—certain words or phrases are *highlighted*: printed in a different *type style* clearly distinguishable from the rest of the text. Highlighting is a technique that should be used sparingly, since overuse diminishes its effectiveness and makes a book's pages look cluttered.

For fairly subtle emphasis (as with paragraph titles; cf. "On the Length of Sections" in Sec. 4.3.3), the best choice is normally *italic* type. **Boldface** type is more dramatic; the reader's attention is automatically attracted because there is an interruption in the page's otherwise uniform print density. For precisely this reason the frequent use of boldface is undesirable: it is simply too distracting. On the other hand, the prominence of boldface makes it a logical (though not essential) choice for *paragraph titles*, which are thereby effectively distinguished from other material highlighted more subtly by italics. Boldface is also useful for alerting the reader to *warnings* (in a laboratory text, for example).

Footnotes, legends to figures, and portions of tables are often set in a miniature typeface called brevier.* Its use in the body of a text is rare, however, and should be specified only with the concurrence of the publisher. Scholarly works sometimes incorporate one additional typeface: SMALL CAPITALS. Here the lower case letters look like capitals that have been reduced in height. The

**a**
```
One may want to include in
final copy recommendations
for the use of special type.
```

**b**
One may want to *include* in final copy **recommendations** for the use of SMALL TYPE.

*Fig. 4-2.* (a) Example of a typescript annotated with *typeface* designations. (b) The corresponding printed text.

---

* Brevier is usually understood to be what printers call an "8-point" type, where the *point* (pt.) is the traditional unit of type size (1 pt. = 0.351 mm in English-speaking countries).

principal use of small capitals is to distinguish personal names, including those of authors whose works are cited. In the past, especially in German publications, highlighting was sometimes accomplished by introducing e x t r a s p a c e between the characters of a word or phrase. This practice has been nearly abandoned, however, because many find it unattractive and because it can make the text more difficult to read.

The standard way of indicating what parts of a manuscript are to be set in special type is by *underscoring*: e.g., a normal (straight line) underscore for text to be italicized, a wavy underscore for boldface, and a double underscore for small capitals. Fig. 4.2 shows an example of a typescript marked in this way. One cautionary note is necessary, however: in some countries printers use different conventions—a wavy underscore might indicate italics, while a normal underscore would specify boldface. One's editor should always be consulted in case of doubt. Special *markup* guidelines always apply to structural formula numbers (boldface, cf. Sec. 3.4.2) and mathematical or physical quantities (italics, cf. sections 3.5.4 and 7.1.3).

The author is not expected to supply typeface designations for *headings* (apart from paragraph titles). The choice of heading style and size is a responsibility of the publisher's *production manager* and depends both on the level of the heading and on aesthetic considerations. Aesthetics also play a role in selection of the typeface for the book as a whole and in *page layout* decisions. These are clearly matters requiring the expertise of a professional.

With the introduction of typewriters featuring interchangeable typefaces, and especially since the advent of word processors, more and more authors find themselves in a position to prepare final copy that already contains italics and even boldface, and underlining begins to seem archaic. Publishers usually have no objection to an author's exploiting this flexibility, but it is still wise to ask.

## *Miscellaneous*

Special marking of one other kind is required in final copy: the printer needs to be alerted to the *first reference* to every figure and table. This is done by drawing a bold, color-coded arrow in the margin next to the line containing the appropriate in-text reference. Red arrows might be used as alerts for *figures* and blue arrows for *tables*. Guided by these arrows, the person responsible for page layout can ensure that the text will be "broken" in such a way that figures and tables are placed as close as possible to the first references to them (cf. Sec. 4.4.2). It is not necessary to interrupt the text with an insert such as:

((Fig. 1)).

To do so could, in fact, be misleading, because limitations in space might well require the figure to appear elsewhere.

The arrangement of chemical formulas is a matter deserving careful thought, just as in the case of journal articles (cf. Sec. 3.4.2). Moreover, as was previously pointed out, there is a fundamental difference between formulas that can be typeset and those that must be drawn (in whole or in part). Actually, book authors should at least try to persuade their publishers to redraw *all* formulas. This produces a more attractive book, one with an appealing uniformity of appearance—but it is also more costly.

Structural formulas, figures, and tables should all be kept separate from the text manuscript for the reasons discussed in sections 3.4.2 through 3.4.4.

Headings also warrant special comment, particularly since their proper preparation entails strict observance of standard *capitalization* conventions. In English, the rules specify that for *chapter* and *section headings*—at least down to headings of the third order—one should capitalize the first letters of all important words.* The first letter of every first word is also capitalized regardless of its "importance". When page proofs of the book arrive (Sec. 4.4.2), the author may discover that a decision has been reached to set first-order chapter titles entirely in capitals, or that a combination of capitals and small capitals has been used. If so, these instructions will have originated with the production manager or the *book designer.*† Since an author has no way of knowing what style will be selected, headings should appear in the typescript with first-letter capitals only. Neither chapter titles nor subheadings should ever be underlined in a typescript. The same advice applies to *table titles* and *column headings* (in what is called the *overhead key line,* cf. Sec. 10.2.1). In the former, *all* important words are sometimes capitalized in the final printed version, but only the first word of the latter. *Figure captions* are often treated differently from table headings, with only the first letter of the first word capitalized. The distinction seems unnecessary, and we are inclined to question its wisdom (cf. Sec. 10.2.1).

One further object of special concern at this stage is preparation of the list of references. The task is a tedious one, but it is important that it be executed with care and proper attention to consistency. A reference list is valuable only if the reader can be sure of its reliability—and some readers *will* want to make

---

* All words except articles, conjunctions and prepositions are "important words" in this context.

† The book designer also selects the size of the page and the typeface and type sizes. Often the person responsible for the production of the book is in charge of design.

extensive use of references. Strategies and techniques for preparation of the list are discussed in Chapter 11.

At this point all that is missing from the final copy (as far as the author is concerned) is the *Table of Contents*. This should be prepared last. Cynical editors often claim they have never seen a book manuscript with a table of contents that corresponds totally and in detail to the headings in the text itself. Waiting until the end to prepare the contents page is the author's best hope for presenting an exception.* When this point is reached the manuscript can be regarded as complete—for the time being; however, we shall see presently that the author's work is not yet finished.

Next, the entire manuscript should be carefully assembled, packaged, and shipped (insured!) to the publisher. It is important to recall that authors usually have a contractual obligation to retain in their possession one complete *duplicate* (security) *copy* of any manuscript submitted, including all extra material such as figures and formulas. When the publisher has received the manuscript and verified that it is intact the author will be sent an official letter of acknowledgment. At this point the process of transforming the manuscript into a book moves out of the author's hands for a time.

## 4.4    The Typesetting Stage

### 4.4.1    Galley Proofs

#### Copy Editing

As soon as the final copy of a book manuscript reaches the publisher it is carefully examined by the editor-in-chief to make sure it meets all of the firm's expectations, particularly with respect to content, style, and form. This first review is likely to be quite critical, and it is entirely possible that the author will be asked to make *revisions*. Even a manuscript that passes this preliminary test will need to undergo changes, however. A certain amount of in-house editing is required on virtually every manuscript. This may only be light *copy editing*,

---

* Actually, this is another responsibility that can sometimes be entrusted to a word processor, at least if it is a sophisticated one.

involving minor changes in *grammatical usage, spelling, punctuation,* or *style* (the discussion of capitalization near the end of the preceding section provides a case in point).

Sometimes, however, an editor will feel it necessary to order more extensive editing. A typical example is a manuscript in English written by someone with a different native language. It may be that when the sample chapter was submitted (cf. Sec. 4.2.2) it was agreed that a certain amount of *language polishing* would eventually be undertaken. Occasionally an editor decides to alter a manuscript in the interest of *conformity* (e.g., to make it more in keeping with other volumes in a series). Changes in this category typically involve *nomenclature, terminology,* or the use of *symbols.* If the book is a *translation,* an outside expert may be asked to verify that technical idioms are properly rendered.

Whether or not measures such as these are taken depends to some extent on the availability of appropriate expertise, but the strictness of the firm's publishing standards can also be a factor. On the other hand, the decision may be tempered somewhat if the editor is conscious of an unusual degree of sensitivity on the author's part. Whatever is done will be both time-consuming and costly, so the editor will try to keep it to a minimum. Moreover, any major changes will probably need to be discussed with the author. The editor may even consider it advisable to return the entire edited manuscript to the author for approval—another inconvenience and loss of time.

The manuscript is then transferred to the *production department,* where it is prepared for *typesetting.* The typesetter may begin by producing a few *sample pages* for critical review by both editor and author.

*Proofreading*

Just like a journal article, a typeset book must pass through two successive stages of *correction*: "galley correction" and "page correction". *Proofreading* the galleys is the author's next major job. *Galley proofs* will arrive in due course, probably in small batches mailed out by either the production manager or someone at the printing plant. These will be accompanied by the (edited) manuscript from which they were prepared. Authors are sometimes disturbed by the physical appearance of galley proofs (and even the later page proofs), but it is important to realize that proofs are not at all representative of the paper or print quality that will characterize the final book. Galley proofs differ from the finished product in substantive ways as well. For example, they usually lack figures and legends, and headings are likely to be set temporarily in the same typeface as the text.

Every single word in the galley proofs must be read ("proofread") with extreme care. The entire text should be compared word for word with the submitted manuscript. No footnote, no entry in a table, not even a line in a figure, should escape this thorough scrutiny (cf. Sec. 3.5.6). Even though the book may contain several million printed characters,* every one should be correct.

Unlike typescript corrections, all necessary *galley corrections* are made in the *margins*, using pen. Typesetting errors should be indicated with a color different from the one used for showing requested modifications. It is essential that corrections be made exclusively with conventional proofreader's symbols (cf. the discussion of journal article galley correction in Sec. 3.5.6 and also Appendix I). Only if the proofs are so marked is it safe to assume that a typesetter will recognize what changes are to be made and where. No corrections or comments should be placed within the body of the text; any that are would tend to be ignored (and there is virtually no space available within the text in any case). Notice that use of a *pen* has been specified. Each correction is a kind of "document", and it deserves more permanence than a pencil would provide.

Sometimes *in-house corrections* (*printer's corrections*) will already be present in the galleys when they are sent to the author. If so, ink of a different color should be used for any additional changes.

The author is expected to read the proofs immediately upon their arrival. Prompt action is more important than might be apparent. Should it be found, for example, that some symbol has consistently been printed incorrectly, swift notification may make it possible to prevent repetition of the error in chapters not yet typeset.

Authors' changes at the proof correction stage should be restricted to true *errors*. It is too late for major alterations in content or style because of the large amounts of text that would need to be reset. Extra typesetting is expensive, and the resulting costs would be reflected in the selling price of the book, further jeopardizing sales. Most publishing contracts contain a clause under which the author can be held personally responsible for a portion of any costs incurred as

---

* This book, for example, requires slightly more than 2 "megabytes" of storage space on a word processor, equivalent to more than two million characters. This number includes space occupied by information (*codes*) establishing formats, defining typefaces, etc.

a result of *last minute changes*. This financial responsibility is usually limited to amounts exceeding 10% of the original typesetting costs, but the 10% latitude must *not* be taken to mean that one is free to change 10% of the text. Changes are far more expensive than initial typesetting, especially if illustrations are involved.

Admittedly, it is still technically feasible to augment text that has reached the galley stage—to incorporate newly published research results, for example. Authors are cautioned to resist the temptation, however. If an *insertion* (or *deletion*) is absolutely necessary, an effort should be made to *delete* (or *insert*) an equivalent amount of text nearby. In this way, only a few words, or at most a few lines, would need to be reset. It is also best to try to plan additions so that they occur at the *ends* of paragraphs, or as *new* paragraphs, since this expedient also helps to minimize new typesetting and its associated costs.

Usually publishers furnish their authors with two copies of all galley proofs. One copy is to be corrected, signed, dated, and promptly returned with the original manuscript. The other should also be annotated with all corrections, but it should then be retained as insurance against loss of the master copy.*

## 4.4.2   Page Proofs

In the next step, correction orders marked in the galley proofs are carried out at the printing house. After this, all tables, figures, and footnotes are incorporated into the typeset text and the copy is broken into pages. Headings are set in final type form, pages are numbered, and *column titles* [†] are added.

A set of *page proofs* is now prepared and sent to the author for approval, this time together with the corrected galleys. In the case of multi-author works, it is the *editor* who usually accepts sole responsibility for checking page proofs.

---

* This is a rule to which many authors fail to adhere, presumably in order to save time. However, it is at their own peril. An enormous amount of time would be lost if something unforeseen should happen to the master copy, which may contain hundreds of corrections! Some authors with access to economical copiers prefer copying the sheets showing corrections, or even the entire set, rather than inserting all corrections twice by hand.

[†] A column title (or "running head") is a type of heading that appears either at the top or bottom of each page in a scholarly book. Typically, the title of the current chapter, occasionally abbreviated, is used as the column title on all left-hand pages (those with even numbers), while the right-hand side (every odd-numbered page) shows the title of the current major subsection. Column titles are designed to help orient the reader, and their presence is considered essential in any book cross-referenced by section.

One might expect that proofreading the page proofs would entail little more than verifying the proper incorporation of all specified galley corrections. This is not strictly the case, however. There is a very real possibility that the typesetter assigned to carry out the galley corrections has inadvertently introduced *new* errors into lines that were previously correct. Fortunately, this problem is less common than it once was. Older technology required the resetting of a considerable amount of type in the course of carrying out corrections—often involving all the lines between the error and the end of the paragraph. Since the introduction of computer-aided typesetting facilities it is possible simply to rearrange the text following the actual site of a repair, often automatically, and the error potential is greatly decreased.

The technique is not flawless, however; new errors can still appear, especially incorrect end-of-line *hyphenation*. Words are often divided at the end of a text line in order to optimize spacing, and this division takes place under the control of an automatic line-breaking routine. The algorithms used are surprisingly powerful, but sometimes they lead to mistakes. It is therefore advisable to check the *surroundings* of all requested changes, not just the corrections themselves.

Besides looking for remaining typographical errors, the proofreader must be on the alert for *missing* text. A paragraph could have been "lost" when the pages were broken, for example, or a figure might have been left out. Similarly, one should verify that everything is in its proper place, particularly the "special" parts: headings, formulas, figures, legends, tables, and footnotes. Page numbers and column titles are obviously new, and these always require careful verification. Finally, any remaining *blockades* (such as an "xxx" in a cross-reference, signifying the need to insert a page number) should be "lifted" by supplying the missing information.

Sometimes a correction will have been improperly carried out, most often one which was difficult to show clearly (e.g., in an *equation*). Should such a mistake be found, it is a good idea, in addition to supplying the usual proof-correction marking, to rewrite the entire corrected segment in the margin, either drawing a circle around it or surrounding it with a loop open to the edge (cf. Appendix I). This extra precaution will give the typesetter the clearest possible guidance. It is also wise to request a revised proof (*revision*) of any critical pages to be sure that they have finally been prepared properly.

When the author has made absolutely certain that everything is correct, an *imprimatur* (lat.: "let it be printed") can be added to the page proofs, signifying permission for publication to proceed. This permission may be qualified in an accompanying letter by noting that approval is conditional and subject to execution of final corrections on specific pages. Often the first page of a set of proofs carries a stamp that reads "o.k. as corrected" (or the equivalent), the signing of which is tantamount to issuing blanket approval for publication with corrections as shown.

## 4.5    Completing the Book

### 4.5.1    Indexes

At this point the author has only one major task remaining: preparation of the book's *index* (or *indexes*). The term "index" is usually taken to mean *subject index*, an alphabetical listing of *index terms* (important words or concepts dealt with in the book), each accompanied by one or more page references. Its purpose is to guide the reader interested in a particular subject to all those places where the subject is treated. Scientific books often contain other kinds of indexes as well—an *author index*, for example, or a *molecular formula index*.

Publishers of technical books have a saying that "a book is only as good as its index", and there is much truth in it. Index preparation is a more formidable task than might at first be apparent. It cannot be done haphazardly if the result is to be the valuable tool readers expect. Publishers and editors have given much thought to the best approach to index preparation, and their wisdom is summarized in the detailed working guidelines provided in Appendix F.

Of the two possible approaches, the first is the most straightforward: manual preparation. This is the traditional method, described in detail in Appendix F.2. Many authors will find it their only option, either because the publisher is unequipped to offer an alternative or because the small size of the planned index prevents a more automated approach from being cost effective.

The alternative for large works is new and extremely attractive, since it involves "subcontracting" much of the labor to a computer. Computer-aided indexing can take a variety of forms, and publishers are experimenting with a number of promising techniques. The least complicated, and probably the most common, requires the author to inform the typesetter what words in a book are to be keyed to the index, and the typesetter in turn gives the information to a computer. Details can be found in Appendix F.3.

Sometimes an even more automated procedure is implemented, one which bypasses the role of the typesetter altogether. The typesetting process itself is invariably computer-controlled, and in principle an index could be generated automatically if one were able to decide in advance what entries it should contain. For example, if the typesetting computer were to discover that a word about to be typeset appears in a predetermined list of index terms, it could take note of the current page number and prepare an appropriate record in a special part of its memory. At some later time the stored information could then be called forth by a different computer program and organized as an index.

The possibilities do not end here. An author preparing a book manuscript on a word processor could agree to mark all potential index terms as they occurred during writing. This would require nothing more than the introduction of appropriate *code symbols*. Later, in the course of typesetting (as pagination is taking place), the publisher's computer would collect all marked items and assemble and alphabetize them. The result would be an index prepared without the need for ever devising a master list of index terms.

In fairness, it should be pointed out that, at least so far, every attempt to automate the process of index preparation has been imperfect, and each current approach still requires a certain amount of manual processing. This is not surprising; it is most unlikely that a computer could be programmed to recognize every possible variant* of a listed index term and to group all coded references properly. Much less would a computer be able to judge sensibly whether a particular word in a particular context is *worth* indexing. Above all, a computer would be at a complete loss trying to deal with the problem of an inappropriately large number of page references associated with a given index term. The result would be little more than a useless, overgrown—but in some respects still incomplete!—catalog of symbols. In other words, even today one can claim that indexing is an art in its own right and a worthy occupation for professionals. Nevertheless, technological help in index preparation is

---

* As a simple example, consider the fact that plural or genitive forms would need to be recognized and then transposed into the corresponding nominative singular. This is not nearly so straightforward as it might appear. In any event it would surely take a human indexer to extract from the phrase "to read the galley proof" the index term "proofreading"—that would be asking too much of a computer!

becoming increasingly available, and the possibilities should definitely be a subject for early discussion with one's publisher. Appendix F should also be consulted for additional guidance.

An index must be proofread just like any other piece of typeset copy, although this task is often undertaken by staff at the publishing house as a way of saving time. Index preparation for a multi-author work is usually the responsibility of the editor.

## 4.5.2    Preliminary Material

### General Observations; Half Title and Series Title Pages

Only a few minor tasks remain before the author's work is finished. However, explaining them requires the introduction of certain terminology that will probably be new for many readers.

Preceding the actual text of a book is a series of *preliminary pages* (the *preliminary material*), also known as *prelims* or *front matter*. These are usually prepared and typeset last, and are generally paginated with Roman numerals to distinguish them from the text proper (which generally begins on the page identified with the Arabic numeral "1"). A book description in a *publisher's catalog* or other promotional material (such as a *prospectus*, or an *advertisement*) often contains such information as:

> viii, 224 pages.

This signifies that the book has 224 pages of text and eight pages of front matter.

Numbering the introductory part separately has the advantage that it permits additions or deletions right up to the time of publication, since corresponding changes in the Roman pagination have no effect on the rest of the book. Thus, preparation of a *dedication* or a *foreword* can be delayed until the last minute if necessary.

All of the first four introductory pages, the *title pages*, have definite functions. Page "i"* is the first right-hand page (apart from the unnumbered *endpaper* or *paste-down*, a double page whose left side is glued to the cover), and it is known in the trade as the *bastard title*. Its alternative designation, *half*

---

* Many publishers, especially in Europe, use capitals for Roman pagination; e.g., "I" rather than the "i" commonly found in English-language texts.

*title*, while less colorful, is somewhat more descriptive: page i is devoted to an unembellished and concise identification of the book. It contains only the major title (without subtitles; cf. the following section) and a list of the authors' last names, e.g.:

> O'Hara/Benwood
> *Chemistry for Lawyers*

Page ii, the *half title reverse side*, is often left blank. In the case of series works, however, this page commonly functions as a *series title page*, identifying the corresponding series and series editor, and sometimes listing the titles and numbers of volumes that have already appeared.

## Main Title Page

Next comes the *main title page,* page iii. This is the actual *title page* in the narrower sense. It is a display page, occasionally quite elaborate, and it bears the full official title of the book (more precisely: its *subject title*), complete with any subtitle and, where applicable, the book's volume number in a designated series. The subject title shown on page iii is always understood to be the official *name* of the book. In addition, the main title page includes the complete name(s) of the author(s), the name and location (or perhaps only the logo) of the publisher, and, occasionally, the year of publication.

It is becoming increasingly standard for the title page to begin with the name of the author, followed by the subject title, although the reverse order is also found. Authors' given names (at least their first names) are usually fully written out. If the number of authors is too great to permit identification of each by name on the main title page, as in the case of contributors to a *multi-author work*, the name of the editor (or *volume editor*) may be substituted, preceded by the words "Edited by". In this case, a *list of contributors* (preferably one containing affiliations or addresses) appears on a separate introductory page. In addition, all authors of individual chapters or contributions are usually identified in the table of contents.

*Conference proceedings* are typical examples of multi-author works. Their title pages usually include the date, title, and location of the conference from which the book originates. The organizers or sponsors of the conference may also be listed (e.g., a scientific society or a research institute).

Occasionally, additional information is required on a title page. If a book is a *translation*, for example, the translator's name may appear, preceded by the phrase "Translated by". New editions of previously published works are also indicated as such by an appropriate title page entry; e.g.:

Second, completely revised edition.

## *The Subject Title*

Before continuing the discussion of a book's front matter and related topics, a brief digression is in order to consider the matter of how and when a book receives its title. It is not uncommon to delay the final wording of a title until work has begun on the preliminary pages. With scholarly works it is especially important that the name chosen accurately characterizes the contents. The wording must not be misleading, nor should it be likely to raise false expectations. Moreover, a *subtitle* is often appended to indicate a book's scope or to identify particular characteristics or emphases in its coverage. A subtitle may also be used to suggest the nature of the intended *target audience*; e.g., describing a given work as a textbook, handbook, or the like. This can be an effective way of ensuring that a book will be recognized for what it is by those who ought to be interested in it. A title should not characterize the target audience too narrowly, however, since this might unduly limit sales. Alternatively, a subtitle can be used to imply a book's *level of sophistication* or point up some unique feature, e.g.:

*Nuclear Magnetic Resonance Spectroscopy: An Introduction*

*Reaction Mechanisms in Organic Chemistry:*
*A Seminar Book with Practice Problems*

In the above examples the main titles clearly establish the subject, while the subtitles suggest the author's approach.

A title also determines where a book will be entered in a *library catalog*, and careful choice of words can lead to optimal placement. For example, the first of the titles cited above would automatically be catalogued under "nuclear", and perhaps "spectroscopy", but it would not be uselessly alphabetized under "introduction", as would be the case if the book were entitled *Introduction to Nuclear Magnetic Resonance Spectroscopy.*

When the matter of a title first arose (in Sec. 4.2.4), it was dismissed with the comment that in the early stages of writing only a "working title" is required. There are several reasons for delaying the final choice of a title, reasons beyond those cited with respect to a thesis (cf. Sec. 2.2.2). Perhaps the most important is the possibility that a similar book may have appeared on the market since the original publishing agreement was signed. This might well necessitate finding a new way to designate this particular work in order to make its unique features clear and to prevent confusion with the unforeseen competitor.

### Imprint Page

The reverse side (*verso*) of the title page, page iv, is called the *imprint page* (*impressum*) or *copyright page*; its purpose is to supply the reader with additional bibliographic and other information. Many technical book publishers have adopted the convention of beginning this page with the author's address (or authors' addresses)—generally a business address—so the reader knows in what institution or firm the book originated and where the author can be reached (cf. the "corresponding author" of a journal article, Sec. 3.3.2).

Further down the page one nearly always finds a standardized block of information known as a *Cataloging-in-Publication* (CIP) entry (cf. Appendix E.2). These first appeared in the mid-1950s as a direct result of the development of "new publications rapid announcement services" by national libraries such as the Library of Congress (Washington, D.C.), the British Library (London), and the Deutsche Bibliothek (Frankfurt). Services of this kind provide librarians and book dealers with weekly listings of all titles newly announced by member publishers. The lists are distributed weeks before the books themselves become available, and all titles included are automatically stored in national library documentation systems. From here, most of them make their way into published catalogs such as *Books in Print* (United States), *British Books in Print* (U.K.), or *Verzeichnis lieferbarer Bücher* (Germany, FRG). Cooperating publishers who routinely submit new book notifications to the appropriate national library are thereby assured that their publications will be documented, but in return they are obliged to display the corresponding official CIP records on the imprint pages of their books. One component of the CIP record is a unique *International Standard Book Number* (ISBN) which helps assure the book of unambiguous worldwide identification (cf. Appendix E).

The imprint page also contains a brief copyright statement (cf. Appendix C). This consists of the copyright symbol ©, the name of the copyright holder (the party holding legal title to the text—usually the publisher), and the year of first and/or most recent publication. Sometimes copyright notices are expanded to describe some of the specific protections of copyrighting: e.g., restrictions on reproduction or on incorporation of the contents into computer-based data collections. The publisher may also add a disclaimer to provide protection from possible misuse by readers of any *trademarks* appearing in the book (even if they are not explicitly or uniformly so identified). Other information on the imprint page may include:

- the title and publisher of an original edition from which the book was translated;
- years of publication of previous editions of the book;
- marketing territory restrictions (in the case of co-publications or licensed publication);
- the number of figures and tables in the book;
- the names of the senior editor and production manager;
- identification of the typesetter, printer, and bookbinder;
- acknowledgment of the designer of the cover jacket or cover and, where appropriate, the source of cover illustrations; and
- the style of type and printing technique used.

It is the publisher who has the responsibility for preparing copy for the four introductory pages described so far. The author will receive proofs to check, however, and these should be examined carefully—even though the publisher will already have proofread them—to be sure that no mistakes have been made. Errors in the opening pages would be particularly embarrassing. Proper attention should also be paid to the selected type styles and sizes. Subtitles, for example, must be set in somewhat smaller type than that used for the main title. If the subject title is too long to fit on a single line it should be divided in a logical way, not merely on the basis of visual aesthetics (although an author's opinion on this point may differ from that of the publisher). All references to authors and translators should be checked carefully for spelling and to make sure they conform to any contractual agreements.

### Preface, Contents

The four title pages are normally followed by a *Preface* and the book's *Table of Contents*, usually simply called the *Contents*.* Both begin on right-hand pages. The author supplies the corresponding manuscripts, although these are likely to undergo modification to bring them into conformity with the

---

* British and American publishing houses often use a different order: contents first, then dedication, foreword, preface, and acknowledgments (if any).

publisher's notions of appropriate style and content. As always, proofs will be prepared, and these will be sent to the author for checking and approval, usually at the same time as the title page proofs. This will most likely be the point at which page numbers are inserted in the table of contents,* a job that can be done only by referring to the final page proofs of the full text. Afterwards, a second set of proofs of the table of contents must be prepared and checked to make sure that no errors have crept in.

*Miscellaneous*

The table of contents may be preceded or followed by such special sections as a *Dedication,* a *Foreword,* a *List of Contributors,* or a *List of Symbols.* The foreword is a short text about the book written by someone other than the author (the editor of the series to which the book belongs, for instance). If provided by a well-known expert in the field, it can be especially useful for publicity purposes.

## 4.5.3   *Cover, Dust Jacket, Promotional Material*

Design details for the cover are usually left largely up to the publisher. The author should be consulted about the design, however, as well as about the *cover type* and the form of *binding*: whether the book will be produced as a *paperback* or as a *hardcover* edition, and whether the pages are to be sewn together or not. A paperback book (also called a *paperbound* or *softcover* edition) is one whose cover is made of a flexible material. The individual pages of such a book are fastened directly to the cover by a layer of adhesive in a process called (perhaps somewhat ironically) *perfect binding.* As a last step the cover is usually "varnished" to enhance its durability. Hardbound books, on the other hand, are more complicated. They usually have a "hollow spine"; that is, their front and back covers are attached to the ends of a piece of gauze, to the midsection of which the pages are glued after first being sewn together in sets *(signatures).* The covers themselves are made of stiff cardboard, hence the expression hardcover (or *hardback*) edition. The two covers, often with a spine insert, are clad with some type of protective material (e.g., linen) before being assembled into a single unit.† A sewn binding and a stiff cover provide greater

---

* No mention is made in the table of contents of introductory pages preceding it.
† This is often known as a "smyth-sewn binding".

protection for a book's pages than the method used for paperbacks, but at a higher cost.

Hardcover books are often further protected by a removable paper *jacket* (or *dust jacket*). This can become the subject of extensive design considerations, as can the cover, especially for a paperback. Both lend themselves well to imaginative layouts and multicolor printing. A *designer* employed by the publisher (or specially hired for the occasion) submits tentative design proposals for the cover and/or jacket of the book, although the author is usually encouraged to offer suggestions, including ideas for possible illustrative material. An author's proposals may be adopted, but they are likely to be extensively reworked by design artists at the publishing house. The danger of course exists that such reworking will lead to a wholly inappropriate result, perhaps one that would mislead potential customers. Should this occur the author need not hesitate to voice disapproval. Authors in any case have a responsibility to check all written portions of a proposed cover to ensure accuracy, as well as to verify that the subject title has been displayed exactly as it appears on the title page*.

The back side and the flaps of the jacket (or the back cover of a paperback or certain types of hardcover editions) are commonly used for providing information about a book. They thus serve as advertising space, sometimes not only for the book itself but also for related titles from the same publisher. The author may be asked to make suggestions for such promotional text (called *blurbs* by those in the business), and later to check it for accuracy, but the actual copy will be developed by the publisher.

Cover text often finds its way into brochures, advertisements, and other book publicity. In a sense, a book jacket or a paperback cover can itself be regarded as *promotional material*, because attractive "clothing" helps a book stand out on a dealer's shelves (ample reason for the author to have a say in its design!).

Shortly before publication, authors often receive an *Author's Questionnaire* from the publisher's marketing department.[†] If so, it should not be taken

---

* The title also appears on the *spine* of the book, but often without any subtitle. Interestingly, English-language publishers orient the spine title so that a vertical book (e.g., in a bookcase) must be viewed from the *left*, while in other countries the opposite orientation is customary. One might argue that the former convention is more practical, since it results in a legible title for a book placed face-up on the desk or a shelf.

[†] Information may also be requested of the author by ISBN and the editors of volumes such as *Books in Print*, who maintain extensive bibliographic files on authors known to be currently active.

lightly even if the questions appear to be ones long-since discussed and laid to rest. The purpose is to give the author one last opportunity to make comments that will influence the firm's marketing and advertising strategy. One is encouraged, for example, to:

● describe once again the target market for the book;
● provide a brief description of the contents, this time not only for the benefit of interested specialists but also for librarians and book dealers;
● compare the book with its possible competitors;
● list periodicals that might publish reviews;
● identify colleagues who should receive complimentary copies; and
● suggest professional meetings where the book might be effectively displayed.*

Most publishers are grateful for help and suggestions, here as throughout the process. They realize that no one knows a book as well as its author, who can now relax and look forward to the pleasure of seeing the fruits—including monetary ones—of a long and prodigious effort!

---

* Some publishing houses solicit this information from authors much earlier using a *project evaluation form*, with the help of which they compile a *basic facts sheet* on the book.

# Part II
# Scientific Writing:
# Materials, Tools, and Methods

# 5    *From Manuscript to Document*

## 5.1    *Introduction*

In this chapter we explore in some depth the matter of converting the polished *manuscript* of a report or thesis into a *finished document* suitable for sharing with others. Sections 5.2 and 5.3 address the major problem: preparing a manuscript whose professional *appearance* is commensurate with the high quality of the contents (assured by adherance to the guidelines in chapters 1 through 4). Here we are concerned primarily with final documents, not copy for a publisher. Nevertheless, many of the suggestions are also applicable to the preparation of *camera-ready copy* (cf. Sec. 3.4.6), as well as copy that will be *typeset*, but in both cases there are usually special rules to observe, particularly with respect to figures, tables, and footnotes. Relevant portions of chapters 3 and 4 should therefore be consulted as required (esp. sections 3.4 and 4.3.7). The format of a *thesis* is also subject to numerous restrictions (cf. Sec. 2.3.2), and university guidelines must always be given strict precedence.

A final manuscript is expected to be virtually perfect;* that is, it must be *complete, attractive,* and *error-free.* All of the "special elements" must be incorporated, such as tables and figures, and here we will assume that these are available in the form suggested in subsequent chapters. It is impossible to overstress the importance of neatness, consistency, and attention to details. Changing or correcting a final copy is both difficult and time-consuming, so every effort should be made to minimize the necessity for changes. Any that are required should be carried out according to the guidelines in Sec. 5.4.

*Duplication* of a report or thesis is normally accomplished either with a photocopier or by the photo-offset process, as discussed in Sec. 5.5. It is the resulting *duplicates* that should be *bound* or otherwise assembled for

---

\* To be more precise, it must lead to *photocopies* that are flawless. In practice, this means that the manuscript itself may contain certain imperfections, but these must be invisible to a photocopier (e.g., pasted corrections, cf. Sec. 5.4).

distribution (Sec. 5.5), while the *original* should be retained in single-sheet form, carefully preserved in case additional copies are later required.*

# 5.2    *Typewriter or Word Processor?*

## 5.2.1    *Typewriters*

The first decision to be reached is whether the final manuscript will be prepared with a typewriter or a word processor (cf. Sec. 4.3.5). Although we begin with a discussion of the procedure to be followed using a typewriter, this choice becomes less desirable with each passing year. Most of the objections once raised with respect to word processors (inordinate expense, inconvenient and limited software, poor print quality) have long since been addressed, and the advantages of using a word processor are formidable.† Today even typewriters are evolving in ways that make them increasingly like "mini word processors".

Nevertheless, some manuscripts are still typed in the traditional way, and if this is the route to be followed then *all* the text should be typed, including tables, figure captions, footnotes, references, and the table of contents. High-quality white bond paper produces the best results. The paper should have a weight of at least 80 g/m2, roughly equivalent to what in the United States is known as "20-pound bond".‡ Standard paper size for reports in the United

---

\* An exception to this rule would be a thesis prepared for a university that requires the submission of bound *originals*.

† A far-reaching further development of word processing is what has recently come to be known as *desktop* or *personal publishing* (cf. the last footnote in Sec. 4.3.5).

‡ Paper in the "20-pound" category has a weight of 74 g/m2. The archaic measuring system used in the United States is described in detail in *The Chicago Manual of Style* (1982). This source also contains information about U.S. and international paper size standards, as well as descriptions of various types of paper. *Rag content* is often used as a measure of paper quality. For example, some universities specify that theses must be prepared using paper with a minimum rag content of 25%.

States is 8 1/2 in. x 11 in. (ca. 216 mm x 279 mm), but nearly everywhere else "A4" paper (ca. 210 mm x 297 mm) is preferred.*

If possible, one should avoid using a *standard* (or *mechanical*) typewriter, because the resulting copy always has an uneven appearance. In contrast, a properly adjusted *electric* typewriter produces sharp uniform copy, particularly if one employs a single-use (carbon or film) ribbon. Moreover, most electric typewriters are now equipped with interchangeable *type elements* or *daisy wheels*, making it possible to introduce supplementary type faces (e.g., italics; cf. Sec. 4.3.7, "Choice of Type"). The choice between *pica* and the smaller *elite* type is a matter of individual preference. Elite has the advantage of producing a more compact manuscript, but pica is better if the manuscript is likely to be photoreduced. Type elements containing special *symbols* are also available. Their frequent use is inconvenient—but preferable to introducing all symbols by hand, particularly for a manuscript containing large numbers of mathematical formulas or physical equations.† Most sophisticated typewriters also provide a *correction* feature far superior to the alternative of correction fluid or tape (cf. Sec. 5.4) because it completely *removes* errors rather than simply covering them. On high-quality paper such corrections are virtually invisible.

## 5.2.2    Word Processors

### The Nature and Operation of a Word Processing Station

The advent of inexpensive *microcomputers* has made it practical (and desirable!) to replace typewriters with *word processors*—not only in offices but also in the home. Today, the price of a minimal word processing system barely exceeds that of a state-of-the-art electric typewriter.

---

* The "A" system of paper designation originated as a DIN Standard, but it has since been adopted by ISO for international usage. A4 paper is converted to the next smaller standard size (A5) by cutting its length in half. Each of the resulting pieces has exactly the same shape as the original, but only one-half the area. Repetition of the procedure again leaves the shape unchanged, but the size (comparable to a postcard) becomes A6. The process actually begins with a rectangular sheet defined as A0 and having the dimensions 841 mm x 1189 mm, with an area of 1 m². For any "A" format paper, the ratio of width to height is 1 : $\sqrt{2}$.

† Accessories also permit the typing of special symbols with a standard typewriter (cf. Sec. 8.1.3).

A typical *word processing station* consists of a *keyboard* similar to that of a typewriter, together with some type of *screen*, usually a *video monitor*. These elements are connected to a *computer* (sometimes housed in the same case as the keyboard), a *magnetic storage device* (usually one employing *diskettes*, perhaps supported by a larger *hard disk*), and a *printer*. In an office, all of these may be centralized for simultaneous access by several word processing stations. The computer is at the heart of the system, since it controls all the other devices and ensures that the operator's instructions are carried out.

In principle, preparing text on a word processor is equivalent to working with an electric typewriter. The major difference is that the entire text is "typed" before anything is printed, though previously entered material can at any time be displayed on the screen. As discussed below, this difference provides numerous advantages without necessarily introducing serious complications. Indeed, word processors can play a significant role at a very early stage in the writing process. For example, the first draft of this book was prepared with a word processor, one so small it could be carried easily in a briefcase. All text was stored temporarily on ordinary tape cassettes and later transferred for editing purposes to a more powerful desk-top microcomputer. This system in turn provided access to a wide range of typefaces, and it possessed outstanding graphics capability, permitting the ready incorporation of illustrations and formulas. Camera-ready copy was then prepared with the same system, taking advantage of sophisticated (but easy to use) *page makeup software* and a *laser printer* (see below).*

## Basic Advantages

What makes a word processor so valuable is the ease with which one can alter magnetically stored information—in this case an emerging manuscript, subject to repeated correction, abridgment, augmentation, and even reorganization. The removal of *errors* is a trivial matter. One simply calls the offending piece of text to the screen, moves an electronic "cursor" to the proper place, and

---

* The "briefcase computer" was a NEC model PC 8201-A. All subsequent development was carried out using Apple Macintosh™ equipment (512 K and Plus), partly supported by a Hyperdrive™ 20 hard disk (General Computer Corp.). Primary software included Microsoft Word™ 1.05 (Microsoft Corp., Redmont, Washington, USA; the dramatically improved version 3.0 had not yet been released) and PageMaker™ 1.2 (Aldus Corp., Seattle, Washington, USA). Camera-ready copy was produced on an Apple LaserWriter™ Plus.

executes a few straightforward keystrokes (essentially typing that which should have appeared in the first place). *Insertions, deletions,* and even *typeface changes* can all be similarly and instantaneously accomplished. Thus, if one wishes to insert an additional sentence, the cursor is placed at the desired point and the new sentence is simply typed; all surrounding text automatically adjusts itself to compensate for the change. A portion of text can be *italicized* by first "selecting" it (i.e., defining a set of boundaries) and then executing from the keyboard (or with a "rolling push-button" called a *mouse*) a single "change" command.

The powers of a word processor become particularly apparent when the need arises to construct a *table* (cf. Sec. 10.2.3). Proper tab settings are the key to success, and these can be altered repeatedly at any time, permitting nearly effortless optimization of column spacing. No matter how one modifies any piece of text the computer attends to all necessary spacing adjustments— everywhere in the document. Thus, the introduction of new text causes instantaneous revision of subsequent line and page breaks; no intervention is required on the part of the typist. The basic page format of a stored document can also be modified at will, and even the introduction of complex graphics into a manuscript can be a simple matter—if one has access to the proper type of system.

Contrary to what some believe, the decision to use a word processor need not bring about a revolution in one's work habits. Many writers who are firmly committed to the new technology continue to rely on traditional methods of revising their manuscripts. There is nothing to prevent one from working at leisure with a printed (or "hard") copy of the text, perhaps in a comfortable armchair, psychologically removed from the original creative act. After a standard pencil reworking of such a *print-out* one simply enters the desired changes into the computer and prints a new copy—clean, free of markings, and ready for another editing cycle. On the other hand, obvious problems—typographical errors, awkward phrases, poor grammar—can be corrected from the screen as soon as they are detected (cf. the first footnote in Sec. 4.3.6).

## *Choice of System*

Probably the most difficult part of making the transition to computerized writing is deciding what system to use. The number of available options is staggering, and unbiased advice is scarce, partly because of the astonishingly rapid pace of technical advancement.* The serious author would do well to study the matter rather carefully before making a commitment, because the required investment is certainly not trivial, and changing systems is both costly and inconvenient. For example, text prepared and stored on one type of computer is often virtually inaccessible to another. There are even limitations on the compatibility of different word processing programs for the same computer.

The most important challenge in selecting a system is ensuring that it is actually capable of accomplishing all that one requires. Thus, a student planning to prepare a thesis on a word processor should verify that the proposed system (hardware *and* software, see below) can produce text consistent in every way with university guidelines. This injunction applies not only to the quality of the print; it also means that both computer and text editing program must be capable of meeting all format requirements, including margin placement (on all sides), type size, pagination, and footnote style.

The science writer should be particularly concerned with the extent to which a given word processor can handle *subscripts* and *superscripts*, unusual *symbols*, mathematical *formulas*, and perhaps even diverse types of *graphic material*. Such capabilities are obviously not essential—one can always introduce the needed elements by hand at the end—but they are certainly welcome. However, this suggestion places significant constraints on potential systems, constraints affecting both *hardware* (computers, printers, etc.) and *software* (text editing and other programs, usually supplied separately in the form of diskettes).

### Hardware

Three major aspects of hardware are subject to variation: the computer, the storage device(s), and the printer.† The *computer* used for word processing may be either a large *mainframe* system, installed at a remote location and accessed through a *terminal*, or a smaller *personal computer*. Both types can be satisfactory, although the personal computer offers the advantage of portability, and the speed with which it operates is unaffected by the activities of others. Portability is particularly valuable to the part-time writer who may

---

* The best sources of current information are the many serial publications found today on nearly every newsstand.
† Other possible hardware acquisitions might include a *modem* (modulator/demodulator) for connecting one's system to others, and an *image scanner* for digitizing graphic copy.

prefer to do some work at home or during a vacation. The author who travels extensively may also want to give serious consideration to a *briefcase computer*, preferably one whose text files can easily be "uploaded" into a more sophisticated system.

It would be a mistake to choose a word-processing computer—or any important tool—solely on the basis of economy. Far more important in the long run are questions of reliability, convenience, service, and software support. The availability of *accessories* is also a factor, but sheer *size* is of secondary importance. Any computer containing the equivalent of 256 or 512 K ("kilobytes") of RAM (*random access memory*, used for temporary internal storage) should provide reasonable performance for most word-processing applications.

The ideal permanent medium for document storage is the *diskette*. (These are also called *floppy disks* or simply *floppies*.) Tape cassettes are less convenient because the associated access times are excessive. Especially recommended are the newer 3 1/2 in. diskettes encased in hard plastic mounts, since these require few special precautions with respect to storage or mailing. The double-sided version of a diskette of this type can hold 800 K of information, corresponding to more than 400 typed pages. A word processing system should normally be equipped with two diskette drives (or one diskette in conjunction with a *hard disk*, a much faster storage device with space equivalent to 10 or more floppies). One storage unit is required for system software, while the other assures ample space for document storage.

The *printers* supplied with most inexpensive personal computers are of the *dot-matrix* type. Characters are formed by combining a limited number of dots, which can be arranged only in certain ways. Early printers of this type were notorious for producing a characteristic "computer script" because the available dot patterns were quite restricted. Recent improvements have made it feasible to increase the dot density and to enlarge the "matrix" of dots (a 7 x 8 matrix is typical for a normal character set), at the same time eliminating the vertical restrictions that once prevented the proper printing of "tails" below the text line (the "descenders" required for correct presentation of letters such as "g" and "y").

A word processing system with an inadequate printer can usually be up-graded through the purchase of a so-called *letter-quality* printer. Some resemble expensive typewriters, but many dot-matrix printers also provide at least "near-letter quality", with the advantage of certain important features absent from other letter-quality machines. These include the ability to produce very acceptable subscripts and superscripts, as well as underlining, alternate type fonts (e.g., italics and boldface) and sizes, and sometimes graphics—although this capability is of little value unless it can be readily accessed by the associated word processing software.

One of the most important advances in output technology has been development of the device known as the *laser printer*. Printers of this type also operate on the dot-matrix principle, but their dot density is so high that discontinuities are difficult to perceive with the naked eye.* Laser printers with varying degrees of sophistication are available from several suppliers. All are still quite expensive, but the price is expected to decrease markedly, and the copy these machines produce is superb.†

Apart from print quality there are other printer characteristics worth considering. *Noise level* is certainly one, but perhaps the most important is convenience and flexibility with respect to *paper handling*. Automatic paper feed is almost essential; few writers are likely to tolerate the need to introduce each sheet singly and at the computer's request. The most common solution is a "tractor-feed" mechanism, requiring "fan-fold" paper. Even better is a reliable single-sheet feed device, since this eliminates the need to buy special paper, as well as the nuisance of separating pages and tearing off marginal perforations.

## *Software*

Software considerations are of four kinds: availability for a given computer, flexibility, ease of use, and cost.‡ Not every word-processing program is available for every computer, and with some computer types the choice is extremely limited. Assuming one does have a choice, however, it is probably best to select a program offering a wide range of features, such as unlimited document length, simultaneous access to more that one document (for easy "importation" of text), search and replace capability, automatic footnoting, spelling verification, and "WYSIWYG".§ Learning to use all the features may

---

* Such results are made possible by the remarkable extent to which a laser beam can be focused and directed.

† The present book can serve as evidence. As noted earlier, it was prepared almost entirely from camera-ready copy generated by a laser printer. Pages were initially prepared in A4 format for subsequent photoreduction by 20%.

‡ It is also worthwhile to investigate *compatibility* with respect to *auxilliary software* (e.g., database managers, outlining programs, graphic generators, and page makeup routines).

§ "What You See Is What You Get"; i.e., the screen version of a document bears a very close resemblance to that ultimately printed.

be a formidable challenge, but not every technique need be mastered at the beginning. Any features that are absent, however, will *always* be absent!

The availability of a wide range of "bells and whistles" should not be the only criterion. Serious attention should also be given to the way the features are accessed. Most authors are better served by "user-friendly" software employing efficient, lucid *menus* than by software that requires the memorization of two-letter acronyms or obscure *function key* commands. Cost is also a factor, but once again economic considerations should not be allowed to dominate, even though beginners are often appalled to discover how expensive it can be to acquire all the programming capabilities dealers delight in demonstrating. Indeed, many computer owners ultimately invest as much (or more) in their software as in the initial purchase of hardware!

### The Benefits of a "Magnetic Manuscript"

It is not our intent to delve further into the myriad technical details of computer-based text editing systems. In any case, the whole field is undergoing such rapid development that any recommendations we might make for specific systems would quickly be outdated. Instead, we conclude with a few general observations about the present and future status of computerized manuscripts, recognizing that capabilities today beyond the financial reach of many are tomorrow likely to be affordable.

The "computerized typewriter" has barely reached adolescence, but its potential impact is already clearly apparent. *Secretaries* with access to word processors commonly experience an increased sense of personal worth as they master a new piece of technology and become expert at polishing the documents they produce. In addition, their work output increases dramatically. Moreover, *writers* still accustomed to giving dictation or preparing hand-written drafts no longer hesitate to modify a first typed version because they know that a corrected copy can be prepared with relatively little effort. Those authors who own word processors report genuine pleasure at finding they can produce entire documents, from first draft to finished copy, without any help from a secretary.

But these psychological consequences represent only one aspect of the growing role of computers in writing. Perhaps more dramatic is the fact that many publishers are now in a position to arrange for *phototypesetting* directly from their authors' original diskettes (cf. Sec. 4.3.7, "New Demands and New Promises"). The realization of a way to avoid a "retyping" step is significant because it eliminates the introduction of errors and at the same time significantly reduces the labor costs associated with publication. Depending on the type of manuscript involved, and on the nature of the envisioned publication, the financial advantage may amount to between 10% and 60% of production costs (including preparation of offset plates, but excluding costs related to printing and binding). This approach is certain to become more general, a trend most authors welcome (partly because it greatly reduces the burden of proofreading).

Unfortunately, authors cannot yet assume their own diskettes will be appropriate for a given publisher's automated typesetting equipment; sufficient *standardization* of the technology has not yet occurred. Writers who use word processors should determine as early as possible whether their systems have characteristics matching the requirements of their publishers. Even if compatibility is assured, certain technical problems are likely to remain. In particular, it will probably be necessary for the author to introduce a host of special commands (*codes*) into each diskette-based manuscript to ensure proper handling of such features as indentations, superscripts or subscripts, and alternate type faces. Dodd (1986) provides specific information regarding preparation and submission of "magnetic manuscripts" for publication in ACS journals, and book publishers also issue guidelines to their authors.*

Numerous opportunities exist for exploiting the ease with which magnetically stored text can be manipulated. For example, a computer-based manuscript can easily be checked for *spelling errors* by utilizing a dictionary file, preferably one that can be supplemented with an appropriate list of technical terms. Moreover, a single command can change the spelling of a particular word everywhere it occurs in a manuscript, allowing one to accommodate the whims of any particular editor (e.g., "sulfur" instead of "sulphur"). Word division (*hyphenation*) at line ends can also be left to the discretion of a computer, and certain types of *grammatical errors* are amenable to automated detection. With certain systems it is possible to create automatically a *table of contents* for a document, complete with page numbers—or even an *index*, provided one first specifies the appropriate index terms (see Appendix F.3 for details). All that is required to take advantage of these and other possibilities is a combination of imagination and the proper software.

---

* Some authors have reluctantly concluded that the extra work presently involved is so considerable as to make the advantages of questionable value. However, this is almost certainly a temporary situation.

One additional advantage of using a word processor deserves special mention: since information on a magnetic disk is present in highly "concentrated" form, storage becomes a trivial consideration relative to the problem of finding space for a huge stack of typing paper, and a diskette is easy and inexpensive to mail. Furthermore, mailing can often be circumvented entirely in the case of computer-based information, because an equivalent set of electronic impulses is readily transmitted from one location to another over the public telephone lines. Transmission occurs at the speed of light and requires only that the sending and receiving computers be equipped with suitable modems. Some authors already use this means to deliver complete "manuscripts" to their publishers, and preliminary versions of the present book were passed back and forth across the Atlantic by telephone from one co-author's desk to another.*

## 5.3    Format

### 5.3.1    Margins and Line Spacings

We come now to the choice of *layout* (page arrangement) for the document one proposes to prepare with a typewriter or word processor. In some cases an author has little freedom in the matter because a particular format is specified by others. This is particularly true for a thesis, as repeatedly noted in Chapter 2. Publishers are also quite prescriptive with respect to the form of submitted manuscripts (cf. Sections 3.4 and 4.3.7), but copy for publication purposes is not our immediate concern. The following guidelines should be regarded only as suggestions, but they may prove helpful in the absence of any external constraints. They are in any case consistent both with common practice and with the recommendations found in numerous style guides.

Rules for *pagination* tend to be rather flexible. In general, page numbers are set about 15 mm below the top edge of the paper, separated from the *first line*

---

* One interesting challenge, however, is deciding how to indicate to the receiving party what *changes* the sender has made in a computer-based draft document. This is a subject treated at length by Roth (1984).

of text by two or three full spaces. Numbers are usually omitted from the first text page (which functions as "page 1"), as well as from preliminary pages (e.g., title page, preface, perhaps an abstract). The *last line* of text should appear about 30 mm above the bottom of the page. If a manuscript is to be bound, a *left margin* of at least 35-40 mm is required; otherwise, 30 mm is sufficient. The *right margin* may be as narrow as 10 mm.* If these guidelines are followed, each text line will be about 16 cm long.

Irregularity of the right-hand margin can be minimized by judicious *hyphenation* of long words. A word processor provides the more attractive alternative of *justified* margins, but even here hyphenation can be useful: it may in some lines prevent inordinately large spaces from appearing between words. Hyphenation can also reduce the number of pages required for a document—especially in the sciences, where long words abound.† *Careless* hyphenation is a frequent source of errors, however, particularly in English. A dictionary should always be consulted to verify the location of breaks between syllables.

In the interest of legibility, typewritten text should always be double-spaced (or perhaps one-and-one-half-spaced).‡ Double-spacing has the extra advantage that it maintains adequate separation between *indices* (subscripts and superscripts, a common feature in scientific text) and adjacent text lines.§ Narrower line spacing is permitted for subordinate material (such as captions, legends, footnotes, long quotations and, occasionally, the abstract), which is then readily distinguishable from the body of the text (cf. Fig. 5-1). Manuscripts intended for typesetting are double-spaced in their entirety, including all "special parts", thereby leaving room for subsequent editing.

---

* It should be stressed that these recommendations, like others in the chapter, do not apply to manuscripts that will be *typeset*. Typesetting is usually preceded by editing, so wider margins are needed—especially on the right—to leave room for editorial comments. As noted in Sec. 4.3.7 under the heading "How Clean is Clean?", some publishers provide authors with special manuscript paper on which margins are explicitly indicated.

† In the previous section it was pointed out that the word processor itself may provide help in this regard. Many advanced systems include automatic hyphenation capability, although it is often applicable only to relatively common words.

‡ "Double-spaced" is understood to mean double *line* spacing. On most typewriters, moving the paper up by two *notches* (typewriter spacings) changes the type position by only *one* text line. "Double-spacing", therefore, corresponds to four typewriter spacings.

§ Word processing systems sometimes provide automatic compensation for the presence of indices in the form of variable line-spacing, but this solution is of questionable value since it gives the page an irregular appearance.

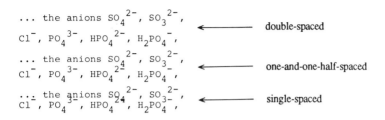

*Fig. 5-1.* Excerpt from a typewritten manuscript showing various line-spacings.

## 5.3.2    Headings, Paragraphs, and Lists

Chapter and section *headings* should be separated from text above and below by at least two full typewriter spacings. If headings and their associated numbers (Sec. 2.2.5, "Structure and Form") cannot be distinguished with the use of a different type face or size they should always be underlined.* Headings should be set either flush with the left margin or centered over the text, and the choice should be the same throughout a manuscript. This rule also applies to *table titles* and *figure captions* (cf. Sections 10.2.1 and 9.1.1, respectively).

In general, text is most readable if *paragraphs* are restricted in length to no more than half a page (cf. Sec. 1.4.2 and the discussion entitled "On the Length of Sections" in Sec. 4.3.3), and a full page should be regarded as a maximum.[†]

It is often advantageous to treat successive short *interrelated* text items (words, phrases—even paragraphs) as *lists* (cf. Sec. 10.3), in which individual components are labeled by number or letter; i.e.,

1. 2. 3.    or    a) b) c)[‡]

In this way the existence of a *relationship* is brought forcefully to the reader's

---

\* This is not true for manuscripts that will be typeset. A printer would interpret underlining as a signal for the use of italics—or perhaps boldface (cf. Sec. 4.3.7, "Choice of Type").

† This rule refers to paragraphs consisting *exclusively* of text; the inclusion of tables or figures can result in acceptable paragraphs that appear longer.

‡ Notice the convention of following *numbers* by periods and *letters* by a closing parenthesis, a practice that derives from an old printer's rule. Alternative forms such as 1), 2), 3) ..., or a., b., c., ..., are discouraged, and one should certainly avoid an *accumulation* of symbols, such as 1.), 2.), 3.), ....

attention without the need for extra headings. Numerical designations must be handled carefully if the manuscript also contains numbered subsections (as recommended in Sec. 2.2.5, "Structure and Form"); otherwise confusion might result. If the *sequence* of listed items plays no major role, and if no reference will be made to particular items within a list, then a neutral symbol is just as effective, such as a dash or a *bullet* (the large dot occurring frequently in this book).

Lists incorporated into the text should not be set apart by indenting. Superfluous indentation disturbs the uniform appearance of a page, and it wastes valuable space. Nevertheless, clarity requires that *formulas* and *equations* always be indented (*displayed*, cf. Sec. 8.1.1).

### 5.3.3    Notes

In scientific writing it is the custom that *substantive notes* containing peripheral explanatory material are printed as *footnotes* (cf. Sections 2.2.12 and 3.4.5). The corresponding *note references* within the text are generally *non-numerical symbols*; e.g.,

$$*, \dagger, \ddagger, \S, \#, \text{etc.*}$$

Footnotes are separated from text on the same page by a solid line (typically ca. 5 cm long) extending to the right from the left margin. Use of symbols rather than numbers for substantive note references prevents any confusion with numbered *source citations* (cf. Sec. 11.3). If for some reason substantive notes must be numbered (e.g., because they will appear at the *end* of the document and there are too many to permit convenient differentiation by symbol), then citation reference numbers are distinguished by enclosing them in *square brackets* set on the text line.[†] Punctuation marks (with the exception of the dash) always *precede* note symbols. Explanatory notes to *tables* are generally indicated by superscript lower case letters (e.g., ...[b]), and the notes themselves are placed directly under the corresponding table (for a typewritten example see Table 10-4b in Fig. 10-3).

---

[*] The order shown here is that recommended in the *The Chicago Manual of Style* (1982).
[†] Substantive notes and references are occasionally combined and numbered consecutively throughout a document. Such combined notes can either be presented as footnotes or collected at the end. Certain journals follow this practice, as described in Sec. 11.3.1.

Notes should always be single-spaced, except in copy to be typeset. Publishers usually require that notes be typed separately from the text manuscript, although some specify their inclusion directly *within* the text manuscript. In this case each note immediately follows the line containing the corresponding note reference, separated from surrounding text by a pair of solid lines extending the full distance between the margins.

Reference marks should be strictly avoided in titles and headings, a prohibition that applies to both substantive notes and literature citations. It is always possible to formulate the opening sentence of the subsequent text in such a way that it can incorporate any note reference one might otherwise be tempted to associate with the preceding heading.

## 5.4     Proofreading and Correction

After the final manuscript has been prepared, the next step is a thorough search for errors. The task confronting the writer (*not* a friend or secretary who may have been responsible for the typing) closely resembles the *proofreading* discussed in Sections 3.5.6 and 4.4.1. These sections should be reviewed prior to beginning the search for errors, even though in this case no actual (printed) "proofs" are involved.

No typist is perfect, and the claim that a fresh manuscript contains no mistakes is almost certainly a sign of a careless *reader*. Sufficient time must be allowed for study of the entire document, *word by word*. A dictionary should be kept handy, because the careful reader will certainly experience uncertainty about the spelling of at least a few words, or the proper placement of an occasional hyphen. Proofreading is slow, intellectually demanding work. We take the time to stress it—at the risk of irritating some of our readers—only because our experience as report *recipients* has made us painfully aware of how seldom the job is done properly. Mistakes are more serious than many seem to realize. Errors in a manuscript of course increase the likelihood of misunderstandings, but they also jeopardize a writer's reputation in general, because they cast doubt on the care he or she applies to *other* tasks. An equally important (though more subtle) threat lies in the fact that mistakes are an annoying source of distraction for the serious reader. Each recognized mistake draws attention away from the subject of the document, essentially lowering the "signal to noise ratio".

The most effective means of error *correction* depends on the nature of the problem. If one is dealing with a document prepared on a word processor, the answer is obvious: enter the changes, and then print the document—or a portion of it—again. Typewritten documents present a greater challenge. Changes can be attempted directly on the offending pages, preferably with the original typewriter's correction feature, but the results are unlikely to be satisfactory because it is extremely difficult to realign the paper with a sufficient degree of accuracy. Some typists are tempted to resort to the use of either correction fluid or correction paper, but the temptation should be resisted. Both techniques are intended to "cover up" errors rather than remove them. Correction paper is simply ineffective; correction fluid is quite opaque, but it leaves an irregular and impermanent surface.

Corrections on typewritten pages are best achieved by typing the proper characters on a separate sheet of paper, cutting out the appropriate strip, and gluing or taping it over the error. If this is properly done, photocopies of the original will show no trace of "patching". The most suitable adhesives are rubber cement and non-glossy transparent tape. The same technique even permits one to cut pages apart and reorganize them. This is also a common way to introduce formulas, equations, figures, and tables into text prepared with either a typewriter or a word processor.

## 5.5    Duplication and Assembly

In Sec. 5.1 it was noted that only *duplicate copies* of an original document should be permitted to leave an author's hands. Technology has dramatically changed the way duplicates are made. Onion-skin carbon copies have thankfully become a thing of the past, due to the nearly universal availability of xerographic dry-copiers. These permit the rapid preparation of an unlimited number of *photocopies*, all of equal quality, on paper comparable to that used for typing the original. Such copiers can even be equipped to sort (*collate*) multiple copies of lengthy documents, and some can produce copies of a *size* different from the original. A reduction to roughly 70% scale is particularly advantageous because it converts standard typed characters to ones approximately the size found in typical printed text. Such copies remain clearly legible, but they take up less space than the original.

Proper attention must be given to the *quality* of the copies produced by a given copier—and on a given day. If the nearest machine is producing unsatisfactory results (e.g., the copies are too light, or they contain unsightly streaks) it is well worth the trouble to look elsewhere, perhaps even to engage the services of a professional copy shop willing to guarantee its product.

Standard photocopying is economically feasible only for relatively small jobs. If the required number of copies of a given page exceeds 20 or 30 it would be wise to inquire at local printshops about the availability and cost of other duplicating processes. Most universities offer a variety of copying options through their own in-house printshops.

Finally, the copies of a report—complete with title page and perhaps a blank cover sheet at the end—must be *assembled* for distribution. A good report warrants better treatment than stapling or the attachment of a paper clip. Stationery dealers offer a wide range of report binders—some of which, unfortunately, are worthless. Binders that rely on frictional force supplied by a plastic spine may look attractive, but too often they allow pages to slip out, and the problem intensifies as the plastic ages. It is safer to rely on the type of binding that requires holes punched in the edge of the document. The best of this kind must be attached with the aid of special machines, requiring that one seek the help of a professional. Whether or not this is warranted depends on the importance of the report—and the extent to which the recipient may deserve (and appreciate) the extra effort.

# 6 Chemical Nomenclature

## 6.1 Historical Perspective

In the course of time, nearly every discipline acquires its own peculiar system for naming, symbolizing, and coding the objects of its investigations. Chemistry is no exception. In this chapter we offer a brief introduction to the structure and current rules of *chemical nomenclature*, with particular emphasis on the underlying systematics.

Many of the chemical elements and a host of individual compounds have been known since antiquity. Names were assigned to these long ago, of course, but the historical names rarely provide chemists with any insight into the nature of the substances themselves. The European alchemists and iatrochemists of the 16th and 17th centuries did little to improve the situation, but they can be credited with first introducing the practice of representing chemical substances *symbolically*, e.g.:

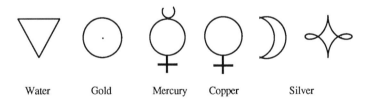

| Water | Gold | Mercury | Copper | Silver |

Unfortunately, the philosophical traditions within which these practitioners worked encouraged them to cloak most of their efforts in secrecy. Moreover, different symbols were employed in different cultures and epochs (as suggested by the fact that two symbols are shown above for silver*), and alchemists were notoriously inconsistent in their use of specific symbols.

---

\* The second of these is familiar as the logo of VCH Verlagsgesellschaft and VCH Publishers.

Eventually the confusion and ambiguity associated with chemical names and symbols was largely dispelled. Much of the credit for this advance is due to early efforts by a few outstanding individuals, first among whom was Guyton de Morveau (*Méthode de nomenclature chimique*, 1787). Others following in the same tradition included Lavoisier, Berzelius, Liebig, Mendeleev, and Meyer.

Chemical nomenclature has always been complicated somewhat by the existence of two apparently distinct branches of the discipline, *inorganic* and *organic*. Each has its own organizational scheme, and the two were long preoccupied with different types of chemical problems. Not surprisingly, two different approaches to nomenclature also developed. Today, the lines of demarcation separating the two subdisciplines are no longer so distinct. In part this is a consequence of recent interest in so-called interdisciplinary (or "transdisciplinary") research, conducted under headings such as *bioinorganic* and *organometallic* chemistry. Nevertheless, two somewhat independent systems of nomenclature remain, a circumstance reflected in the structure of this chapter.

However, one particular development has had a major unifying influence on naming throughout the discipline. Responsibility for coordinating all practices concerning chemical nomenclature is now in the hands of a single international body, the International Union of Pure and Applied Chemistry (IUPAC).

# 6.2    IUPAC Rules for Inorganic Nomenclature

## 6.2.1    The Origin of the Rules

Present-day chemical nomenclature had its formal origin in 1921 with the establishment of a Commission on Nomenclature of Inorganic Chemistry by the International Union of Chemistry (IUC, a predecessor of IUPAC). This group undertook to conceive, discuss, and draft a comprehensive set of "rules" for the naming of all inorganic compounds. A major challenge recognized from the beginning was the need for chemical names that could easily be adapted to all the world's major languages.

Many meetings were required before the Commission finally issued a definitive report: *Nomenclature of Inorganic Chemistry* (IUPAC, 1971).* The system described in this document (called the "Red Book") is generally regarded as the best approach currently available for naming inorganic compounds, but it is nonetheless assumed that certain specific features may eventually need to be altered or even discarded.

The principal motivation for devising rules for naming compounds is the desire to provide chemists with words or sets of words that are *unique* to particular substances. However, an ideal name is more than unique: it also conveys *information* about the substance it represents. A name should at least imply a particular *empirical formula*, but it is desirable that it also reveal significant *structural features* of the compound in question. One skilled in using the rules of nomenclature should be able to envision a specific molecular structure solely on the basis of a compound's name, even in the case of an unfamiliar substance.[†] These criteria were applied during the development of the IUPAC system of nomenclature described briefly[‡] in the following pages (for more detail see, e.g., Cahn and Dermer, 1979; Hellwinkel, 1974).

## 6.2.2 Writing Chemical Formulas

### Elements

The first problem confronting IUPAC was that of providing all the chemical *elements* with unique names, and also with universally acceptable *symbols*

---

[*] This report was an outgrowth of the so-called "1940 Rules", and was preceded by an earlier edition (the "1957 Rules", published in 1959).

[†] This goal has indeed been met by IUPAC. Unfortunately, however, the converse does not always apply. It can often be shown that a given structure is consistent with various names, all of which are in general accord with IUPAC rules. Thus, the rules of nomenclature are not "strict" in the sense that a mathematician might expect, particularly the rules applied to organic compounds (Sec. 6.3). In many ways this is actually an advantage, but it poses a serious problem for a comprehensive indexing service like Chemical Abstracts. Indeed, CA is forced to adapt the IUPAC rules somewhat so they become strict, thereby ensuring consistent use of a single unique name for each known compound (cf. the comments at the end of Sec. 6.2.4, but also Sec. 6.4). CA adheres firmly to whatever names it has chosen, although changes in practice are sometimes introduced at the beginning of a new period of literature coverage.

[‡] The terminology and many of the examples employed throughout the chapter are drawn directly from appropriate IUPAC publications.

*Table 6-1.* Official IUPAC English-language list of the chemical elements.

| Name | Symbol | Atomic number | Name | Symbol | Atomic number |
|---|---|---|---|---|---|
| Actinium | Ac | 89 | Iridium | Ir | 77 |
| Aluminium | Al | 13 | Iron (Ferrum) | Fe | 26 |
| Americium | Am | 95 | Krypton | Kr | 36 |
| Antimony | Sb | 51 | Lanthanum | La | 57 |
| Argon | Ar | 18 | Lawrencium | Lr | 103 |
| Arsenic | As | 33 | Lead (Plumbum) | Pb | 82 |
| Astatine | At | 85 | Lithium | Li | 3 |
| Barium | Ba | 56 | Lutetium | Lu | 71 |
| Berkelium | Bk | 97 | Magnesium | Mg | 12 |
| Beryllium | Be | 4 | Manganese | Mn | 25 |
| Bismuth | Bi | 83 | Mendelevium | Md | 101 |
| Boron | B | 5 | Mercury | Hg | 80 |
| Bromine | Br | 35 | Molybdenum | Mo | 42 |
| Cadmium | Cd | 48 | Neodymium | Nd | 60 |
| Caesium | Cs | 55 | Neon | Ne | 10 |
| Calcium | Ca | 20 | Neptunium | Np | 93 |
| Californium | Cf | 98 | Nickel (Niccolum) | Ni | 28 |
| Carbon | C | 6 | Niobium | Nb | 41 |
| Cerium | Ce | 58 | Nitrogen | N | 7 |
| Chlorine | Cl | 17 | Nobelium | No | 102 |
| Chromium | Cr | 24 | Osmium | Os | 76 |
| Cobalt | Co | 27 | Oxygen | O | 8 |
| Copper (Cuprum) | Cu | 29 | Palladium | Pd | 46 |
| Curium | Cm | 96 | Phosphorus | P | 15 |
| Dysprosium | Dy | 66 | Platinum | Pt | 78 |
| Einsteinium | Es | 99 | Plutonium | Pu | 94 |
| Erbium | Er | 68 | Polonium | Po | 84 |
| Europium | Eu | 63 | Potassium | K | 19 |
| Fermium | Fm | 100 | Praseodymium | Pr | 59 |
| Fluorine | F | 9 | Promethium | Pm | 61 |
| Francium | Fr | 87 | Protactinium | Pa | 91 |
| Gadolinium | Gd | 64 | Radium | Ra | 88 |
| Gallium | Ga | 31 | Radon | Rn | 86 |
| Germanium | Ge | 32 | Rhenium | Re | 75 |
| Gold (Aurum) | Au | 79 | Rhodium | Rh | 45 |
| Hafnium | Hf | 72 | Rubidium | Rb | 37 |
| Helium | He | 2 | Ruthenium | Ru | 44 |
| Holmium | Ho | 67 | Samarium | Sm | 62 |
| Hydrogen | H | 1 | Scandium | Sc | 21 |
| Indium | In | 49 | Selenium | Se | 34 |
| Iodine | I | 53 | Silicon | Si | 14 |

*Table 6-1.* Continued.

| Name | Symbol | Atomic number | Name | Symbol | Atomic number |
|---|---|---|---|---|---|
| Silver (Argentum) | Ag | 47 | Tin (Stannum) | Sn | 50 |
| Sodium | Na | 11 | Titanium | Ti | 22 |
| Strontium | Sr | 38 | Tungsten (Wolfram) | W | 74 |
| Sulphur | S | 16 | Uranium | U | 92 |
| Tantalum | Ta | 73 | Vanadium | V | 23 |
| Technetium | Tc | 43 | Xenon | Xe | 54 |
| Tellurium | Te | 52 | Ytterbium | Yb | 70 |
| Terbium | Tb | 65 | Yttrium | Y | 39 |
| Thallium | Tl | 81 | Zinc | Zn | 30 |
| Thorium | Th | 90 | Zirconium | Zr | 40 |
| Thulium | Tm | 69 | | | |

(*element symbols*).* Since inorganic chemistry is the branch most closely associated with the breadth of the periodic table, it was in the context of inorganic nomenclature that IUPAC established and disseminated rules for the elements.

The current IUPAC list of elements for use in English-speaking countries is contained in Table 6-1. Even a cursory inspection will quickly reveal that not all the names listed are "systematic". IUPAC accepted as inevitable the persistence of many traditional names, despite the fact that "common" or *trivial names* vary considerably from one language to another, as the following examples illustrate:†

| Element | English | German | French |
|---|---|---|---|
| N | nitrogen | Stickstoff | azote |
| Fe | iron | Eisen | fer |
| Cu | copper | Kupfer | cuivre |
| Au | gold | Gold | or |
| Pb | lead | Blei | plomb |

* The challenge is similar to that which IUPAC accepted with respect to quantities and units (cf. Chapter 7).
† Slight differences even exist among lists of elements appearing within the English-speaking realm. American chemists nearly always employ the spellings "cesium", "sulfur" and "aluminum", while their British counterparts prefer "caesium", "sulphur" and "aluminium" (cf. the disagreement with respect to proper spelling of the units for length and volume noted in a footnote to Sec. 7.3.2). We have adopted the American forms in the interest of consistency.

The problem is alleviated somewhat by the fact that many substances *containing* these elements are named in an internationally uniform way since for most elements there is also a *systematic name*. (Those differing from the English names are shown in parentheses in Table 6-1.) Thus, one speaks of a "ferrate", "cuprate", or "aurate"—not of an "ironate", "copperate", or "goldate". The elements sulfur and nitrogen lack official systematic names, but names for sulfur compounds commonly include the syllable "thio-" (from Greek, $\theta\varepsilon\iota o\nu$ "brimstone"), and nitrogen compounds are regularly distinguished by the syllable "azo-" (from French *azote*).

Hydrogen is the only element for which individual *isotopes* have been given special names: "deuterium" (symbol D) and "tritium" (T), the isotopes with relative atomic masses two and three, respectively. Isotopes of all other elements are designated by appending a hyphen and the appropriate mass number to the name of the corresponding element, as in "carbon-13".

Provision has already been made for future additions to the list of elements. For example, any name assigned to a metallic element should end in the syllable "-ium", consistent with the way most existing metals are named (e.g., sodium, uranium). Elements with atomic numbers beyond 105 have been given *tentative names* reflecting the corresponding atomic number. The provisional symbols for such elements are unique in that they consist of three letters rather than one or two. Thus, element 107 is "unnilseptium", with the symbol "Uns".

A complete *element symbol* consists of one or more letters (the first of which is capitalized) surrounded by a set of numbers: the element's *mass number* (as a left upper index), *atomic number* (left lower index), and *ionic charge* (right upper index). The right lower position is reserved for a *multiplier*, indicating how many such atoms are present in a given molecular array. Thus, the complex symbol

$$^{15}_{7}N^{3-}_{2}$$

represents a species comprised of two nitrogen atoms (atomic number 7), each with the mass number 15, and three extra electrons. The atomic number is actually redundant, and in most cases it can be omitted. The absence of a specification of mass number in an element symbol is taken to signify an isotopic composition equivalent to that found in nature. Notice in the example that the Arabic numeral specifying the magnitude of the charge *precedes* the corresponding sign. The numeral "1" is usually omitted in this context. Thus, simple species like the magnesium cation and the chloride and peroxide anions are written:

$$Mg^{2+} (\text{not } Mg^{+2} \text{ or } Mg^{++}); \quad Cl^- (\text{not } Cl^{1-}); \quad O^{2-} (\text{not } O^{-2})$$

Certain elements exist in various gaseous, liquid, or solid *modifications* known as *allotropes*, usually comprised of different numbers of atoms. These are named with the use of *multiplying affixes** based on Greek numbers (cf. Sec. 6.2.3, "Binary Compounds and Other Simple Substances"). Occasionally such names are further elaborated with *structural affixes*. The following examples are illustrative:

| Symbol | Trivial name (English) | Systematic name |
|---|---|---|
| H | atomic hydrogen | monohydrogen |
| $O_2$ | (molecular) oxygen | dioxygen |
| $O_3$ | ozone | trioxygen |
| $P_4$ | white (yellow) phosphorus | *tetrahedro*-tetraphosphorus |
| $S_8$ | $\mu$-sulfur | *catena*-polysulfur, polysulfur |

Certain *groups* of elements have been provided with *collective names*. These include the *halogens* (F, Cl, Br, I, At), the *chalcogens* (O, S, Se, Te, Po),[†] the *alkali metals* (Li, Na, K, Rb, Cs, Fr), the *alkaline-earth metals* (Be, Mg, Ca, Sr, Ba, Ra), the *noble gases* (He, Ne, Ar, Kr, Xe, Rn), the *lanthanides* (elements 57 ... 71, La through Lu), and the *actinides*[‡] (elements 89 ... 103, Ac through Lr).

## Compounds

*Compounds* are most concisely represented by collections of element symbols. The proper *sequence* of the symbols depends somewhat on the circumstances, as will be apparent from the following discussion. Indices are normally omitted, with the exception of *subscripts* to indicate more than one atom of a given type. If subscripts are used to group all atoms of each element, what results is a *molecular formula*. Such a formula accurately reports the

---

* Affix" is a rarely encountered term covering letter combinations used in any of three ways: as *prefixes*, *suffixes*, or *infixes*. Its use here is a signal that the syllables in question are sometimes preceded by other syllables (other affixes), and therefore they cannot be classed strictly as prefixes.

† Collective names also exist for *compounds* of the first two sets of elements. One can thus refer to halogenides (or halides) and chalcogenides.

‡ The last two groups have been named as they would be in the United States. IUPAC prefers the terminology "lanthanoids" and "actinoids".

*stoichiometry* of a compound, but it provides no direct information about *structure*.\* More informative *structural formulas* (*compound symbols*) can be constructed by selective placement of at least some of the symbols.† Numerous examples appear in this chapter and in Chapter 8. Discussion of the *naming* of compounds is deferred to Sec. 6.2.3.

*Binary compounds*—molecular entities comprised of two elements—are generally written with the more *electropositive* constituent first. For example, one should always write "KCl" and "MgS", not "ClK" or "SMg". For binary compounds between non-metals the proper sequence of elemental symbols is

Rn, Xe, Kr, B, Si, C, Sb, As, P, N, H, Te, Se, S,
At, I, Br, Cl, O, F,

resulting in formulas such as $XeF_4$, $H_2S$, $SF_6$, $Cl_2O$, and $OF_2$. In the case of *intermetallic compounds* the element symbols are arranged in *alphabetical order* (e.g., $CuSn_2$, $SnTa_5$). Deviations are allowed, however, and this is useful if one wishes to suggest the presence of ionic character (e.g., $Na_3Bi_5$), or to underscore relationships or analogies between substances (e.g., $Cu_5Zn_8$ and $Cu_5Cd_8$).

*Isotopic labeling* can be specified in a formula by introducing the proper left upper index. An analogous symbolic notation is also employed when one wishes to incorporate reference to a particular isotope into the *name* for a compound. In this case the letters comprising the isotope symbol are set in italics, and the entire symbol is enclosed in brackets and appended to the appropriate word; e.g:

$^{32}PCl_5$   phosphorus[$^{32}P$] pentachloride
$^2H^{36}Cl$   deuterium chloride[$^{36}Cl$]
$^{15}N^3H_3$   ammonia[$^{15}N$, $^3H$]

---

\* The element symbols are usually listed in alphabetical order; however, in view of the prominence of the elements carbon and hydrogen in organic chemistry, the symbols C and H are frequently placed first. Sodium acetate might therefore be symbolized $C_2H_3NaO_2$. In certain cases an even simpler formula may be appropriate, one that shows only *relative* stoichiometry. A formula of this kind is called an *empirical formula*. For example, ethane, $C_2H_6$, has the empirical formula $CH_3$.

† An example of a structural formula of sodium acetate would be $CH_3COONa$, or $[CH_3COO]^- Na^+$, or

$$H_3C—C{<}^O_O{-}\ Na^+.$$

According to IUPAC, a structural formula is a description of "the sequence and spatial arrangement of the atoms in a molecule" (*J. Amer. Chem. Soc.* **1960**, *82*, 5517-5584).

In some cases, an entire atomic grouping must be singled out and added to a formula (or even a name) in order to establish clearly the *location* of a particular isotopic label; e.g.:

$NO_2{}^{15}NH_2$  nitramide[$^{15}NH_2$].

Inorganic compounds comprised of *more than two elements* are treated much like binary systems. Electropositive elements are listed first, followed by the electronegative elements. Both sets are usually arranged alphabetically. However, in the case of small molecules (or ions) in which the atoms form a *chain* it is customary to write the formula with the elements arranged in an order consistent with the bonding pattern, starting from the most electropositive end; e.g.:

HOCN (cyanic acid)
HONC (fulminic acid)
$NCS^-$ (not $CNS^-$ or $SCN^-$)

This represents a first step in the direction of the *structural formulas* mentioned previously. If two or more different atoms or groups are attached to a single *central atom*, the symbol of the central atom is placed first, followed by the symbols of the remaining atoms or *coordination groups* (*ligands*), arranged in alphabetical order; e.g., $PBrCl_2$, $SbCl_2F$, $PO(OCN)_3$. As shown in the last example, the presence of several identical *groups* of atoms is indicated by enclosing the appropriate symbols in parentheses or square brackets and appending the proper subscript multiplier. Parentheses should never be "nested" in a formula. Instead, one employs the combination of parentheses enclosed in square brackets, illustrated by the following example of a substance containing a complex cation:

$[Cu(NH_3)_4]Cl_2$

## A Few Special Situations

*Water of crystallization* and other loosely attached (solvent) molecules are denoted in a special way. Their multipliers are placed *on* the text line and *preceding* the corresponding portion of the formula; e.g.:

$Na_2SO_4 \cdot 10\,H_2O$

Note also the use of *centered dots* to separate the individual components of the solvate. Solvated species are also *named* in special ways (cf. Sec. 6.2.3, "Binary Compounds and Other Simple Cases").

The formula for a *free radical* should always show the presence of the unshared electron. This is done by placing a *dot* either above the atom on which the electron is formally localized or centered at the end of the formula; e.g.,

$$H_3C\cdot \, , \, HO\cdot \, , \, \overset{.}{N}H_3^+$$

Finally, crystalline phases of *variable composition* are represented by either of two special notations depending on the intended degree of precision. Thus, a symbol like (Cu,Ni) can be used to denote mixtures of copper and nickel over the complete range from one pure substance to the other. Likewise, K(Br,Cl) implies the range from pure KBr to pure KCl. A more detailed formula would indicate atom-for-atom substitution of element A by element B; e.g.:

$$A_{m+z}B_{n-z}C_p$$

A particular composition range is specified by supplying values for the variables. For example:

$$KBr_xCl_{1-x}, \; 0.2 < x < 0.8$$

*Oxidation Numbers*

The *oxidation number* of an atom in a compound is the charge that atom would bear if the electrons comprising each bond were assigned to the more electronegative of the two connected atoms. Oxidation numbers often provide insight into chemical behavior, and for this reason they are occasionally incorporated into atom and compound symbols. Roman numerals set in parentheses are used for this purpose. Oxidation number zero is represented by a cipher. The following examples illustrate the principle:

| Chemical formula | Hypothetical composition | Symbolic representation |
| --- | --- | --- |
| $CH_4$ | one $C^{4-}$ and four $H^+$ ions | C(-IV), H(I) |
| $NF_4^+$ | one $N^{5+}$ and four $F^-$ ions | N(V), F(-I) |
| $[Ni(CO)_4]$ | one uncharged Ni atom and four uncharged CO molecules | Ni(0), C(II), O(-II) |

Oxidation numbers also play a role in the naming of compounds when the *Stock system* is used (cf. "Coordination Compounds" in Sec. 6.2.4).

## 6.2.3 The Naming of Inorganic Compounds

### Binary Compounds and Other Simple Cases

The IUPAC *names* for inorganic compounds, like chemical formulas, indicate what *constituents* are present and in what *proportions*. For a binary compound, the name begins with the name of the more electropositive of the elements. This is followed by a word derived from the *systematic* name for the electronegative element, adapted so that it ends in the syllable "-ide". The proportions of the two elements are specified by *multiplying affixes* representing the various integers. The first twelve of these affixes are

> mono, di, tri, tetra, penta, hexa, hepta, octa, nona (or ennea), deca, undeca (or hendeca), dodeca.

The appropriate affix is placed directly *before* the word for the element to which it applies (i.e., as a *prefix*), without interruption by either a space or a hyphen. The affix "mono" is often omitted; it is simply taken for granted in the absence of any other numerical indicator. Even "di" and "tri" are frequently dropped in the case of simple salts where no ambiguity can arise. The following examples illustrate the use (or omission!) of multiplying affixes:

| | |
|---|---|
| $CaCl_2$ | calcium chloride |
| $Na_2S$ | sodium sulfide |
| $AlCl_3$ | aluminum trichloride |
| | (the "tri" would here be considered optional) |
| $AgBr$ | silver bromide |
| $KI_3$ | potassium triiodide |
| $Li_2Pb$ | dilithium plumbide |
| $CO_2$ | carbon dioxide |
| $N_4S_3$ | tetranitrogen trisulfide |

The rules of *alphabetization* of compound names (e.g., for inclusion in an index) require that numerical affixes be ignored. For this reason, compound names appearing in an alphabetical context often have their prefixes italicized and set apart by hyphens. Thus, $U_3O_8$ might be found listed in a manufacturer's catalog as "*tri*-uranium octaoxide".

The same affixes used in a slightly different way sometimes convey other types of information, particularly in the case of more complex entities. Thus, an affix can indicate extent of substitution   (e.g., $SiCl_3H$, trichlorosilane; $PO_2S_2^{3-}$, dithiophosphate ion), number of identical coordination groups (e.g., $[CoCl_2(NH_3)_4]^+$, tetraamminedichlorocobalt(III) ion), number of identical central atoms in a condensed acid or its ions (e.g., $H_2Cr_2O_7$, dichromic acid; $H_2S_3O_{10}$, trisulfuric acid; $P_3O_{10}^{5-}$, triphosphate ion), or number of atoms of some skeletal element in a particular type of molecular or ionic array (e.g., $B_{10}H_{14}$, decaborane(14); $Si_3H_8$, trisilane).

In addition to the multiplying affixes, a set of so-called *multiplicative* affixes has also been defined. These have the form bis, tris, tetrakis, pentakis, etc. The function of a multiplicative affix is to indicate the extent of a set of identical subunits, each member of which is connected to some central atom in the same way. For example, $P(C_{10}H_{21})_3$ is called "tris(decyl)phosphine". The purpose of having two different types of affixes is to avoid potential ambiguity, such as that which might exist between the compound just cited (containing three "decyl" groups, cf. Sec. 6.3.2, "Alkanes, Cycloalkanes, and Alkyl Radicals") and the substance tridecylphosphine, $PH_2(C_{13}H_{27})$.

*Arabic numerals* also play a role as affixes. Their assignment is to serve as *locants*, specifying particular atoms within an atomic chain, ring, or cluster at which substitution, replacement, or addition has occurred; e.g.:

$H_3Si$-$SiHCl$-$SiH_2$-$SiH_2$-$SiH_3$
2-chloropentasilane

Arabic numerals are also used as *multipliers* within names, but only for indicating molecular proportions in solvates. For example, $CaCl_2 \cdot 6\ H_2O$ is sometimes referred to as "calcium chloride-water (1/6)" or "calcium chloride 6-water", although the name "calcium chloride hexahydrate" is also permitted.

*More Complex Entities*

Compounds containing two or more independent electronegative constituents are named by citing portions of the corresponding (systematic) elemental names in *alphabetical order*. The last member of such a set is given the usual compound ending "-ide", but all others are terminated with "-o". Thus, PbClF would be given the name "lead chlorofluoride".

A few complex cations and anions are treated for naming purposes as if they were elemental units, leading to the term *pseudo-binary* systems; e.g.:

| | |
|---|---|
| $NH_4Cl$ | ammonium chloride |
| NaCN | sodium cyanide |

Many polyatomic anions are named by first selecting the atom (or group of atoms) that is most *characteristic* or *central* with respect to the whole. All components are cited in the name, but the central or characteristic portion is cited last, and its name is modified to end in the syllable "-ate"; e.g.:

| | |
|---|---|
| $BrO^-$ | oxobromate |
| $ICl_4^-$ | tetrachloroiodate |
| $SCN^-$ | thiocyanate |

Such polyatomic groups are known as *complexes,** and the atoms, radicals, or molecules bound to the central (or characteristic) atom are called *ligands*.

A similar approach can be seen in an *abbreviated naming sytem* often applied to anions derived from oxoacids. This is illustrated by the name "trisodium phosphate" for $Na_3[PO_4]$, in which the presence of oxygen must be inferred. This type of name has the serious disadvantage that it fails to specify stoichiometry. A partial remedy is provided by the "rule" that if the characteristic element shows a "lower than normal" oxidation number the terminating syllable is changed to "-ite". For example, $KClO_3$ is designated as "potassium chlorate", while $KClO_2$ is "potassium chlorite", because the anions contain $Cl(V)$ and $Cl(III)$, respectively.[†] Such abbreviated names are convenient, but the longer systematic names are more informative, as illustrated by the following examples:

| Chemical formula | Abbreviated name | Systematic name |
|---|---|---|
| $Na_2[SO_4]$ | sodium sulfate | disodium tetraoxosulfate |
| $Na_2[SO_3]$ | sodium sulfite | disodium trioxosulfate |

Compound names must sometimes be supplemented to include *stereochemical information* in order to permit a distinction beetween *stereoisomers*. This is also accomplished by means of prefixes, including

$(R)$-, $(S)$-, $(E)$-, $(Z)$-,[‡] *meso-*, *asym-*, and *cyclo-*.

Prefixes of this type are always *italicized* and connected to the remainder of the

---

* Sometimes it is the substances containing such groups (i.e., in conjunction with other groups or ions) that are called "complexes" (cf. Sec. 6.2.4).
† The tradition from which this distinction derives was actually quite elaborate. A name with the suffix "-ate" could be further modified by the prefix "per-" to show an exceptionally high oxidation state, while an exceptionally low oxidation state was indicated by use of the prefix "hypo-" in conjunction with the suffix "-ite"; e.g., $KClO_4$ contains $Cl(VII)$ and had the name "potassium perchlorate", while the $Cl(I)$ in $KClO$ dictated the name "potassium hypochlorite". Today such nomenclature is regarded as obsolete.
‡ The first four examples are replacements for the less definitive (and thus less desirable) D-, L-, trans-, and cis-.

name by a hyphen. Italics are also used when an element *symbol* is included in a name. Often this is done to provide a *locant*—designating, for example, which atom in a heteroatomic chain or ring bears a substituent (e.g., $CH_3ONH_2$, *O*-methylhydroxylamine), or which atom of a ligand is coordinated to the central atom. Italicized element symbols can also be used to specify the presence of a bond between two metal atoms, to identify the point of attachment of one group to another, or to define the nature of a particular isotope present in a labeled compound; e.g.:

$[(OC)_3Fe(C_2H_5S)_2Fe(CO)_3]$
   bis($\mu$-ethylthio)bis(triscarbonyliron)(*Fe-Fe*)
$CH_3ONH_2 \cdot BH_3$   *O*-methylhydroxylamine (*N-B*)borane
$H_2{}^{18}O$   water[$^{18}O$]

Sometimes symbols or lower case letters are used as locants—to specify the spatial position of a particular group relative to the central atom in a coordination compound, for instance. These are also printed in italics as shown in the first example above, where "$\mu$-" indicates a bridging group.

## 6.2.4     Additional "Systematic" Aspects of Inorganic Nomenclature

### Coordination Compounds

Current IUPAC recommendations permit the limited use of several additional systems for naming inorganic compounds. Each of these systems has the advantage of emphasizing some particular aspect of compound structure. For example, the *Stock system* is widely used in conjunction with compounds of elements whose oxidation numbers vary (e.g., the transition metals). Oxidation states are made explicit (cf. "Oxidation Numbers" in Sec. 6.2.2), often at the expense of a clear statement of stoichiometric proportions. However, the latter can usually be inferred (after a bit of reflection) from the information provided. The relevant oxidation numbers are specified as Roman numerals set in parentheses immediately following the names of the corresponding elements.* e.g.:

---

* Unfortunately, the first two compounds continue in many quarters to be referred to as "ferric chloride" and "ferrous chloride". This archaic way of specifying oxidation states is no longer acceptable, and its use should be strongly discouraged.

| Molecular formula | Stock name | Standard systematic name |
|---|---|---|
| $FeCl_3$ | iron(III) chloride or ferrum(III) chloride | iron trichloride |
| $FeCl_2$ | iron(II) chloride or ferrum(II) chloride | iron dichloride |
| $Pb_3O_4$ (2 $Pb^{2+}$, $Pb^{4+}$, 4 $O^{2-}$) | | |
| | dilead(II) lead(IV) oxide | trilead tetraoxide |
| $Na_2[Fe(CO)_4]$ | sodium tetracarbonylferrate(-II) | disodium tetracarbonylferrate |

When *coordination compounds* or complex ions are named by this system, ligands are cited (in alphabetical order) *before* the central atom. Usually the ligand names are given the ending "-o", although there are exceptions, such as "ammine", "aqua", and "carbonyl", as shown in the following examples:

| | |
|---|---|
| $K_4[Ni(CN)_4]$ | potassium tetracyanoniccolate(0)* |
| $Na_2[Fe(CO)_4]$ | sodium tetracarbonylferrate(-II) |
| $[CoCl(NH_3)_5]^{2+}$ | pentaamminechlorocobalt(III) ion |
| $[Cu(H_2O)_2(CO)_2]^+$ | diaquadicarbonylcopper(I) ion |

The *Ewing-Bassett system* is another approach to naming polyatomic species, one which emphasizes the *charge* residing on a complex or unfamiliar ion. An Arabic numeral followed by the appropriate sign is placed in parentheses immediately following the name of the relevant ion; e.g.:

| | |
|---|---|
| $UO_2SO_4$ | uranyl(2+) sulfate |
| $K_4[Fe(CN)_6]$ | potassium hexacyanoferrate(4–) |

While there are other systems for naming particular types of compounds (such as those of boron, both alone and in combination with nitrogen) we regard these as too specialized to warrant consideration here.

## *Species Containing Hydrogen*

The developers of the IUPAC rules agreed to retain the traditional names assigned to the *volatile binary hydrides*. These are based on the systematic name of the corresponding characteristic element, which is shortened and terminated with the syllable "ane". For example, $GeH_4$ is known as "germane" rather than the more obvious alternative "germanium tetrahydride". Similarly,

---

* Notice the use of the recommended (but rarely employed) systematic spelling for the element nickel.

$Si_3H_8$ is "trisilane"; $PbH_4$ is "plumbane"; and $P_2H_4$ is "diphosphane".* In deference to custom it was also agreed that the hydrides containing a single atom of the Group V elements P, As, and Sb might continue to be called "phosphine", "arsine", and "stibine" (from "stibium"), respectively. Other recognized exceptions for the names of molecular hydrides are "water", "ammonia", and "hydrazine" for $H_2O$, $NH_3$, and $N_2H_4$, respectively (cf. also the concluding remarks in this section).

One additional series of hydrogen-containing compounds requires special comment: the *acids. Binary* acids are normally referred to by systematic names of the form "hydrogen ...ide"; e.g.:

> $H_2S$      hydrogen sulfide
> HCl      hydrogen chloride

Older names such as "hydrofluoric acid" also remain in common use, but it is intended that these now be used to designate *aqueous solutions* as distinct from pure compounds. Thus, gaseous HF should be referred to only as "hydrogen fluoride".

Many *ternary* acids have retained their historic trivial names, often ones designed to reflect the oxidation state of the central atom (cf. the next to last footnote under "More Complex Entities" in Sec. 6.2.3); e.g.:

> HClO      hypochlorous acid      [CA Reg. No. 7790-92-3][†]
> $HClO_2$      chlorous acid      [CA Reg. No. 13898-47-0]
> $HClO_3$      chloric acid      [CA Reg. No. 7790-93-4]
> $HClO_4$      perchloric acid      [CA Reg. No. 7601-90-3]

Certain acids that differ from one another only in formal water content have similar names distinguished only by unusual prefixes. The most familiar example is $H_3PO_4$ and $HPO_3$, known as "orthophosphoric" and "metaphosphoric" acids, respectively. In both cases the central phosphorus atoms are pentavalent.

## Double Salts and Polymorphs

*Double* or other *multiple salts* are named by citing first the cations and then the anions. Rules of order are basically the same as those applied to coordination compounds; e.g.:

---

* Even the name "methane", $CH_4$, owes something to this practice, a fact that has important consequences for the entire system of organic nomenclature (Sec. 6.3).
† The significance of the information in brackets is clarified in Sec. 6.4.

> CaKPO$_4$    calcium potassium phosphate
> LiNaS      lithium sodium sulfide

Cationic elements are usually listed alphabetically, although an order based on increasing electronegativity is also permitted.

Salts containing acidic hydrogen were formerly subject to special treatment: the presence of hydrogen was signified by the prefix "bi-" (e.g., NaHCO$_3$, "sodium bicarbonate"). Such names are still common in the world of commerce, but they are no longer appropriate in a scientific context. Only the systematic names should be used; e.g.:

> KH$_2$PO$_4$   potassium dihydrogen phosphate
> NaHCO$_3$    sodium hydrogen carbonate

Naturally occurring *minerals* provide many examples of one of the major challenges confronting any system of nomenclature: *polymorphism*. It is quite common for two minerals to have very similar compositions and yet differ markedly in their structures and properties. Perhaps it is not surprising that the names commonly applied to minerals are quite unsystematic. Quartz, tridymite, and cristobalite are typical examples. To a geologist these names refer to three very different materials, yet each has a chemical composition corresponding essentially to the name "silicon dioxide".

Polymorphs of simpler substances are sometimes distinguished by adding arbitrary symbols such as Greek letters (e.g., $\alpha$-iron, ß'-titanium) or Roman numerals (e.g., ice-II) to systematic names. Designations of this type should be regarded only as provisional, however, or as instances of trivial names for materials more properly identified by providing (in parentheses) the specification of a particular crystal system; e.g., zinc sulfide (cub.), whose formula would be written ZnS (cub.).

## Concluding Observations

From the foregoing discussion it will have become apparent that the systematic nomenclature of inorganic chemistry is not quite as "systematic" as one might expect. A writer is often free to use any of several names for a given compound, although there may be a "best" choice dictated by the context in which the name is to appear. It should again be emphasized, however, that the situation is very different for the compiler of a computerized database, or the staff preparing a reference work such as *Gmelin's Handbook of Inorganic Chemistry* or *Chemical Abstracts*. The editors of such works generally insist on the exclusive use of one consistent system of nomenclature, since this is the only way to prevent ambiguity, as well as to limit duplication or dispersal of information (cf. the second footnote in Sec. 6.2.1).

The shortcomings of systematic nomenclature help to explain why it is not a static entity, as demonstrated by the permanent status of the various IUPAC nomenclature commissions. The efforts of IUPAC in this respect have always been complicated by the fact that chemists, like others, tend to be creatures of habit, often clinging to outmoded terminology despite the vigorous protestations of well-intentioned reformers. Recognition of the power of habit has led IUPAC to adopt numerous compromises short of the goals that strict logic might demand. The names of the nitrogen hydrides may serve as a useful concluding example, since they show the current "state of transition" of one aspect of inorganic nomenclature:

| Molecular formula | Systematic name | Name(s) still in common use |
|---|---|---|
| $NH_3$ | azane | ammonia |
| $N_2H_4$ | diazane | hydrazine |
| $N_2H_2$ | diazene | diimide |
| $HN_3$ | hydrogen trinitride | hydrogen azide, hydrazoic acid |

# 6.3 IUPAC Rules for Organic Nomenclature

## 6.3.1 Introduction

The earliest systematic approach to naming the "organic" carbon-containing compounds was actually quite analogous to that used for inorganic substances. Thus, benzoyl chloride was so named because it was regarded as an organic equivalent of the inorganic salt sodium chloride. During the 19th century, however, it became apparent that the number of possible organic compounds was almost inconceivably vast, and a nomenclature was required that would be both more elaborate and more systematic, particularly with respect to compound structure. The new system was inaugurated by an international congress held in Geneva, Switzerland, in 1892 (hence the term *Geneva rules*). Modifications were adopted in 1930 at an IUC meeting in Liège, Belgium, and since 1949 the system has been further developed under the auspices of IUPAC.

The most authoritative and comprehensive description of organic nomenclature presently available (IUPAC, 1979b, known as the "Blue Book") divides the subject into the following subtopics:

A  Hydrocarbons
B  Fundamental Heterocyclic Systems
C  Characteristic Groups Containing C, H, O, N, Hal, S, Se, and Te
D  Organic Compounds Containing Elements That Are Not
   Exclusively C, H, O, N, Hal, S, Se, and Te
E  Stereochemistry
F  Natural Products and Related Compounds; and
H  Isotopically Modified Compounds

We highly recommend this primary source. Other useful guides to organic nomenclature include the books by Cahn and Dermer (1979), and Lees and Smith (1983).

The fundamental principle underlying contemporary organic nomenclature is that the naming of a compound should begin with a description of the corresponding *hydrocarbon* (or *heterocyclic*) *skeleton*. Details are then added in the form of appropriate affixes (cf. Sec. 6.2). The reason for this particular approach is that the skeleton of an organic compound tends to be resistant to change, and frequently remains unaltered in the course of chemical transformations.

Some of the most important strategies and techniques employed in the systematic naming of organic systems are described very briefly in subsequent sections. It should be noted, however, that many traditional names, also known as *trivial* names, have survived the trend toward systematization. Innumerable examples could be cited, such as the familiar names "camphor" and "cholesterol". Trivial names are often indicative of the origin or occurrence of the corresponding substance in nature. For example, the names "malic acid" and "lactic acid" derive from the fact that these compounds were known to be present in apples and milk, respectively (cf. lat. *malum*, apple; *lac*, milk).*
Many trivial names are even acknowledged within the IUPAC system of

---

* As these cases show, trivial names tend to be short and thus easy to use and memorize, the primary reason why they are still in common use.

nomenclature as allowed "exceptions". Nevertheless, the tendency today is to employ systematic names whenever possible, in order to convey the maximum amount of information. Our aim here is to demonstrate very generally how this is done.

## 6.3.2    Classes of Compounds

### What Is a "Class"?

Compounds that share some important structural feature comprise what is known as a *class*. Most classes are defined in terms of one or more specific types of *bonds* linking particular atoms, bonds responsible for the characteristic chemical properties of the class. Simply identifying the class to which a compound belongs produces what is sometimes called a *generic name*.* For example, the compound $H_3C-OH$ is *an alcohol* because it contains an -OH group bonded to a saturated carbon atom.

The most fundamental classes are those open to *hydrocarbons*, compounds consisting only of the elements carbon and hydrogen. By contrast, many other common classes are like the alcohols in that they require the presence of *heteroatoms*. For example, *ketones* and *ethers* share with alcohols a requirement for oxygen. *Amines* and *nitriles* must contain nitrogen, whereas all *thiols* contain sulfur.

For nomenclature purposes, there are three particularly important classes: the acyclic and cyclic *saturated* hydrocarbons (*alkanes*† and *cycloalkanes*), the *aromatic* hydrocarbons (*arenes*), and the *heterocycles*. These furnish the molecular *skeletons* that are the basis for all systematic names. All other compounds are regarded as "derivatives" of more fundamental "parent" structures. Therefore, the first prerequisite for understanding organic chemical nomenclature is a general knowledge of the approach to naming molecular skeletons.

---

* The term "generic" has quite a different meaning in other contexts. In pharmaceutical chemistry, for example, it refers to material of a particular chemical composition, as distinct from a specific brand of that substance.

† An older name for the alkanes is "paraffins". The latter term is occasionally still encountered, especially in industrial or commercial contexts. The *alkenes* (hydrocarbons containing carbon-carbon double bonds) are similarly known as "olefins".

## Alkanes, Cycloalkanes, and Alkyl Radicals

We begin our discussion with *linear alkanes*. Like all alkanes, these have the general formula $C_nH_{2n+2}$, but they contain only *continuous chains* of carbon atoms. The first four members of the series ($n = 1$ through $n = 4$) retain unsystematic traditional names: methane, ethane, propane, and butane. All other linear alkanes are named by combining a Greek or Latin designation for the appropriate number $n$ (cf. the list of multiplying affixes near the beginning of Sec. 6.2.3) with the ending "-ane". Thus, the unbranched compound $C_5H_{12}$ is pentane, $C_6H_{14}$ is hexane, and $C_{20}H_{42}$ is eicosane. *Trivial* names for a few of the *branched* alkanes are similar in that they also reflect the total number of carbon atoms present. For example, there is one *branched* structure with the formula $C_4H_{10}$, and it is sometimes called by the trivial name "isobutane". As explained below, however, the preferred name for this compound is the more systematic "2-methylpropane".

Carbon atoms can also be joined in such a way that a *ring* is produced, giving rise to the term *cycloalkanes*. If only one ring is present the formula must be $C_nH_{2n}$, and such a compound is named according to the size of the ring. Thus, the compound with the formula $C_7H_{14}$ containing a seven-membered ring is "cycloheptane". Other cycloalkanes also correspond to the formula $C_7H_{14}$, however, including one called "methylcyclohexane". The name reveals that this substance must contain a six-membered ring (bearing a one-carbon "substituent").

More complex molecules may contain several rings, and even *bridged rings* are relatively common. A special set of rules known as the *von Baeyer system* is used for naming compounds of the latter type. The resulting names are quite distinctive, but also rather cumbersome. For example, compounds **1** and **2**, commonly known as "norbornane" and "cubane", acquire the systematic names "bicyclo[2.2.1]heptane" and "pentacyclo[4.2.0.0$^{2,5}$.0$^{3,8}$.0$^{4,7}$]octane", respectively.

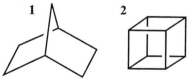

The *radical* counterparts to alkanes play an especially important role in organic nomenclature. Here the term "radical" is used in the special sense of a hypothetical entity obtained from a parent hydrocarbon by mentally abstracting a single hydrogen atom. Names for such radicals are obtained from those of the parent compounds by replacing the hydrocarbon ending "-ane" with "-yl". Thus, the "methyl" radical (or "group") has the composition $CH_3$. The generic name for any radical derived from an alkane is *alkyl*.

Radical names are most commonly invoked for designating *alkyl substituents* introduced into more fundamental structures, such as simple unbranched alkanes (cf. Sec. 6.3.3). The principle is easily demonstrated with the help of a simple acyclic hydrocarbon such as compound **3**, identified previously as "isobutane". Its structure can be seen to consist of a three-carbon "backbone" bearing a one-carbon "branch". To name it, one adds the name for the one-carbon radical ("methyl") to that of the three-carbon alkane ("propane"), prefacing the whole with the number "2" to serve as a "locant" for the radical. The systematic name for **3** is thus "2-methylpropane".

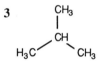

A skeleton can also be modified by introducing one or more sites of carbon-carbon unsaturation. The presence of such multiple bonds is indicated in a systematic name by prefacing the appropriate alkane name with a locant showing where the unsaturation begins, and replacing the parent ending "-ane" with "-ene" (for C=C) or "-yne" (for C≡C). Examples are "2-pentene" and "1-butyne". If both types of unsaturation occur in the same molecule, both must be specified. The syllable "-ene" always precedes "-yne", and the locant for the triple bond is placed in front of the corresponding ending. Here, as in other cases, any "e" that would precede a written or spoken vowel (ignoring locants) is omitted. Application of these principles results in names such as "1-buten-3-yne".

As suggested by the examples above, numerical locants are commonly used to specify where alterations have been made in a parent structure. The numbering process always begins with a search for the *principal chain* (or ring) in a molecule, and this requires the use of a set of rules. One is a *seniority rule* that makes it possible to decide where numbering should begin (e.g., in the case of an acyclic system, at which end of the principal chain; cf. Sec. 6.3.3). A distinctive feature of the numbering rules is the *principle of lowest numbers*, according to which the set of locants "1,6,7" would be preferred over a potential alternative such as "2,3,8".

## Arenes

The second fundamental class of hydrocarbons for naming purposes is the *aromatic* group, also called the *arenes*. All of its members are cyclic. The category includes monocyclic substances like benzene, but also a wide variety of "*ortho*-fused" and "*ortho*- and *peri*-fused"* polycyclic hydrocarbons. All arenes contain the maximum possible number of non-cumulative double bonds. The layman would surely find it odd that a group of compounds defined in terms of a particular mix of topology and unsaturation should occupy so fundamental a position in organic nomenclature. In fact, however, this combination of structural features is extremely prevalent, and the compounds in question exhibit unique properties that give them a special place in the world of organic chemistry.

Most of the simple members of the aromatic class have trivial names. For example, the name "anthracene" is associated with the system comprised of three linearly fused six-membered (benzene) rings. Heptalene (**4**) is an example of an aromatic compound containing two *ortho*-fused seven-membered rings. Compound **5** can be regarded as the result of fusion (with loss of several atoms) of two molecules of *indene* (**6**), and this observation serves as the basis for its name: "indeno[1,2-*a*]indene".† Notice that all three of these compounds indeed contain the required maximum possible number of non-cumulative double bonds.

**4**　　　　　　　**5**　　　　　　　　　　　**6**

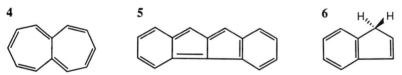

A substance that can be regarded as a partially or fully hydrogenated *derivative* of an aromatic system is often named accordingly. That is, the name

---

* A compound is described as "*ortho*-fused" if it contains two rings that have in common two (but *only* two) adjacent atoms. An "*ortho*- and *peri*-fused" compound is one in which at least one ring shares two atoms with each of two or more rings, so that all are part of a contiguous series.

† Numbers and italicized letters in square brackets are locants to express the nature of the ring fusion, both here and in the von Baeyer names illustrated previously. Such locants are determined in complex ways, and even the numbering of polycyclic and heterocyclic systems is a process difficult to summarize. The *relative* position of two groups on an aromatic ring is indicated by the prefixes *ortho*-, *meta*-, and *para*- (abbreviated *o*-, *m*-, and *p*-) for 1,2-, 1,3-, and 1,4- positioning, respectively. The reader interested in more details should consult one of the definitive works on nomenclature listed in Sec. 6.3.1.

of the corresponding arene is preceded by a prefix such as "dihydro-" or "tetrahydro-", denoting the addition of two or four hydrogen atoms, respectively. In the case of complete saturation, the prefix "perhydro-" is used.

### Heterocycles

The third important class of skeletal compounds is the *heterocycles*, cyclic systems whose rings contain elements other than carbon. Many compounds in this class have trivial names, such as "pyridine" and "oxazole", and these are most commonly used. Elaborate systematic nomenclature also exists for describing such molecules, leading to names such as "1,2,4-triazine" and "imidazo[2,1-*b*]thiazole" for structures **7** and **8**. The systematic rules for naming fused (polycyclic) heterocycles are especially complicated, and it is usually more convenient to use "replacement nomenclature" (cf. Sec. 6.3.4). The latter approach is in fact the only one available for naming bridged heterocycles.

        **7**                     **8**

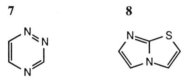

From the foregoing discussion it will be apparent that systematic names for organic compounds are usually comprised of several fragments—syllables, numerals, letters, and occasionally other symbols, all working together to produce an unambiguous description of the structure of a molecule. So far we have been concerned mainly with molecular skeletons. In the following sections we examine some of the rules for specifying details on the periphery of a skeleton.

## 6.3.3    Substitutive Nomenclature

*Substitutive nomenclature* is the most important system for generating an unlimited number of *derived names* from a limited set of *parent names*. The name of a parent compound becomes a *base* or *root* from which other names are formed. For example, the name "propanol" results from carrying out a substitution on the word "propane", just as the name for chlorobenzene is derived from the name of the parent compound benzene.

Substitutive nomenclature always implies the replacement of *hydrogen* by some other atom or group. The principle was actually introduced briefly in the previous section (under "Alkanes, Cycloalkanes, and Alkyl Radicals"), where skeletal branches were treated as "alkyl substituents". The same procedure is used for describing the presence of almost any appendage, including ones with heteroatoms (e.g, $-Cl$, $-OH$, $-NO_2$, $-OCH_3$, etc.). The envisioned replacement of a hydrogen atom is strictly a formality; in no sense does it imply that the compound in question could actually be obtained in the laboratory by carrying out a "substitution reaction" on the parent hydrocarbon (although this may occasionally be the case).

Substitution is denoted by adding an appropriate affix (either a prefix or a suffix, cf. Sec. 6.3.1) to the name of a parent compound (sometimes after the deletion of a terminal "e"). The name "ethanol" (from ethane) is an example in which a suffix has been employed, and "chlorobenzene" illustrates the use of a prefix. In the first case, one hydrogen atom in the parent compound has been replaced by the hydroxyl radical $-OH$ ($C_2H_6 \rightarrow C_2H_5OH$), while in the other the substituent is a Cl atom ($C_6H_6 \rightarrow C_6H_5Cl$).

The choice between using a prefix or a suffix is not arbitrary. For example, certain substituents are always designated by *prefixes*. These include -F ("fluoro-"), $-Cl$ ("chloro-"), $-Br$ ("bromo-"), $-I$ ("iodo-"), $-N_3$ ("azido-"), and $-NO_2$ ("nitro-"). Hydrocarbon radicals (alkyl groups) also fall into this category.

Certain other substituents have been assigned two affixes: one for use as a prefix, the other as a suffix. A few of these groups are listed in Table 6-2, which also contains the corresponding affixes. When only one of these occurs

*Table 6-2.* Selected organic functional groups, together with some of the ways they may be designated.

| Class | Formula | Prefix | Suffix |
|---|---|---|---|
| Carboxylic acids | $-COOH$ | carboxy- | -carboxylic acid (-oic acid) |
| Amides | $-CONH_2$ | carbamoyl- | -carboxamide (-amide) |
| Nitriles | $-C\equiv N$ | cyano- | -carbonitrile (-nitrile) |
| Aldehydes | -CHO | formyl- | -carbaldehyde (-al) |
| Ketones | $>C=O$ | oxo- | -one |
| Alcohols | $-OH$ | hydroxy- | -ol |
| Thiols | $-SH$ | mercapto- | -thiol |
| Amines | $-NH_2, -NR_2$ | amino- | -amine |

in a molecule the *suffix* form is always employed. If several such substituents are present, one is cited as a suffix and all others are specified in prefix form. Multiple prefixes are arranged in alphabetical order. The group designated as a suffix is referred to as the *principal* (or *senior*) group, and its location in the molecule is often critical in determining the overall pattern of numbering applicable to that compound. Before one attempts to assign a systematic name it is therefore essential to establish the identity of the principal group. A decision is reached with the help of a *seniority rule* that ranks all possible substituents. The rule is necessarily rather complex, and for our purposes it is sufficient to note that the carboxyl group –COOH (along with its "derivatives", including esters, acid halides, and amides) appears high on the list, and the aldehyde, ketone, alcohol, thiol, and amine groups follow in that order, as indicated in Table 6-2. Thus, the compound $HSCH_2CH_2OH$ is correctly named "2-mercaptoethanol",* not "2-hydroxyethylthiol", because –OH is "senior" to –SH.

The entry labeled "ketones" in Table 6-2 requires a word of explanation. The structural element >C=O is actually the *carbonyl group*, and it is also found in a number of other functional groups. Occasionally the word "carbonyl" itself appears as part of a prefix, as in the case of the substituent $–C(O)OCH_3$ ("methoxycarbonyl"). In other circumstances the moiety "=O" is treated as an *oxo* substituent doubly-bonded to one of the members of a carbon skeleton.

## 6.3.4 Other Types of Organic Nomenclature

The principles outlined so far suffice for the naming of most organic compounds but sometimes other approaches are useful or even necessary. The following is a sample of the most important alternative and supplementary systems of organic nomenclature.

---

* The compound therefore is regarded as a derivative of ethanol. Some reference works use this fact as a basis for alphabetization, listing the compound as "ethanol, 2-mercapto". The latter is an example of an *inverted* name, and it has the advantage of ensuring that the compound in question will be listed in close proximity to its "parent", ethanol. *Chemical Abstracts* makes extensive use of this procedure. Cyclic "parents" as defined by CAS are listed in a *Ring Systems Handbook*. Useful sources of inverted systematic names include the CAS *Registry Handbook—Common Names* and the chemical substance section of the *CA Index Guide*.

*Radicofunctional nomenclature* can occasionally serve as a convenient alternative to the substitutive system already described, particularly when one wishes to emphasize *relationships* within a family of compounds. In this system, suffixes are circumvented, and the *functional class name* by which one wishes to characterize a compound is expressed as a separate word. This is preceded by a name for the *radical* that remains after removal of the functional group. For example, the systematic name corresponding to the formula $CH_3OH$ is methanol, but the substance can also be called "methyl alcohol". The functional group –OH is singled out, emphasizing the fact that the compound belongs to the generic class of *alcohols*. What remains is the radical $CH_3$ (methyl).

*Replacement nomenclature* (also known as *a-nomenclature*) signifies the presence of *heteroatoms* within what would otherwise be a familiar carbon skeleton. The names assigned to such compounds suggest that the heteroatoms are "replacements" for carbon. The nature of the heteroatom is clarified by the use of an abbreviated form of the element's systematic name. All such forms end in the letter "a", which explains the alternate name for the system. Its application is illustrated by the name "2-oxa-7,9-dithia-4,5-diazadecane" for the structure

$$CH_3-S-CH_2-S-CH_2-N=N-CH_2-O-CH_3.$$

Replacement notation is particularly common for heterocycles.

*Subtractive nomenclature* also relies on relationships to simplify names. In this case, a substance is described as differing from some well-known reference compound through the *absence* of one or more specific atoms or groups. Thus, "didehydro-L-ascorbic acid" is a molecule similar to L-ascorbic acid but with two fewer hydrogen atoms. Similarly, "deoxyribonucleic acid" (abbreviated DNA) is "deficient" in oxygen atoms, but otherwise identical to ribonucleic acid (RNA). Here the difference is the absence of one oxygen atom in each of many ribose units. The affix "nor-" is used to indicate the absence of one or more *methyl groups* relative to a familiar parent.

*Additive nomenclature* works in the opposite way: it signifies the formal *addition* of some molecule or atom to a reference compound. The name "styrene oxide" is a convenient example; the corresponding structure is derived by adding an oxygen atom across the double bond of styrene, thereby creating a three-membered ring:

Styrene    Styrene oxide

The name "ethylene dibromide" is similarly applied to the compound with the formula $BrCH_2CH_2Br$.

Finally, *conjunctive nomenclature* permits one to name substituted ring systems by combining the names of two separate compounds, the parent cyclic molecule and the parent of the substituent. The implication is that fusion has taken place between the two substances, with concurrent loss of one hydrogen atom from each. For example, "cyclohexaneacetic acid" is a name commonly applied to the compound whose formula is $C_6H_{11}-CH_2COOH$.

## 6.3.5    Nomenclature in Areas Related to Organic Chemistry

For much of its nomenclature, the field of *biochemistry* relies on the rules that apply to organic compounds in general. However, biochemists have also introduced a number of special naming practices, primarily as a result of the incredibly complex nature of many of the compounds they encounter. Moreover, a great many compounds of biological interest were named before their chemical structures were determined, and the corresponding trivial names continue to play an unusually important role. Both factors help to explain why compounds of interest to biochemists often bear names more closely related to biological functions or isolation procedures than to chemical constitution.

*Carbohydrates*, *fats*, and *proteins* (including *enzymes*) are among the substances covered by specialized naming rules. Such rules were first introduced in 1921, and since that time they have undergone extensive modification. These rules were formerly the responsibility of an IUPAC Commission on the Nomenclature of Biological Chemistry, but they are now supervised by a joint commission established by IUPAC and the International Union of Biochemistry (IUB). Other rules have been established for naming such classes of compounds as vitamins, steroids, quinones, and the broad category of "natural products" (IUB, 1979). The interested reader is advised to consult specialized sources for more information.

One unique aspect of nomenclature in the field of biochemistry is the *symbolic notation* long associated with the formulas of *peptides* and *proteins*. These materials can be regarded as condensation products of *amino acids*, whereby the amino group of one amino acid joins the carboxyl group of another. The result is an amide (also known as an "α-peptide bond"). Molecules of this type are invariably represented by a sequence of three-letter *amino acid symbols* rather than by standard chemical formulas. Thus, "Ala-Gly-Ser-Arg" denotes a tetrapeptide derived from alanine, glycine, serine, and arginine, in which the individual amino acids are linked in the order shown. The remaining free carboxyl group is always understood to be in the amino acid listed *last*.

Whenever the terminal syllable "-ase" is encountered in biochemistry it is an unmistakable signal that an *enzyme* is under discussion. Often this suffix is combined with the name (or name fragment) of a substance or type of substance upon which an enzyme *acts*, and the combination is used as a name for the enzyme. Thus, a "maltase" is an enzyme that acts on the sugar maltose, while an "esterase" promotes the cleavage of ester-like linkages. These generic names can be made more specific by coupling them with the name of the organism from which the enzyme is isolated (e.g., "yeast alcohol dehydrogenase"). A more systematic but less transparent approach to enzyme nomenclature is the *Enzyme Classification* (EC) system (IUB, 1979) which associates the cryptic identifier "E.C.5.2", for example, with enzymes of the "*cis-trans*-isomerase" type. The greater specificity implicit in a name like "maleate isomerase" is reflected in the correspondingly more complex designation "E.C.5.2.1.1".

Biochemists are not the only ones who have discovered a need for a different approach to naming certain types of organic systems. The field of *macromolecular* (or *polymer*) *chemistry* also presents a series of challenges. This is an area undergoing especially rapid development, and the same can be said with respect to its nomenclature. IUPAC has established a separate Commission on Macromolecular Nomenclature that has so far distributed recommendations for the naming of single-strand polymers, as well as rules for designating the stereochemical aspects of such systems. Uniform polymer abbreviations have also been published (IUPAC, 1987).

Other IUPAC commissions have issued recommendations of varying complexity for nomenclature and terminology in such areas as colloid and surface chemistry, methods of chemical analysis, spectroscopy, nonproprietary drugs (*International Non-proprietary Names*, INN), and pesticides. All these fall outside our area of immediate concern, although it should be noted that some are sufficiently important to have been incorporated into appropriate national and international standards.

# 6.4    CAS Registry Numbers

We conclude with a brief description of the origin and use of *CAS registry numbers*, although these are generally not considered within the domain of chemical nomenclature. The reason we raise the subject is that registry numbers are employed increasingly as *replacements* for the names of chemical compounds. For example, it is now common to conduct *literature searches* solely on the basis of registry numbers. The reader interested in a detailed discussion of registry numbers should refer to Schulz (1985) or to the definitive publications issued by Chemical Abstracts Service.

Since 1965, every compound newly described in the chemical literature has been assigned a unique CAS registry number. Each number can be relied upon to correspond unambiguously to one particular substance. It is important to realize that a CAS registry number in itself conveys no information whatsoever about the chemical nature of a compound; it is simply a number, issued and administered by CAS. The digits that comprise it are only a reflection of the state of the registry system at the time the substance in question was entered.*

---

* The main impetus for the creation of a system of registry numbers was the need to provide a means for rapid and exhaustive searching of computerized databases. Many urged that a different approach to the problem be explored: one oriented around computer-readable *structural* information. Advocates of this notion argued that its adoption would permit the conduct of more informative searches, such as trying to identify all known compounds possessing a particular structural feature. Probably the most widely discussed system of this type was the so-called *Wiswesser Line-Formula Notation*, described in detail by Smith and Baker (1976). The Wiswesser system is remarkably sophisticated but at the same time quite straightforward. A chemist who chooses to employ it can quickly become expert in "coding" compound structures, thereby producing "names" like "QV4" for pentanoic acid, "ZY" for 2-aminopropane, or "1Y&VY" for diisopropyl ketone. Indeed, computer programs have even been developed that permit the chemist to *draw* the structure of interest and then wait for the computer to supply the corresponding Wiswesser designation. Much ingenuity and foresight were demonstrated by the developers of the Wiswesser system, but it appears almost certain that for the foreseeable future most computer searches will need to be conducted on the basis of registry numbers instead.

An example is the number 7732-18-5, which has been permanently assigned to the compound water—$H_2O$ of normal isotopic composition. All registry numbers have essentially this form, although there may be as many as six digits preceding the first hyphen. Two digits are always found between the hyphens, and the last digit serves only a control function.

The principal virtue of registry numbers is the ease with which they can be housed and searched in a database, but their use also permits a completely unambiguous specification of a particular molecular structure, an attribute that even the most sophisticated naming system cannot yet claim. The CAS *Registry Handbook—Number Section* is the authoritative source that relates these numbers to the corresponding systematic compound names, although many chemists find it easier to extract the numbers directly from a computer database. It is anticipated that two supplementary volumes will need to be added to the *Registry Handbook* each year to list the numbers and names of the roughly 360 000 compounds currently added annually.

Some (about 3%) of the compounds introduced into the system lack definitive structural assignments at the time of their entry. This means that certain files require subsequent elaboration. Moreover, it is occasionally discovered that a compound reported to be "new", and thus assigned a registry number, is in fact a substance already present in the system. Whenever this occurs, one of the registry numbers must be cancelled, a circumstance which can arise for other reasons as well. All changes affecting registry numbers are scrupulously reported in the "Registry Number Update" section of the *Registry Handbook*.

Even though the registry number system is managed by the publishers of *Chemical Abstracts*, not every registered compound will have been reported in the abstracting journal at the time its number is assigned. This is because many publishers (and others as well) apply for CAS registry numbers at a very early date, often before reports of the substance in question appear in print *anywhere*.

Chemists have been remarkably successful in finding concise ways to use words for describing molecules, but the challenge has become progressively larger as the number of known compounds has multiplied—and at an ever-increasing rate. While it would certainly be premature to assert that word-based nomenclature schemes have finally become obsolete, it is reasonable to assume that registry numbers or their equivalent are now a permanent part of the "nomenclature" of chemistry.

# 7 Quantities, Units, and Numbers

## 7.1 Quantities

### 7.1.1 The Meaning of the Term "Quantity"

One of the distinguishing characteristics of most scientific work is a reliance on numerical, or *quantitative*, measurement. Some measurements relate to the properties of *objects*, but others have to do with some aspect of an *event* or a *condition*. Regardless of what is measured, however, and irrespective of the complexity of the measuring operation, the ultimate result is specification of the *value* of some *physical quantity*—so named, perhaps, as a tribute to physics, the "mother" of all sciences (although the adjective "physical" is often omitted in references to quantities). Since quantitative data derived from measurements play such a major role in science, it should come as no surprise that much attention has been directed to seeking the best ways to present such information. It will simplify our task of discussing how data should be treated in a manuscript if we begin by reviewing the nature and significance of quantities in general.

To report or utilize a "quantity" is to make a statement about an amount or intensity of something: a 10-kilogram *mass*, a 30-degree Celsius *temperature*, or a 50-hertz *frequency*, for example. A measurement has been performed, or at least implied. But what kind of measurement? Measurement in what *dimension*? *Length* is probably the most commonly encountered dimension. In its abstract sense—the way we here use the term—the *dimension* "length" is what one invokes in measuring any physical *distance*, including those referred to as "width", "thickness", and "height". *Time* is also a dimension in this sense, as is *mass*. Before further exploring the concept of dimensions, however, it is convenient to establish the fact that each dimension serves as the basis for a type of *quantity*. Thus, any actual measurement in the "length dimension" results in a quantitative statement of *a* length. We are no longer treating length as an abstract concept; instead we are contemplating the *quantity* "length", and in order to express it we are forced to establish a *reference value* in the form of a *unit*.* Assuming we agree to use the "meter" as that standard, we can describe a *quantity* length (symbolized *l*, see below)—a particular measurement associated with the *dimension* "length" (**L**)—as having the *value* "three meters" (3 m).

Quantities can be of many types, but all share the property that they can be assigned numerical values. Certain quantities are unusual in that they can be expressed *simply* as numbers. Examples include the quantities "refractive index" and "mass fraction". Such "unitless"[†] quantities are rare, however. In general, a specification of a quantity only carries meaning if it consists of the *product* of a number *and* a unit, as in the case of the length cited above, or the various examples earlier in the discussion. This principle can be stated more concisely in the form of a general equation:

$$\text{Quantity} = \text{Numerical value} \times \text{Unit} \qquad (7\text{-}1)$$

Further examination reveals that quantities fall into two basic categories: the values of some (*extensive* quantities) are dependent on the "size" (extent) of the object or event being measured, while in other cases (*intensive* quantities) values are associated with intrinsic properties of the subject of investigation. The quantity "mass", for example, is extensive; its value varies with an object's size. "Temperature", on the other hand, is "size-independent", so it must be an intensive quantity.

Measurements are useful only to the extent that they are reproducible and subject to comparison, preferably without regard to disciplinary or geographical boundaries. It is for this reason that interdisciplinary and international *standards* of measurement have been established. Achieving this goal has required the broadest possible agreement on the nature of quantities themselves and on the units to be employed in their measurement. This subject

---

* In other words, a quantity "quantifies" the corresponding dimension, which by its very nature, is only a qualitative concept.

† Strictly speaking, even quantities of this type generally have a unit associated with them, the unit "one", because their values arise from the division of two quantities sharing identical physical units. For example, a refractive index is obtained by comparing the propagation velocities of light in two adjacent media, where each velocity alone might be expressed in terms such as m s$^{-1}$. But m s$^{-1}$/(m s$^{-1}$) = 1, so the units "cancel". The expressions "dimensionless" or "without dimensions" are sometimes applied to such quantities, but these should be avoided. They are clearly misleading, but perhaps worse, they utilize the word "dimension" in a (vague) sense quite different from the technical meaning of the word introduced previously.

is examined in more detail in Sec. 7.3, but it is important to note here that one fundamental consideration is the extent to which different quantities (e.g., mass and length) can or should be interrelated. It has long been recognized that there are advantages to be gained by minimizing the number of totally independent quantities and units. For example, while it is obviously possible to analyze the ability of a machine to do work by counting the number of horses the device could replace (i.e., by specifying the corresponding "horsepower"), this approach is of limited value. Only if one further defines the capability of a "standard horse"—in terms involving the transport of a given (defined) mass under a particular set of circumstances (probably involving both time and distance)—can the measurement be used for purposes of comparison. As a result of this further definition, however, the horse becomes irrelevant, and the unit "horsepower" is reduced to an unnecessary and, for most purposes, inconvenient anachronism; the quantity to which it once related would be expressed today in terms of the more widely applicable *watt*.

## 7.1.2 Base Quantities and Derived Quantities

Various attempts have been made in the past to define the smallest and most convenient set of mutually independent quantities. From such a set of *base quantities* it would be possible in principle to derive all other required quantities simply by using multiplication and division to generate appropriate combinations.* The potential number of such secondary or *derived* quantities would be limitless, but the most important ones could easily be identified and granted "official" status as a way of ensuring their universal recognition and consistent use.

There are many possible approaches to defining a set of base quantities, and it was not easy to choose the best. Indeed, controversy has even existed over the *number* of base quantities required to ensure that all of science is adequately served. Our attention here is restricted to the single system—summarized in Table 7-1 and described in Sec. 7.3.1—that has recently been granted exclusive status within science generally and even by most governments (cf.

---

* Quantities obtained by use of operations other than multiplication and division are technically known as *functions*. Examples familiar to the chemist include the Gibbs free energy ($G$, also called *Gibbs function*) and pH, where $G = H - TS$ and pH = $-\log_{10}[c(H^+)/(\text{mol } L^{-1})]$. Here, the mathematical operations involved are subtraction and conversion to a logarithm, respectively.

Table 7-1. Base quantities and units in the SI.

| Base quantity (dimension) | Quantity symbol | Dimension symbol | Corresponding base unit | Unit symbol |
|---|---|---|---|---|
| length | $l$ | **L** | meter | m |
| mass | $m$ | **M** | kilogram | kg |
| time | $t$ | **T** | second | s |
| electric current | $I$ | **I** | ampere | A |
| thermodynamic temperature | $T$ | **Θ** | kelvin | K |
| amount of substance | $n$ | **N** | mole | mol |
| luminous intensity | $I, I_v$ | **J** | candela | cd |

Standard *ISO 31/0-1981* and IUPAC, 1979a; Standards *ISO 31/1* to *31/13* are also relevant in this context).

While the number of base quantities is small, hundreds of derived quantities are in common use, and not only among scientists. *Density* is a typical example; the quantity "density" is defined as the quotient obtained when the base quantity *mass* is divided by another quantity, *volume*, where the values of both apply to the same sample. Density so defined is thus a technical term with an unambiguous meaning.* Due to the efforts of the international scientific community similar "official" definitions exist for a great many derived quantities, and each has its own *name*—the latter in various versions applicable to different languages. These are the only names that should be employed when quantities are mentioned in formal scientific reports.

Adherence to standardized *terminology* is extremely important. Modern science draws much of its strength from the fact that it is international and interdisciplinary in character. Scientists unable or unwilling to agree on a single vocabulary would quickly find themselves in the same dilemma as the builders of the tower of Babel. However, standardization with respect to quantities goes beyond names. Scientists make extensive use of *equations* (cf. Chapter 8) and other forms of shorthand based on symbolic notation. For this reason it has also been necessary to adopt uniform *symbols* for the most important quantities

---

* Observe that the definition has been given strictly in terms of the mathematical operation "division": density *is* mass divided by volume. It is often *interpreted* as (average) mass *per* (average) unit volume, but this is an interpretation, not a definition, and there is a growing tendency to avoid such terminology. An analogous argument applies to units.

*Table 7-2.* Supplementary and derived SI units having special names and symbols approved by CGPM, listed in terms of the quantities to which they apply.

| Quantity[a] | Quantity Symbol | Name of derived SI unit | SI unit symbol | Expression in terms of SI base units | Expression in terms of other units |
|---|---|---|---|---|---|
| activity[b] | $A$ | becquerel | Bq | $s^{-1}$ | |
| quantity of electricity, electric charge | $Q$ | coulomb | C | s A | |
| Celsius temperature | $\Theta\,(\vartheta)$ | degree Celsius[c] | °C | K | |
| capacitance | $C$ | farad | F | $m^{-2}\,kg^{-1}\,s^4\,A^2$ | C/V |
| inductance | $L$ | henry | H | $m^2\,kg\,s^{-2}\,A^{-2}$ | Wb/A |
| frequency | $\nu, f$ | hertz | Hz | $s^{-1}$ | |
| energy, work, quantity of heat | $E, W$ | joule | J | $m^2\,kg\,s^{-2}$ | N m, W s |
| luminous flux | $\Phi$ | lumen | lm | cd sr | |
| illuminance | $E$ | lux | lx | $cd\,sr\,m^{-2}$ | $lm/m^2$ |
| force | $F$ | newton | N | $m\,kg\,s^{-2}$ | |
| resistance | $R$ | ohm | $\Omega$ | $m^2\,kg\,s^{-3}\,A^{-2}$ | V/A |
| pressure | $p$ | pascal | Pa | $m^{-1}\,kg\,s^{-2}$ | $N/m^2$ |
| plane angle | $\alpha, \beta, \gamma$ | radian | rad | | |
| conductance | $G$ | siemens | S | $m^{-2}\,kg^{-1}\,s^3\,A^2$ | A/V |
| dose equivalent | $H$ | sievert | Sv | $m^2\,s^{-2}$ | J/kg |
| solid angle | $\Omega$ | steradian | sr | | |
| magnetic flux density | $B$ | tesla | T | $kg\,s^{-2}\,A^{-1}$ | $Wb/m^2$ |
| electric potential | $U$ | volt | V | $m^2\,kg\,s^{-3}\,A^{-1}$ | W/A |
| power | $P$ | watt | W | $m^2\,kg\,s^{-3}$ | J/s |
| magnetic flux | $\Phi$ | weber | Wb | $m^2\,kg\,s^{-2}\,A^{-1}$ | V s |

[a] Arranged in alphabetical order of the corresponding *unit symbols*.
[b] Activity of a radioactive substance, also called "radioactivity".
[c] Strictly speaking, the degree Celsius is not a derived unit, but it may be treated as such for our present purpose; cf. eq. (7-3) and the related discussion in Sec. 7.3.1.

(both base and derived), again on an international basis and this time without regard to language. Most of the currently approved quantity symbols are capital or lower case letters of the Latin alphabet, but Greek letters also appear. A few symbols consisting of two letters continue to be used,* although this practice clearly violates the spirit of the adopted guidelines (cf. Rule 1.5 in the familiar "Green Book": IUPAC, 1979a). *Superscripts* (including *primes*, *double primes*, and the like) or *subscripts* (*subindices*) are often added to quantity symbols when further differentiation is required (an example in the case of heat capacity $C$ appears near the end of this section; others will be found in Sec. 7.2).

Table 7-1 contains the accepted English names and corresponding symbols for the seven official base quantities. It should be noted that each is related to one of seven *base dimensions* within the framework of which measurements are conducted. Table 7-2 provides information regarding a number of derived quantities commonly used in chemistry and related disciplines. These tables also list the *units* to be employed whenever the corresponding quantities are measured and reported; units themselves are the subject of detailed discussion in subsequent sections of this chapter.

The scientific community is relatively forgiving when its members are found guilty of minor breaches of the international recommendations on quantities and quantity symbols. Tolerance in this regard probably does little harm so long as clarity is maintained. However, much stricter discipline is imposed with respect to units. Improper use of units and their needless proliferation seriously impede communication and the free interchange of results. The implied demand for rigor need not be seen as burdensome: the number of units to be mastered is really quite limited, since most quantities are measured in terms of the units associated with the seven base quantities. Density, for example, is usually specified in grams divided† by cubic centimeters (g cm$^{-3}$) rather than in terms of a separate "density unit".

Adherence to the proper use of units is actually a matter of *law* in many countries; their governments literally forbid the public use of any units other than those with international sanction. Indeed, mankind at long last appears to be on the road to adopting a single uniform basis for measurements, not only in the abstract realm of science but also in commerce.

---

* The latter are limited to certain quantities with the unit "one" used in the study of transport phenomena. All such two-letter symbols should be enclosed in parentheses to prevent confusion; e.g., (*Re*) Reynolds number, (*Nu*) Nusselt number.
† See the second footnote in this section regarding terminology.

The situation with respect to quantities is rather different. No limit can be imposed on the number of "valid" quantities, and new ones will undoubtedly continue to be introduced as science pursues its relentless journey into the unknown. Total and inflexible standardization of quantities is clearly an unrealistic objective, but there is certainly reason to strive for uniform practice regarding the quantities most often encountered.

### 7.1.3    Symbols and Typography

It was noted earlier that symbols play an important role in communicating information related to quantities. Unfortunately, the common alphabets do not provide a sufficient number of letters to permit exclusive assignment of concise alphabetical symbols to all the defined quantities, and units also require symbols. Many letters have therefore been pressed into service more than once. Depending on the context, for example, $M$ may represent such diverse quantities as molar mass, the magnetic quantum number, the moment of force, magnetization, or luminous or radiant excitance. Obviously there is potential

| | |
|---|---|
| $a$ | surface area divided by volume of packed bed, $m^{-1}$ |
| $c$ | adsorbate concentration in bulk fluid phase, $mol\ m^{-3}$ |
| $c_F$ | specific heat of fluid, $J\ kg^{-1}\ K^{-1}$ |
| $D_{ax}$ | axial fluid dispersion coefficient, $m^2\ s^{-1}$ |
| $D_p$ | adsorbent particle diameter, m |
| $h_p$ | particle-to-fluid heat transfer coefficient, $W\ m^{-2}\ K^{-1}$ |
| $-\ \Delta H$ | heat of adsorption, $J\ mol^{-1}$ |
| $k_F$ | thermal conductivity of fluid, $W\ m^{-1}\ K^{-1}$ |
| $(Nu)$ | $=\ h_p D_p/k_F$, Nusselt number |
| $R$ | particle radius, m |
| $R_C$ | column radius, m |
| $T_A^*$ | defined in eq. (11 a), K s |
| $t$ | time, s |
| $\alpha_{ax}$ | axial fluid thermal dispersion coefficient, $m^2\ s^{-1}$ |
| $\varepsilon$ | root-mean-square error, defined in eq. (23) |
| $\tau^*$ | time required for tail of thermal wave to vanish, s |

*Fig. 7-1.* Example of a list of symbols.

for ambiguity, and authors are strongly advised to record all symbols appearing in a particular report or paper in a *List of Symbols* containing clear, concise identifications. An example of such a list in typewritten form is shown in Fig. 7-1.*

Official international and national guidelines specify that symbols indicating quantities should appear whenever possible in *italics* (italic or sloping type; cf. Sec. 3.5.4). This rule applies not only to letters in the Latin alphabet, but to Greek letters and other symbols as well. In a typescript the necessary distinction may be made by underlining, a practice consistent with standard conventions in the printing industry (cf. Sec. 4.3.7, "Choice of Type"). Unit symbols, on the other hand, should never be printed in italics, nor should the symbols for the chemical elements, but rather in ordinary roman ("upright") type. These rules should be carefully followed in all scientific text; i.e., not only within the narrative portions but also in figures (cf. Sec. 9.2.3, "Letters and Symbols") and tables.

Insistence on the use of italics for quantity symbols may be perceived as a nuisance, and it is certainly costly for publishers, but the potential benefit to readers is significant in terms of enhanced clarity. One might also argue that since physical quantities are the "atoms" of which scientific information is comprised it is appropriate that they be distinguished from background text and from other symbols, such as those for numbers and units.

One interesting consequence of the italics rule is that it doubles the arsenal of available symbols. For example, the two symbols "m" and "$m$" have different meanings: the former has been assigned to a base unit (meter) and the latter to a base quantity (mass). Some authors find the italic convention troublesome, but its application is really very simple: any symbol capable, in principle, of assuming a variety of values (and associated, perhaps, with a unit)—i.e., any *variable*—should always be printed in italics.† Everything else is set in roman type, including mathematical functions such as "ln" or "sin" and the differential operator "d".

The *natural constants* (or *fundamental constants*) are a special exception to the above rule; their symbols are always printed in italics even though they

---

* Note that we have deliberately *avoided* writing a sign of equality between a symbol and its definition. The reason is that in most cases what appears is a *representation* followed by an *explanation*, not an *equality* in the mathematical sense. The *sign of equality* (=) should be reserved for use in the latter context, illustrated in Fig. 7-1 by the entry for "Nusselt number" (cf. also the discussion in Sec. 7.6).

† The rule also applies to "variables" in the general mathematical sense of the term.

presumably represent constants. Thus, the gas constant is represented by $R$ and the Planck quantum by $h$. One other exception is the mathematical symbol "$f$" (for "function"), also printed in italic type presumably because $f$ may refer to any kind of dependency.

The italicization rule for symbols applies to subindices as well as to parent symbols. $C_p$, for example, is the symbol for heat capacity at constant pressure. "Pressure" is a quantity, so the corresponding subscript is set in italics. On the other hand, if "$C_B$" were intended to represent the heat capacity of substance B, a roman subscript would be correct.

It is perhaps appropriate to add a word here about the special typographic conventions applicable to *vectors* and *dimensions*. It was formerly the custom to indicate a vector (i.e., a quantity directed in space) by placing an arrow above the corresponding quantity symbol, but this practice is no longer recommended. Instead, the quantity symbol is set in boldface italic type (e.g. force $\boldsymbol{F}$, electric field strength $\boldsymbol{E}$). In *dimensional notation*, a derived quantity is conceived of in terms of a corresponding *derived dimension*. This derived dimension is itself a result of combining an appropriate set of base dimensions. The process is symbolized by the use of *dimensional symbols* (cf. Table 7-1) joined by mathematical operators. It is recommended that dimensional symbols be printed in upright "sans-serif" type,* if possible in *boldface*. Thus, for the dimension "force":

$$\dim F = \mathbf{L\ M\ T^{-2}}$$

# 7.2    *Quantitative Expressions*

## 7.2.1    *General Remarks*

One of the most distinctive characteristics of the physical sciences is the fact that they purport to be "exact". Rules such as those discussed above certainly

---

\* "Sans-serif" is a printer's term to describe a particular type *style*. Sans-serif letters are devoid of all embellishment, such as the extra lines often found at the top and the bottom of a capital "I" or at the bottom of a letter like "m".

appear to reflect such exactness. Nevertheless, closer inspection reveals that scientists are not always as painstaking and precise in their speech and writing as one would (or should!) expect. Consider, for example, the following specifications, typical of many that appear in print:

3 kg sulfuric acid

1500 r.p.m.

$10^{12}$ neutrons/s cm$^2$

The expressions certainly look scientific, but in fact they are *colloquialisms* masquerading as scientific statements, usages more closely allied to conversation in the everyday world.

The problem may be more readily appreciated if we first examine a more blatant example:

"3 kg sulfuric acid/kg sodium hydroxide".

Here we see two items connected by a division sign (i.e., incorporated into a fraction). Since the expression is taken from a scientific manuscript, should one conclude that sulfuric acid is to be divided by sodium hydroxide? Unfortunately, even textbooks are rife with similar careless statements.

The logical absurdity posed by the last example actually begins with a lack of precision in the first of the expressions above: "3 kg sulfuric acid". When written this way, "sulfuric acid" appears to have the status of a unit: to be a part of a product of units, as in the expression "kg m" (cf. Sec. 7.3.3); or perhaps it somehow qualifies the unit that precedes it. Obviously neither of these can be the case, since "sulfuric acid" is clearly not a unit, nor does it specify a particular kind of kilogram. In fact, the words "sulfuric acid" are intended to specify the nature of a certain portion of material whose mass is 3 kg. More correct forms of expression for a 3 kg portion of sulfuric acid are:*

$$m_{H_2SO_4} = 3 \text{ kg} \quad \text{or} \quad m(H_2SO_4) = 3 \text{ kg}$$

Similarly, one should write

$$f = 1500/\text{min}$$
$$\phi = 10^{12}/(\text{s cm}^2) \quad \text{or} \quad 10^{12} \text{ s}^{-1} \text{cm}^{-2}$$

---

* The second notation (that using parentheses) is a common way of specifying "chemical" quantities; it has the advantage that it avoids the typing and typesetting problems caused by "subscripts to subscripts".

for the other examples, which were attempts to specify a particular frequency of rotation and a neutron flux density, respectively.

We can summarize by saying that there is cause for concern whenever one finds numbers, symbols of quantities and units, and mathematical operators mixed together with words. Granted, an expression like "the mass fraction of cantharidin, with respect to dry weight, was 20 mg/kg", or even "... was $\omega$ = 20 mg/kg", may be cumbersome, but there is virtue in forcing oneself to formulate expressions this way: the effort provides practice in thinking clearly about quantities and their meanings. A more pragmatic rationale is that proper formulation leads to greater clarity in stoichiometric and other *calculations*. The advantages become particularly apparent when the need arises to use a computer for performing calculations: computers are notoriously intolerant of data that have been incorrectly treated.

If a situation should arise in which precise symbolic formulation is truly impractical there is always one straightforward and technically correct alternative: revert to the use of *words*. As a case in point, consider the need to specify "mass of fluorine incorporated into plastic film (in grams divided by square meters), divided by the amount-of-substance concentration (in moles divided by liters) of reagent solution". (We challenge the reader to express this relationship conveniently using symbols!)

## 7.2.2    Dual Notation

Symbols can be used to express quantitative relationships in either of two ways, the distinction being one of completeness. Each notational mode has its place, in deference to which the term *dual notation* has been coined (IUPAC, 1979a). If one wishes to express a *general* relationship between several quantities, then a highly compressed notation is best; e.g., the equilibrium constant for a system of several components $i = 1, 2, ...$ is

$$K_c = \prod_i (c_i)^{n_i}$$

An equation of this type is straightforward, and it is relatively easy to remember. The number of symbols has been kept to an absolute minimum, and the whole is written as compactly as possible.

If, on the other hand, one's primary goal is the elimination of all ambiguity in a *specific* case, then a more elaborate notation is called for; e.g.:

$V(K_2SO_4, 0.1 \text{ mol dm}^{-3} \text{ in } H_2O, 25 \text{ °C}) = 48 \text{ cm}^3$

$\alpha(589.3 \text{ nm}, 20 \text{ °C}, \text{sucrose}, 10 \text{ g dm}^{-3} \text{ in } H_2O, 10 \text{ cm}) = 66.47°$

The expanded notation is useful for specifying the meaning of *numerical values*, although it could prove awkward in a table unless incorporated into a table footnote. The examples above should be regarded as extremes. In practice one usually employs "hybrid" formulations that fall somewhere between the two implied limits.

## 7.2.3    Expressing Quantities

The mathematical expression defining a quantity was introduced as eq. (7-1). The same equation can also be written in symbolic terms:

$$G = \{G\} \times [G] \qquad (7\text{-}2)$$

Here, $G$ by itself signifies the *value* of some particular quantity.* The *braces (curly brackets)* { } denote "numeric value associated with" (or "number applicable to"), while the *square brackets* [ ] stand for "units of". This notation makes it apparent that it would be incorrect to write a unit symbol [the last term on the right in eq. (7-2)] immediately after a quantity symbol; this would place it on the "wrong side" of the equation. For example, the following are improper specifications of a unit for measure of electrical energy:[†]

$$W \text{ kW} \cdot \text{h} \quad \text{or} \quad W \text{ [kW} \cdot \text{h]}$$

Some authors reading these lines will undoubtedly be dismayed to discover that they have been deprived of their favorite means of indicating units to readers of their papers, but there are other ways of easily providing the same information. For example, unit symbols can be enclosed in *parentheses* following a quantity symbol, separated from it by a space. Alternatively, a comma (cf. Fig. 7-1) or the preposition "in" can be inserted between the two:

$$p \text{ (N/m}^2) \quad \text{or} \quad p, \text{N/m}^2 \quad \text{or} \quad p \text{ in N/m}^2$$

---

\* The reader is forewarned that the terms "quantity" and "value of quantity" are often used interchangeably, even in official documents (e.g., IUPAC 1979a).

[†] Erroneous notations similar to those shown are particularly common in tables and figures. For examples of correct unit specifications see Fig. 9-10 and Fig. 10-1.

The last two representations have the advantage over the first that they are more obviously *explanations*, so they cannot be confused with instructions for multiplication.

Probably the best way to specify the set of units one has employed is to show a *division* of the appropriate quantity by its units, which leads to a value whose "dimension" is "one", i.e., to a number. For example, a set of data referring to *pressure* might appear as numbers under the heading $p/(\text{N m}^{-2})$. That the quotient from such an operation must be a number becomes apparent when eq. (7-2) is solved for $\{G\}$. Authors are increasingly adopting this notation, especially for labeling *data columns* (numbers) in tables and for specifying *scales* in figures. Examples can be found in Sections 9.2.3 and 10.2.

# 7.3    *Units*

## 7.3.1    *SI Units*

*Units* are the *standards* or *measures* needed for making numerical comparisons —as amounts, sizes, or intensities—in terms of physical quantities. The process of measurement itself consists of determining how many *unit amounts* of a physical quantity are associated with a given "sample".

All the quantities of interest to scientists can be dealt with using only a very small number of units, although many quantities demand the use of several units in combination. Interestingly, modern science was long plagued by a bitter debate over precisely how many fundamental and independent units should be recognized, and what they should be. Until recently, three closely related unit systems remained strong contenders. Each was referred to by the initial letters of its most important, or base, units. These were the familar

> CGS (from centimeter, gram, second),
> MKS (from meter, kilogram, second), and
> MKSA (from meter, kilogram, second, ampere)

systems. The controversy was finally resolved in 1960 as scientists from throughout the world pledged themselves to adhere to a single, consistent system of units, the *Système International d'Unités* (SI). The SI—to say "the

SI *system*" would be redundant—was modified in 1971 (above all by inclusion of the mole as a seventh base unit; see below), and it is now generally regarded as the best solution currently available to a complex problem. There is little doubt that its adoption will ultimately lead to a marked improvement in communication across a host of cultural and disciplinary boundaries.

The SI is comprised of seven *base units*. Each of the seven originates in one of the seven *base quantities*: *length, mass, time, electric current, thermodynamic temperature, amount of substance*, and *luminous intensity*. The corresponding base units are the *meter*, the *kilogram*, the *second*, the *ampere*, the *kelvin*, the *mole*, and the *candela*. *Unit symbols* for the base units are listed in Table 7-1.

It would be inappropriate here to delve into the precise definitions of all the base units, although one, the mole, is discussed in some detail in Sec. 7.4. Instead, we simply note that the SI essentially absorbed the MKS and MKSA systems, and that the units of length and mass of the CGS system are smaller than their SI counterparts by two and three orders of magnitude, respectively.*

Units were also specified in the SI for a number of derived quantities, but these *derived units* are all products of two or more base units, each to the appropriate (positive or negative) power. A few derived units are considered sufficiently important to merit their own names and symbols,[†] as illustrated in Table 7-2. Two, the radian and the steradian, appear to have no associated dimensions because the corresponding product of base units to the correct powers happens to be unity. These belong to a special category known as *supplementary units*.

The *degree Celsius* (°C, also called *degree centigrade*) represents a peculiar case. In a sense it is really not a separate unit at all, since it is equal in size to the kelvin (K). It owes its existence to the unusual nature of *temperature*, the quantity with which it is associated. Temperature is an intensive quantity, and only rarely is one concerned with systems in which temperature is totally "absent" (i.e., systems at "absolute zero", where the magnitude of the temperature is zero kelvin). For most purposes it is more convenient to

---

* It is interesting to note that the base unit for mass is the *kilo*gram rather than the gram, i.e., that it contains a multiplying prefix (cf. Sec. 7.3.3). This is a result of the long-standing relationship between the kilogram and the meter (in the MKS and MKSA systems).

[†] Occasionally a distinction is made between those SI-related units that have acquired official status (and a name of their own) and other derived units. The former are "derived units" in the narrower sense, while strict terminology would designate the latter as *compound units*.

compare the temperature of a given system with a "standard" (or "zero point") easily appreciated in terms of everyday experience: the temperature at which water freezes, approximatively 273.15 K. This reference point is assigned an alternative value, 0 °C. Temperatures below the (carefully defined) freezing point of water thus become negative quantities when specified in the unit "°C". The relationship between the kelvin and the degree Celsius is symbolized by the expression

$$T/K = \Theta \,/°C + T_0\,/K = \Theta\,/°C + 273.15 \tag{7-3}$$

where $T$ is some *absolute temperature*, $\Theta$ is the corresponding *Celsius temperature*, and $T_0$ is the absolute temperature of the zero point on the Celsius scale. Note that the symbols corresponding to these two units are correctly written "K" (not "°K") and "°C" (not simply " ° ").

SI unit recommendations have met with surprisingly rapid acceptance, at least in official circles. They were first issued in 1960 by the 11th international Conférence Générale des Poids et Mesures (CGPM) and quickly made their way through task groups and committees of a great many international and national scientific societies, including the International Union of Pure and Applied Physics, IUPAP, and the International Union of Pure and Applied Chemistry, IUPAC (IUPAP, 1978; IUPAC, 1979a). They also became a part of various international and national *standards* (e.g., Standard *ISO 31/0-1981*; Standard *ANSI/IEEE 268-1982*; Standard *BS 3763: 1976*; Standard *ISO 1000, 1981*; Standard *DIN 1301, Teil 1, 1985*). The SI recommendations have even been incorporated into *legislation* in a number of countries, and new editions of virtually all scientific reference books (e.g., the *CRC Handbook of Chemistry and Physics*) feature data adjusted to conform to SI guidelines. One small dictionary and handbook of quantities and units (Drazil, 1983) is particularly suited to the international scientific community because of its extensive discussion of the terminological aspects of the SI in the languages English, French, and German.

No report of a measurement is complete unless it contains an explicit and unambiguous statement of the corresponding units. Unit *symbols* should always be used when quantities are stated in numerical form ( e.g., "2.4 cm$^2$"). In all other cases, the full name of the unit should be written: "a few square centimeters".

Unit symbols never vary; for example, no distinction is made between singular and plural forms. Moreover, a unit symbol is never followed by a period (except at the end of a sentence) because it is not an *abbreviation*. When printed, unit symbols are always set in ordinary roman type. An extensive list of quantities and units, along with their names and symbols, can be found in Appendix K.

## 7.3.2 Additional Units

A few so-called *additional units*, although not strictly a part of the SI, have received the sanction of the agency that supervises the SI (Comité International des Poids et Mesures, CIPM); these are listed in Table 7-3. Many relate to measurement of space and time. It should be noted that the symbol for *hour* is "h", not "hr", and that for *minute* is "min". The *unit* "second" should always be designated in scientific writing by "s", even though in non-technical prose the *word* "second" might be *abbreviated* "sec.", with the plural form "secs.".*

The unit *liter*† warrants special comment. Two symbols for "liter" are listed in Table 7-3: "l" and "L". The capital letter is now widely regarded as preferable. This constitutes a change, and it results from a suggestion first made by the CIPM and quickly seconded in 1978 by the American Chemical Society. The capital letter has the important advantage that it is unlikely to be mistaken for any other symbol, while the lowercase "l" is easily confused with the numeral "1". It should also be noted that the former distinction between the liter and the cubic decimeter has now disappeared: L (or l) is simply the special symbol for a derived unit identical to $10^{-3}$ m³ (or 1 dm³; cf. Sec. 7.3.3).‡

Many publications still tolerate the use of certain non-SI units (cf. Table 7-3), particularly the *bar*, the *are* (a), the *hectare* (ha), the *röntgen* (or *roentgen*; R), the *curie* (Ci), the *decibel* (dB), and the *diopter* (dpt). "English system" units such as the inch and the pound have no place in scientific writing.

---

* Some have urged that the symbol "a" be adopted for "year" (from lat. *annus*), but no corresponding action has been taken. Consequently, the word "year" can be employed where otherwise a symbol would be appropriate.

† The spelling "liter" used here is that advocated by scientific societies in the United States, despite the fact that the original English-language documents defining the standards employ the British spelling "litre" (cf. IUPAC, 1979a); a similar dichotomy exists with respect to the meter (or "metre").

‡ For applications demanding high precision, some published data related to volumes must be modified by a correction factor. Prior to 1969 the liter was so defined that it was slightly *larger* than the cubic decimeter (equal to 1.000 028 dm³).

*Table 7-3.* Additional units.

| Quantity | Quantity symbol | SI unit | Other units and their names |
|---|---|---|---|
| plane angle | $\alpha, \beta, \gamma$ | rad | degree (°), $1° = \pi$ rad/180<br>minute ('), $1' = 1°/60$<br>second ("), $1'' = 1'/60$ |
| area | $A$ | m² | hectare (ha), 1 ha $= 10^4$ m²<br>are (a), 1 a $= 100$ m² |
| volume | $V$ | m³ | hectoliter (hL), 1 hL $= 10^{-1}$ m³<br>liter (L or l), 1 L $= 10^{-3}$ m³<br>centiliter (cL), 1 cL $= 10^{-5}$ m³<br>milliliter (mL), 1 mL $= 10^{-6}$ m³ |
| time | $t$ | s | day (d), 1 d $= 24$ h<br>hour (h), 1 h $= 60$ min<br>minute (min), 1 min $= 60$ s |
| pressure | $p$ | Pa | bar (bar), 1 bar $= 10^5$ Pa $= 10^2$ kPa<br>$= 1.019\ 72$ kp/cm² |
| activity (of a radionuclide) | $A$ | Bq | Ci (curie), 1 Ci $= 3.7$ x $10^{10}$ Bq |
| exposure to nuclear activity | $X$ | C kg⁻¹ | R (röntgen), 1 R $= 2.58$ x $10^{-4}$ C/kg |
| dimensionless quantities | $n$ | | dB (decibel), $n = 10 \lg(Q_1/Q_2)$, where $n$ is the number of decibels |

One other unit of special interest is the *atomic mass unit,** formerly abbreviated "amu". The atomic mass unit is recognized as an "additional unit" within the SI, but its symbol has been changed to "u". It is defined as 1/12 the mass of an atom of the nuclide $^{12}$C. Thus:

$$1\ u = 1.660\ 565 \times 10^{-27}\ kg$$

The mass of an atom or molecule expressed in units of u has a numerical value identical to the "relative molecular mass" of the substance (cf. Sec. 7.4.3 for further discussion of relative molecular mass).

---

* The term "dalton" appears in some scientific literature as a synonym for "atomic mass unit". This is particularly the case in biochemical publications, and it illustrates an unfortunate tendency toward the proliferation of technical terminology. Another related term (cf. Drazil, 1983) is *atomic mass constant (unified)*, with the symbol $m_u$. In this case the value $1.660\ 565 \times 10^{-27}$ kg is treated not as a *unit*, but as a *natural constant* like $R$ or $h$.

## 7.3.3    Prefixes and Spacing

Scientific units of measure can readily be adapted to suit a wide *range* of measurements by combining them with *multiplying prefixes*. Prefix conventions have also been the subject of international debate, resulting in the official list of allowed prefixes and prefix symbols presented in Table 7-4. A prefix has the effect of increasing or decreasing the fundamental value of a unit by a specific power of ten (i.e., by one or more "orders of magnitude"). Negative powers are all represented by lower case letters, while both lower and upper case letters occur in the symbols for positive powers. In one instance, "micro" ($10^{-6}$), a Greek letter has been employed ($\mu$).

Prefixes are to be used only singly, never in combination. This rule holds true even in the case of mass, which has (for historical reasons) the "kg" as base unit rather than the "g"; e.g.,

$$\text{not} \qquad 1 \text{ pkg} = 10^{-12} \times 10^{3} \text{ g}$$
$$\text{but rather} \quad 1 \text{ ng} = 10^{-9} \text{ g}$$

*Table 7-4.* Multiples and sub-multiples of units.

| Factor by which the unit is multiplied | | SI prefix | Symbol |
|---|---|---|---|
| 1 000 000 000 000 000 000 = | $10^{18}$ | exa | E |
| 1 000 000 000 000 000 = | $10^{15}$ | peta | P |
| 1 000 000 000 000 = | $10^{12}$ | tera | T |
| 1 000 000 000 = | $10^{9}$ | giga | G |
| 1 000 000 = | $10^{6}$ | mega | M |
| 1 000 = | $10^{3}$ | kilo | k |
| 100 = | $10^{2}$ | hecto | h |
| 10 = | $10^{1}$ | deca | da |
| 1 = | $10^{0}$ (unity) | | |
| 0.1 = | $10^{-1}$ | deci | d |
| 0.01 = | $10^{-2}$ | centi | c |
| 0.001 = | $10^{-3}$ | milli | m |
| 0.000 001 = | $10^{-6}$ | micro | $\mu$ |
| 0.000 000 001 = | $10^{-9}$ | nano | n |
| 0.000 000 000 001 = | $10^{-12}$ | pico | p |
| 0.000 000 000 000 001 = | $10^{-15}$ | femto | f |
| 0.000 000 000 000 000 001 = | $10^{-18}$ | atto | a |

A prefix symbol is always written directly in front of the symbol for the corresponding unit; no space is ever left between the two. The entire unit symbol (with prefix) should be separated by one space from the numerical value to which it applies, but the two should appear on the same line of text if at all possible.

Derived units are often written as products and/or quotients of other units. A space must always be left between the symbols for any two units to be understood as multiplied by each other.* The division operator should be represented by a *slash* ("/ "; other names for this symbol include "oblique," "slant," and "solidus"). A group of (multiplied) unit symbols following a slash must be enclosed in parentheses in order to avoid ambiguity. The alternative of employing negative powers eliminates the need for a slash and is often preferable; e.g., $S = 0.133$ J mol$^{-1}$ K$^{-1}$; similarly, "$c$, mol/L" or "$c$/(mol L$^{-1}$)" could be written as "$c$, mol L$^{-1}$" or "$c$ mol$^{-1}$ L" .

The rule requiring a space between unit symbols is not arbitrary: its purpose is to ensure a clear distinction between a product of units, on one hand, and either a unit preceded by a prefix or a unit whose symbol is comprised of two or three letters (e.g., Pa or mol), on the other. As a further safeguard against ambiguity it was agreed that the unit "m" (*meter*) should always appear *last* in any product expression of which it is a part, thereby avoiding any possible confusion with the *prefix* "m" (*milli*); thus, "mN" would signify "millinewton", whereas "N m" is "newton-meter".[†]

---

* No such rigid rule applies with respect to quantities; cf. Sec. 8.1.2. With units it is extremely important, though too often ignored.

† There is no need to invoke this rule for products involving exponentials or expressions containing multiplication signs. Thus, m$^2$ kg s$^{-2}$ is a valid formulation.

## 7.4     *Quantities and Composition in Chemistry*

### 7.4.1     *The Mole*

The SI was the first system of units to dignify the discipline of chemistry with a base unit of its own: the *mole* (unit symbol: mol). Surprisingly, this tribute actually appears to be a source of consternation among some chemists, and many fail to understand its proper use. The source of the problem is not the unit: it is the physical quantity with which the unit is associated, a quantity with the official name *amount of substance* and the symbol $n$ (cf. Sec. 7.3.1).

The name chosen for this quantity is probably an unfortunate one. While the names for the other base quantities (length, mass, time—even luminous intensity) have *a priori* meanings more or less clear to everyone, the expression "amount of substance" apparently does not. The term *sounds*, in fact, more like a vague piece of everyday speech than a precise, carefully defined concept.* Moreover, the significance of the quantity itself is difficult for many to grasp. Even IUPAC documents fail to come to terms adequately with the problem. Rather than attempting a rigorous definition of the quantity "amount of substance" IUPAC (1979a) simply notes the interesting anomaly that the concept of the corresponding unit, the mole, is older than the quantity for which it is a measure: "One of these independent base quantities is of special importance to chemists but until recently had no generally accepted name, although units such as the mole have been used for it."

The *mole*† is officially defined as "the amount of substance of a system which contains as many elementary entities as there are atoms in 0.012

---

* The term adopted in German for "amount of substance", "Stoffmenge", has also been criticized. In this case, the expression at least *sounds* technical (unlike the English version), but unfortunately one of the component nouns, "Menge", had already been appropriated by mathematicians as their designation for what in English is known as a "set".

† The mole is the only one among all SI units, base and derived, whose symbol consists of *three* letters, and it is also the only one whose name (mole) and unit symbol (mol) differ only by one letter. This should not be taken as an excuse for carelessness, however. As with other units, the *name* "mole" should be used when a *word* is appropriate (i.e., in a sentence), and it is perfectly correct to employ a plural form ("several moles"); in conjunction with numerals, however, the official *symbol* for the unit is what is required, as in the phrase "2 mol of $H_2SO_4$". The symbol "mol" should always be used in equations.

kilogram of carbon-12." The actual number of such particles is, of course, the *Avogadro number*, $6.022 \times 10^{23}$, and the *Avogadro constant*, $N_A$, is expressed

$$N_A = 6.022 \times 10^{23} \text{ mol}^{-1}$$

It has been argued by some that the mole is actually nothing more than a counting convention, like the dozen, and that there was no need to begin suddenly treating it as a formal unit. Be that as it may, the mole *is* now a unit—related to a quantity—and it should be dealt with accordingly. As one example of the carelessness disconcertingly prevalent consider the statement "the compound took up two moles of bromine". It is most unlikely that the sample in question really reacted with 320 g of bromine, yet this is what has been stated. In all probability the author actually intended to report that two moles of bromine were consumed *per mole of substrate*. A *proportionate* specification was called for, not a statement of an *absolute* amount. Errors of this kind would never be tolerated in the case of other quantities, such as length or mass!

The principal advantage of the mole is that it facilitates dealing with *sets* of particles and therefore with "equal portions", which may differ considerably in mass and volume. Moreover, the huge factor contained in the Avogadro constant helps bridge the gap between the invisible and unimaginably small world of individual atoms and that of everyday experience. Indeed, chemical calculations would become a nightmare in the absence of a unit such as the mole. Chemists should perhaps take some pride in the fact that their needs were taken so seriously, and that the mole acquired a status equivalent to that of the meter.

## 7.4.2    The Amount of Substance

Returning to the problem of the quantity $n$ ("amount of substance") we might begin by defining it as follows: "$n$ is the magnitude or extent of a *portion* of substance as measured by the number of multiples of $6.022 \times 10^{23}$ *particles* it contains." Notice that defining the amount of substance in this way, and accepting the mole as equivalent to $6.022 \times 10^{23}$ particles, amounts to assuming as truth what was once regarded only as hypothesis: the dual postulates that all matter *consists* of particles, and that the *nature* of the particles can be described. Many decades have passed since the acceptance of these postulates became general, and it seems not unreasonable for them now to be the basis for a unit of measure and a corresponding quantity, the "amount of substance."*

The nature of the particles making up the "portion" referred to above is subject to no restrictions—they may be *atoms, molecules, ions, electrons,* or even (at least for purposes of calculation) *fractions* of particles (*equivalent particles*). Nevertheless, the type of particles counted must always be *specified*. If *non-uniform* matter is to be measured this way it must be explicitly treated as a mixture of components; i.e., each type of particle present must be counted separately.

## 7.4.3    Molar Quantities

Chemists have long recognized the advantages of working with derived quantities obtained by dividing certain other quantities by the corresponding amount of substance. Such derived quantities are invariably of the intensive type, and they have traditionally been named by prefacing the name of the original numerator quantity with the word "molar" (IUPAC, 1979a).[†] For example, the quantity *molar volume* results from dividing the *volume* of a portion of matter by its *amount of substance* (i.e., by the number of particles it contains, measured in moles).[‡] The symbol for the new quantity, $V_m$, is

---

* The quantity "amount of substance" is not merely a special way of treating mass. This is perhaps best illustrated by the fact that photochemists and electrochemists regularly—and quite properly—use expressions like "two millimoles of electrons" (or photons) even though the amount of substance to which they are referring—by no means insignificant—has a mass well below the limits of measurement of the most sensitive balance. The terminology remains a problem, however. Consider the following example. It is quite appropriate to ask either "what *mass* of NaCl will react with ..." or "how many *grams* of NaCl will react with ...". There is also no problem involved in asking "how many *moles* of NaCl ...", but it is most awkward in English to formulate the corresponding statement in terms of a *quantity*. Nevertheless, one should avoid phrases such as "the number of moles" and instead say "the amount of substance". The skeptic should try substituting "number of meters" or "meter number" as synonyms for "length".

[†] The mole is the only unit whose name, when transformed into an adjective ("molar"), is used as part of the name of a derived quantity.

[‡] Another set of intensive quantities is similarly obtained by dividing appropriate extensive quantities by *mass*. These are denoted by preceding the name of the corresponding extensive quantity by the word *specific* (e.g., specific heat of fusion, $\Delta H_f/m$).

simply that of the numerator quantity (volume) elaborated with the roman subscript "m". Similarly, the molar Gibbs free energy is defined as

$$G_m = G/n$$

Units of molar quantities always contain the factor "mol$^{-1}$". Probably the most commonly used of the molar quantities is the *molar mass* $M_m$, the mass of a portion of matter divided by its amount of substance. The corresponding unit is g mol$^{-1}$, and the numerical value obtained is identical to the *relative molecular mass* of the material, $M_r$. The term "molecular weight" was once associated with such a number, but the IUPAC *Manual of Symbols and Terminology for Physicochemical Quantities and Units* no longer recommends this terminology. The same fate has befallen the "atomic weight", a term that has been replaced by "relative atomic mass".

Discussions of solution chemistry make frequent use of the quantity *amount-of-substance concentration,* the amount of substance of a given solution component divided by the total volume of the solution.* If the substance in question is B, then two acceptable symbols for this quantity are $c_B$ and [B],[†] where the unit of measure is mol L$^{-1}$. Older conventions remain rather firmly entrenched, however. Thus, a solution of concentration 0.1 mol L$^{-1}$ is still frequently called a "0.1 M solution" (enunciated "point one molar solution"). In view of the IUPAC definition for the word "molar" such terminology could hardly be called "scientific". "M" in this context must be clearly understood as a form of shorthand notation rather than as a unit symbol if reference to "0.1 M HCl" is to be acceptable. Stoichiometric calculations are possible only with the correct unit: mol L$^{-1}$. Similar arguments apply to so-called "normal" solutions (e.g., "0.1 N KMnO$_4$"), in which "equivalent particles" take the place of molecules.[‡]

The amount-of-substance concentration is only one of several quantities used for describing the *composition* of a mixture or a solution, and archaic (and no longer appropriate) usages accompany others as well. It is not our

---

* Traditionally this was referred to as the "molar concentration", but this usage conflicts with the technical definition presented above for the adjective "molar". The same problem arises when one attempts to characterize a solution in terms of its "molarity".

† Use of the latter symbol should be restricted to "law of mass action" expressions.

‡ The convention of the abbreviation "N" will probably persist for a long time. "2 mL of 0.1 N KMnO$_4$" (where KMnO$_4$ is understood to be an oxidizing agent capable of supplying five electrons per molecule) is certainly shorter and more convenient to write than the correct form "2 mL of a solution, $c(1/5 \text{ KMnO}_4) = 0.1$ mol/L".

intent to range too far into what is properly the content of a general chemistry course, but it may nevertheless be prudent to point out a few other outdated forms still found even in textbooks. Consider, for example, notations such as "vol-%" or "weight-%" (wt.-%). Admittedly these are intelligible, but they clearly have much in common with colloquialisms of the type we condemned in Sec. 7.2.1. In both of the cases cited, names of quantities (volume and "weight", where the latter actually refers to mass) have been joined to a mathematical operator (%)—and the result is alleged to be a unit! This is another striking example of "unscientific" expression. The same can be said regarding a notation such as "%(v/v)". Correct ways of expressing what is actually intended include

$$\varphi = 0.28 = 0.28 \text{ L/L} = 28\%$$
$$\omega = 0.50 = 0.50 \text{ g/g} = 50\%$$

where $\varphi$ is the *volume fraction* and $\omega$ the *mass fraction* of a substance in a mixture (cf. Appendix K). Both quantities have the unit "one". Writing "L/L" or "g/g" as the corresponding "units" may be helpful as a reminder, but this is obviously not necessary, because the process of arriving at the quantities is implied in the definitions of $\varphi$ and $\omega$ (volume of component divided by volume of mixture, and mass of component divided by mass of mixture, respectively). Use of the symbol "%" (for *percentage*) conveys the message "multiply the value shown by 0.01", hence it results in another valid notation. Observe that this unit symbol—in its original meaning a mathematical operator —is unique in that it is *not* separated from the corresponding numerical value by a blank space (cf. Dodd, 1986). Special symbols are used to express mass (or other) fractions of trace substances (cf. Appendix K).

## 7.5    Numbers

Next we consider correct ways of presenting *numbers* and the numerical portions of quantity specifications (for more details see, for example, Leaver and Thomas, 1974). *Single-digit* numbers within a sentence are normally written as words, but numerals are always used for numbers equal to or larger than 10 (unless a number begins a sentence, in which case it is often written as a word):

Fourteen grams of ...; ... was 14 times as large.

Avoiding the need for numbers at the beginning—or end—of a sentence is an even better solution. If two or more numbers appear in the same sentence, and if at least *one* of them is 10 or larger, then numerals should be used for *all* of them, e.g.:

Each of 14 elements (9 nonmetals and 5 metals) was ...

One exception to this rule is a sentence containing a numerical value for a quantity in addition to other numerical information. Here a mixture of words and numbers may well be proper; e.g.:

Each of four animals received doses of 0.5 mg per kilogram of body weight ...

In general, numbers are printed in roman type, although italics can be used in exceptional cases for display purposes. Any number containing more than four digits before or after the decimal point should be broken into *triads* (groups of three digits), with division proceeding in each direction from the decimal point. The triads themselves should be separated from one another by spaces,* not commas;[†] e.g.,

13 296    0.124 35    61 325.014 51    1.020 07 x $10^{-6}$

Four-digit numbers are normally left unbroken:

1000    3.1457

Consistency should be maintained, however. For example, if some numbers in a table need to be written in triads then all should be, including those comprised of only four digits.

Because *billion* is the word for "a thousand million" (prefix *giga*, $10^9$) in the United States, but "a million million" (prefix *tera*, $10^{12}$) in most other countries, billion and related terms such as trillion should be strictly avoided in scientific or technical writing.

---

* In printed work the space should be narrow (approximately the width of the letter "i") and should always be the same even if variable spacing is present between words (as in text that has been "justified").

[†] A comma could be misleading since, unfortunately, the comma is used as the decimal sign in certain countries. It is most remarkable—and also disconcerting—that no attempt has yet been made to obtain international agreement on so elementary a matter.

Standard conventions should be followed for the *rounding* of numbers. These conventions are discussed in many reference works and even in most elementary textbooks, and we refrain from reviewing them here, apart from illustrating the principles with the following examples:

$$2.17 \approx 2.2$$
$$2.15 \approx 2.2$$
$$2.14 \approx 2.1$$
$$3.131\ 59 \approx 3.132 \approx 3.13 \approx 3.1$$

Unit *conversions* should be made with careful attention to the relationship between the known accuracy of the data and what is implied by the number of digits shown. The number of digits retained in a conversion should be such that accuracy is neither sacrificed nor exaggerated (i.e., the correct number of *significant digits* should always be reported). For example, a length of 3.0 ft. is equivalent to 91 cm (not 91.44 cm as a calculator might suggest), while "3 ft." is presumed to be less precise and should thus be converted to 90 cm (or better—because less ambiguous—0.9 m). Had the measurement been reported as 12.5 ft. one would assume that the corresponding metric length would be exactly 3.81 m. If, however, the 12.5 ft. length is known to have been obtained by rounding to the nearest 0.5 ft., the conversion should be given as 3.8 m.

Proper conversion is accomplished by *first* multiplying the quantity at hand by the appropriate conversion factor, and *then* rounding to the correct number of significant digits, consistent with the known (or assumed) precision of the data. One should round neither the factor nor the quantity before performing the multiplication, since to do so would reduce the accuracy of the result. The precision should never be assumed to exceed the *accuracy* of measurement, or the *tolerance* (or "margin of error") if the latter is known. Examples:

A stirring rod is said to be 6 in. long. The accuracy of the statement is estimated to be 1/2 in. (± 1/4 in.), or, after conversion, 12.7 mm. The converted value of 6 in. is 152.4 mm. This must be rounded to the nearest 10 mm, or 150 mm, but it would be better to present the result as "15 cm" or "1.5 x 10$^{-1}$ m", because these forms clearly imply two-digit precision.

A test pressure is described as 200 ± 15 lb./in.$^2$ (pounds per square inch, psi). Since one-tenth of the tolerance is ± 1.5 lb./in.$^2$ (± 10.34 kPa), the converted value should be rounded to the

nearest 10 kPa. Conversion leads to
1 378.951 4 ± 103.421 35 kPa, which is presented as
1380 ± 100 kPa.

*Fractions* comprised of numbers are best written in the slash notation, consistent with the suggestion for quotients of units and quantities (see Sections 7.2.3 and 7.3.3). Parentheses should be incorporated where necessary; e.g.,

$$1/2 \quad 83/100 \quad 1/(2\pi)$$

Adherence to the slash convention is especially important if fractions are to appear within a text line. (Cases requiring the use of a horizontal quotient line are treated in Sec. 8.1.1.)

Decimals should be represented by periods set on the text line* ("decimal points"), although publications from some European countries in languages other than English persist in using commas for the purpose (see the second footnote in this section).[†]

# 7.6    *Troublesome Mathematical Symbols*

Our discussion here is restricted to certain specific symbols that are frequently misused. The reader is referred to specialized texts for more information about mathematical symbols in general.

A *range*, in the sense of "from ... to", is commonly indicated by a horizontal line or dash (a hyphen in typed copy, a somewhat longer symbol called an *en dash* in printed works); for example,

$$800\text{-}1000 \text{ bar}$$

Many feel that a different notation is preferable for scientific purposes, since the dash might be mistaken for an instruction to perform a *subtraction*. One obvious possibility is simply to resort to words:

---

* Use of centered dots ("0·5") is now considered obsolete.
[†] Many European journals have recently changed to the use of periods, often in order to simplify the task of producing English translations.

a pressure range of 800 to 1000 bar.

However, a different solution has been proposed in Standard *DIN 1338, Beiblatt 2, 1983*: use of the symbol "...";* i.e.,

800...1000 bar

Another troublesome symbol is that for the concept "approximately equal to". The correct symbol consists of two curved lines ("tildes"), one above the other (≈). A single such line (~) is frequently substituted, but this should be used to convey quite a different meaning: it correctly signifies "proportional to".

Inconsistency is quite common with respect to the symbols for "greater than or equal to" and "less than or equal to". In this case the correct representation is a vertical *combination* of signs: that for either "greater than" or "less than" together with that for "equal to"; i.e.,

$$\geqq \quad \leqq \quad (\text{not: } \geq \quad \leq)$$

IUPAC (1979a) proposes substitution of a different variant:

$$\geqslant \text{ and } \leqslant$$

Unfortunately, this suggestion suffers from the fact that it is virtually impossible to simulate on a typewriter, and even most word processors lack the required capability.

It may seem surprising that one of the most misused symbols of all is "=", the *equal sign* (*sign of equality*). "Equal" is a mathematical concept, and the corresponding symbol should be used only where true equality exists between pairs of mathematical or physical quantities, numbers, or expressions; that is, this symbol should always constitute part of an *equation*. As has already been pointed out (first footnote in Sec. 7.1.3), it is inappropriate to use a sign of equality as shorthand in an explanation (such as in the legend to a figure; see Sec. 7.1 and Fig. 7-1). Most violations of this rule occur in places where no symbol at all is required—a simple space would serve the same purpose, as

---

* The principal objection to this usage is the fact that three dots also have the meaning "and so on, up to", as in a sequence. The dual meaning is unlikely to cause confusion, however, and it therefore seems reasonable to encourage use of the "three-dot" symbol in the interest of clarity. Only in the case of page numbers and literature citations should a range always be indicated by a dash; i.e., pp 14-18, or [3-11].

would a set of parentheses, as in the following expressions, both of which are quite inappropriate as they stand:

> "Phylogeny (= racial history)"
> "V = valve"

In both cases the equal sign is superfluous, but also incorrect. Similarly, one should never write

$$mp = -10\ °C$$

because "mp" is not the symbol for a quantity. The correct representation is simply

$$mp\ -10\ °C$$

On the other hand, this information *could* be represented in the form of a true equation:

$$t_m = -10\ °C$$

Here, $t_m$ is indeed a symbol for a quantity: a particular kind of "temperature"— a "melting temperature", as indicated by the subscript "m". The correct formulation of equations in general is discussed in the following chapter.

# 8 Equations and Formulas

## 8.1 Mathematical Equations

### 8.1.1 Some General Rules

In this chapter we consider the techniques for incorporating mathematical or chemical *equations* or *formulas* in a manuscript. While some of the recommended practices no doubt have their origin in aesthetics, they have been refined according to two principal criteria: clarity for the reader and convenience for the typesetter. We first discuss *mathematical* equations, although certain of the points made also apply to the *chemical* equations discussed in Sec. 8.2.2.

Generally speaking, all equations in a manuscript should be placed on separate lines and *indented*.* In the terminology of copy-editors and typesetters, material treated in this way is *displayed*, a technical term that describes the primary purpose.† A brief explanation or commentary separating two or more equations should also occupy a separate line, but the text should be set flush left.

Authors are often unsure of the correct way to punctuate text containing displayed expressions and equations. In principle, an equation falling at the end of a sentence should be followed by a period, and one in the middle of a sentence might logically be set off by commas. Such commas and periods are usually omitted, however, for two pragmatic reasons: punctuation at the end of a complicated mathematical expression can be difficult to place correctly, and it would probably be overlooked in any case.

Occasionally an equation will prove too long to fit on a single line. If so, it should be divided so each line ends with a mathematical operation symbol,

---

* In manuscripts to be typeset the same treatment is recommended also for long *expressions*, such as $\exp[-E_a/(RT)]$.
† The display technique also serves a practical purpose. It is often extremely complicated, if not impossible, for a printer to "justify" (i.e., terminate exactly at the right margin) a text line containing an equation or a long expression.

preferably a "+" or a "−", and the same symbol should be *repeated* at the beginning of the new line, e.g.:

$$c^4(H_3O^+) + c^3(H_3O^+) \cdot K_{S1} + c^2(H_3O^+) \cdot (K_{S1}K_{S2} - K_W) +$$
$$+ c(H_3O^+) \cdot (-2C \cdot K_{S1} - K_W) - K_{S1}K_W = 0$$

If a series of *transformations* of an equation is to be shown (as in a *derivation*), each new form of the equation should occupy a separate line, and each should begin with an "equals" sign (*sign of equality*). The resulting series of equality symbols should be aligned vertically; e.g.,

$$\Delta G^\circ = \Sigma\,(...) + ... - (...)$$
$$= (... + ... - ... - ...) -$$
$$- T\,(...)$$
$$= -62.6 \text{ kJ mol}^{-1}$$

Alignment of equals signs is recommended even if the equations are separated by short text lines (e.g., a few words of explanation).

All *arithmetic symbols* within an equation should be typed at the same level. This rule applies not only to multiplication, addition, subtraction and equality symbols, but also to the horizontal line used in fractional notation to symbolize division. In typewritten text the latter is most easily created by rolling the paper down one-half space and using the underline key. The x commonly employed as a multiplication symbol, for example in exponential notation (e.g., $3 \times 10^4$), is best simulated in a typescript by a lowercase "x", preceded and followed by a space. *Centered dots* are also frequently employed to indicate multiplication, especially in printed copy, but these are almost impossible to duplicate with a typewriter. Periods are unsatisfactory even if the paper is moved down for proper positioning, because the resulting dot is too small. The best procedure is simply to leave appropriate spaces while typing, and later to add the dots with a pen and black ink. Two or more adjacent *quantity* symbols are *assumed* to require multiplication, as in $pV = nRT$. Products of *units* may be implied in a similar way, although in this case a space should always be left between the individual symbols (cf. sections 7.3.3 and 8.1.2).

Special care should be taken to ensure the proper use of *grouping* symbols, such as parentheses. In particular, one should verify that all *pairs* are complete and that the symbols of *nested groups* appear in the correct sequence: *parentheses* should be enclosed within *brackets*, and brackets within *braces*; e.g.:

$$\{\,[\,(\quad)\,]\,\}$$

Grouping symbols of all types should be drawn large enough to enclose all parts of the expression they are intended to isolate.

The following examples illustrate several of the suggestions made above. All were created with a typewriter; they have been photoreduced to 60% of their original size (i.e., subjected to a 40% linear size reduction):

$$C = F* \cdot \left\{ x^n + \sum_{i=1}^{n} \left[ x^{n-i} \cdot (1 - a) \cdot \prod_{j=1}^{i} k_j \right] \right\}$$

$$\lim_{i \to \infty} \left| \frac{c_i}{c_{i+1}} \right| < 1$$

$$\frac{d^2 \tau}{dx^2} = (\ln 10)^2 \cdot \left[ \frac{k_1}{x} - x + x \cdot k_2 \cdot \frac{x - k_2}{(x + k_2)^2} \right]$$

## 8.1.2    Placement and Spacing

*Subindices* (*subscripts*) and *exponents* (or other *superscripts*) should be set directly adjacent to the symbols to which they apply. If a variable contains both a subscript and a superscript, it is usually considered preferable to type one directly above the other rather than stagger them;* e.g.

$$10^{-3} \quad e^{0,027} \quad y^{1/3} \quad K_S^2 \quad a_1^{-0,5} \quad z_{abs}^2$$

Subscripted exponents should be avoided if possible; they may lead to ambiguity unless they are very carefully prepared. Usually it is not difficult to devise some equally suitable substitute notation; e.g.,

$$\exp(-x_2) \quad \text{rather than} \quad e^{-x_2}$$

Correct *spacing* is also important for maintaining clarity in mathematical expressions. Typewritten or typeset copy should always contain single blank spaces at the following points:

---

* Adherence to this suggestion may be impossible in text prepared on a standard word processor. Subscripts should then *precede* superscripts. Some editors follow a different convention in which vertical alignment is reserved for specifying *limits* (e.g., "$a_1^3$" means "$a$ from one to three"), while *exponents* are placed after the complete expression to which they apply (e.g., "$a_1{}^3$" is the correct form for "$a_1$ cubed").

- before and after operation and relation signs, such as +, −, or =; e.g., "$z = a + 2b$";
- before and after integral ( $\int$ ), summation ( $\Sigma$ ), and product ( $\prod$ ) signs;
- before and after abbreviations for trigonometric and logarithmic functions; e.g., "$2 \tan x$";
- before, but *not* after, a derivative or differential sign (e.g., $dx$, $\partial y$, $\Delta z$);
- between a number and a unit symbol;
- between symbols of units multiplied together (cf. sections 7.3.3 and 8.1.1).

*No* space should be left between a coefficient (or *multiplier*) and the symbol to which it applies (e.g., $4x$), nor is a space appropriate between terms understood to be multiplied together [e.g., $(abc)$].* Also, *no* space is left between the % (*percent*) sign and the number to which it applies.

## 8.1.3    Special Symbols

Producing equations with a typewriter is at best an awkward undertaking. This is particularly true when special symbols are involved, including the letters of the Greek alphabet and such standard mathematical symbols as

$$[ \; ] \; \{ \; \} \; \cap \; \otimes \; \equiv \; \neq \; \Leftrightarrow \; \Rightarrow \; \infty \; \Sigma \; \prod \; \bullet \; \times \; \rightarrow$$
$$\sqrt{} \; \exists \; \forall \; \in \; \notin \; \partial \; \nabla \; \wedge \; \vee \; \approx \; \div \; \cap \; \cup \; \uparrow \; | \; \varnothing$$

Many *word processors* (cf. Sec. 5.2.2) offer much greater flexibility; some systems even permit the user to create customized symbols. Nonetheless, novices should be forewarned that the *availability* of a variety of symbols on a word processing system does not in itself guarantee their easy incorporation into text, nor is all word processing software equally suited to the formatting of complex equations. Anyone proposing to use a word processor for routine preparation of highly mathematical text is strongly advised to shop carefully before deciding on a particular system. Indeed, it would be wise to insist on a "hands-on" demonstration in order to assess the ease with which the desired results can be obtained.

A second factor to consider is the *quality* of the resulting printed symbols, a function not only of the computer and its associated software but also of the printer. A "letter quality" printer (cf. "Hardware" in Sec. 5.2.2) may produce

---

* Actually, high-quality printed text usually does contain very small spaces between the terms of a product, and the same effect can be achieved with certain word processors.

excellent text, but it is likely to be nearly as limited as a typewriter with respect to symbols; dot matrix printers on the other hand are enormously versatile, but their output is not always pleasing to the eye.* A wide range of printed output (including, for example, italicized symbols) should be examined minutely to ensure its suitability for camera-ready copy (cf. Sec. 3.4.6).

In the absence of satisfactory computer-aided techniques for incorporating symbols into manuscripts one is forced to resort to less convenient alternatives. Those most frequently chosen include:

● carefully introducing all symbols with *India ink*, either *free-hand* or (preferably) with the aid of special *stencils* (or *templates*), such as the ones illustrated in Fig. 8-1;

● using adhesive *dry transfer letters* (examples: Letraset, Datak, Cello-Tak, Alfac);

● purchasing an appropriate *typewriter accessory* for the temporary insertion of special characters (example: Typit, a device adaptable to most standard machines). The *type elements* or *daisy wheels* (containing letters, punctuation marks, numerals) in many modern typewriters can be replaced temporarily by ones containing special symbols.

Finally, it should be noted that even certain *common* symbols can pose problems in typewritten equations. An example is the numeral "1". Many typewriters lack a separate "1" key, while others produce a "1" that is virtually indistinguishable from the letter "l".[†] The resulting ambiguity may be sufficiently serious to warrant hand-lettering in certain cases. Similarly, many typewriters lack a separate symbol for "zero". The standard alternative here is to use the capital letter "O", but this may also lead to ambiguity. It is unfortunate that the "O-slash" (Ø) form of "zero"—long familiar to computer programmers—has not established itself as a standard character in scientific writing.[‡]

---

* So-called "laser printers" are a clear exception. The best laser printers will reproduce faithfully any symbol created by a word processor, and the resulting copy is of "near-print" quality (resolution approx. 300 dots per in.[2], as contrasted to approx. 1200 dots per in.[2] for a high quality professional *photo-typesetting machine*) even though it is based on a dot matrix principle.

† It is largely for this reason that capital "L" rather than small "l" is now the recommended symbol for "liter": 1 L = 1 dm[3] (cf. Sec. 7.3.2).

‡ Copy for typesetting should be carefully annotated to distinguish between zero and capital "O" wherever confusion is possible. For example, "30 H" might well be read as "3 OH" in the absence of a clarifying marginal note.

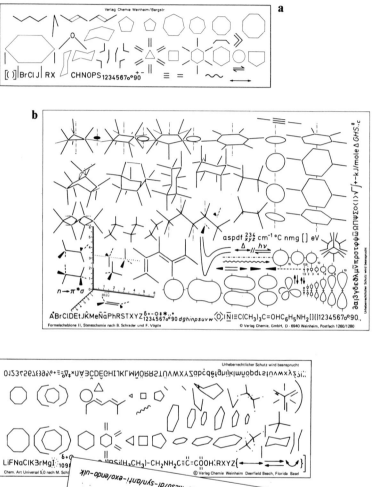

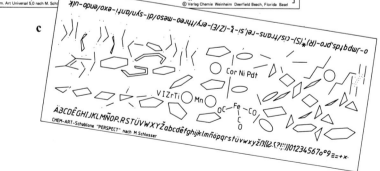

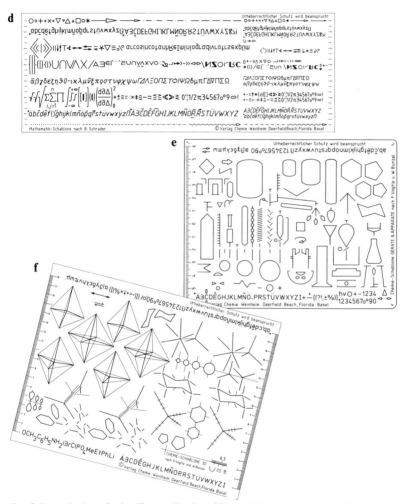

Fig. 8-1. A selection of scientific stencils; those illustrated here have been developed by, and are available from, VCH Verlagsgesellschaft mbH, Weinheim, and VCH Publishers, Inc., New York.—a) *Chemistry Stencil planar* for drawing organic chemical structures (designed by F. Ehrhardt, size 258 mm x 83 mm); b) *Formula Stencil II Stereochemistry* (designed by B. Schrader and F. Vögtle, size 320 mm x 220 mm); c) *Chem-Art universal 5.0 and Chem-Art perspect* chemistry stencils (designed by M. Schlosser); d) *Mathematical Stencil* with 5 mm and 3.5 mm letters and symbols, upright and italic (designed by B. Schrader, size 294 mm x 108 mm); e) *Chemistry Stencil Apparatus and Equipment* (designed by F. Vögtle and W. Bunzel, size 168 mm x 130 mm); f) *Chemistry Stencil 3D for Drawing Stereo Formulae and Stereo Diagrams* (designed by F Vögtle and W. Bunzel, size 256 mm x 193 mm).

## 8.2    Chemical Formulas and Equations

### 8.2.1    Chemical Formulas

*General Considerations*

A *chemical* (or *structural*) *formula* should always be regarded as a single, indivisible entity. In other words, structural formulas must never be divided for continuation on a second line. Considerable thought should be given to the way in which such formulas are typed. Every effort should be made to convey a maximum amount of structural information in the simplest possible way. The following examples illustrate a variety of techniques for producing typewritten structures:

$$H_3C-CH=CH_3 \qquad H_3C-CH=CF_2-CHCl_2 \qquad \begin{matrix} H_3C \\ H_2N \end{matrix}>CH-COOH$$

$$H_5C_6-SO_2-N<\begin{matrix} CH_3 \\ O-Si(CH_3)_3 \end{matrix} \qquad \begin{matrix} H_3CSO_2 \\ H_3C \end{matrix}>N-N=C<\begin{matrix} C_6H_5 \\ H \end{matrix}$$

Note that *atom connectivities* and relevant *spatial relationships* have been made explicit wherever practical. Ideally, single bonds should be represented by a line somewhat longer than a hyphen (an "en dash" in typeset material).

Single bonds are often omitted at places where a practiced chemist would instinctively understand their presence (except in elementary textbooks), permitting simplification of the first two examples to $H_3CCH=CH_2$ and $H_3CCH=CFCHCl_2$. This saves space, as does the use of abbreviations (see below).

More compact (or *condensed*) formulas are occasionally used for the purpose of supplementing and clarifying compound *names*. Such formulas should always be written directly following the names to which they correspond, separated from the text by commas:

... was treated with hexamethyldisilazane, $HN[Si(CH_3)_3]_2$, ...

Word processing equipment and associated software have now advanced to such a degree that it has become quite practical to incorporate very sophisticated *structural formulas* into text manuscripts, and authors are certainly encouraged to avail themselves of such facilities whenever possible.* Nevertheless, situations will continue to arise in which it is more convenient (or even necessary) to prepare formulas in the traditional way—by hand, with appropriate pens and India ink. *Vellum* is the preferred medium for such work, and *stencils* should be employed whenever possible to minimize irregularities. Fig. 8-1 illustrates a number of convenient and readily available scientific stencils of the type to which every chemist/writer should have access. The stencils included are designed to produce formulas with capital letters 5 mm high, a size ideal for copy that will be photoreduced before printing. Photoreduction leads to a significant improvement in appearance, but the *CHEM-ART universal* stencil (Fig. 8-1c) is also available in a smaller size (with 3.5 mm letters) suitable for preparing structures directly in a typewritten manuscript. Even if one's structures are to be photoreduced it is still convenient to have stencils of two sizes available, because the smaller stencils are ideal for preparing such subsidiary structures as those above or below reaction arrows.

Stencils are usually accompanied by detailed directions for their use, including information on the proper choice of pens for various line widths (cf. Sec. 9.2.3, "Drawing the Lines"). With a little practice—and the proper tools (cf. Sec. 9.2.2)—one can generate even rather complex structures quite efficiently. Fig. 8-2 shows a number of structures prepared with the help of the stencils illustrated in Fig. 8-1 a-e. Specific guidance on the use of stencils and other drawing equipment is reserved for Chapter 9 (esp. Sec.9.2.3).

Several of the structural formulas used here as examples contain *abbreviations* in addition to bonds and atomic symbols. The intent is to conserve space and at the same time eliminate the distraction that might accompany excessive detail. Thus, a representation such as "–OSiMe$_3$" is easier to grasp than "–O–Si(CH$_3$)$_3$". However, care is required to ensure that no ambiguity is introduced. Appendix H contains a selection of abbreviations suitable for symbolizing chemical substituents within complex structures.

If the need arises to distinguish between two otherwise identical *substituents* one should always use *superscripts*, not subscripts; i.e., "R[1]" and "R[2]" are the

---

* An example of a suitable software package adaptable to virtually every personal computer is *Liniengraphik: Chemische Strukturen, Diagramme, technische Zeichnungen.* This program is a member of the VISPER-32 software series available from VCH Verlagsgesellschaft mbH, Weinheim (Federal Republic of Germany). An English-language version is currently being prepared. A powerful structural program specifically for the Apple Macintosh computer is *ChemDraw* (Cambridge Scientific Computing, Cambridge, Massachusetts, USA).

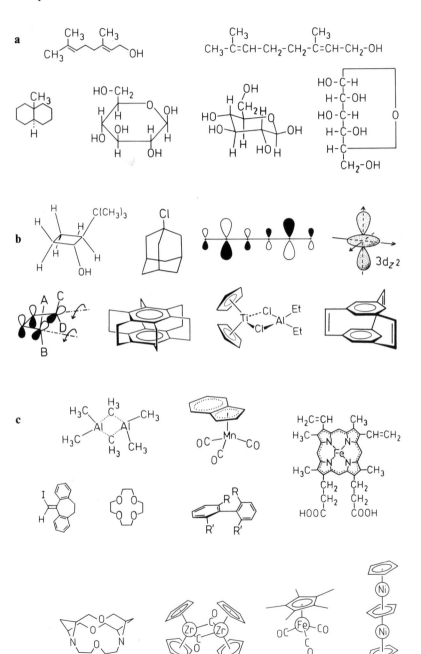

**d** $\quad \alpha_{\alpha\beta\gamma} = \dfrac{2}{\hbar} \sum_j \left[ \dfrac{w_{jn}^2 + w^2}{\hbar \left(w_{jn}^2 - w^2\right)^2} \right]_0^2 A_{\alpha\beta\gamma}$ $\qquad \sqrt[n]{\dfrac{\sqrt[3]{\varepsilon + A_3^*}}{d^2 - \sqrt[5]{\mu\tilde{\nu}}}} \; \sqrt[3]{\psi^2 + C_2^3}$

$$\int_0^{\pi/2} \frac{dx}{\sqrt[3]{\cos x}} \qquad \int_{-\infty}^{+\infty} \psi_j^* \psi_i \, d\tau = \delta_{ji} \qquad \lim_{x \to x_0} \frac{f'(x)}{g'(x)} \qquad D_{\infty h} \quad \Sigma_g^+$$

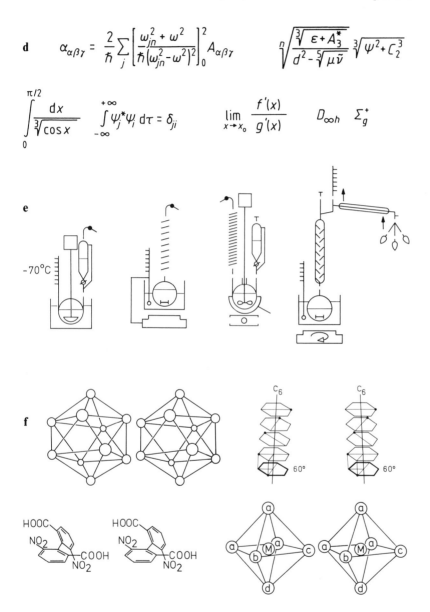

*Fig. 8-2.* Various structural formulas and mathematical expressions prepared with the aid of the stencils illustrated in Fig. 8-1 a-f (reduced linearly to 60%).

proper designations, *not* "$R_1$" and "$R_2$". According to standard rules of nomenclature (cf. Sec. 6.2.2), "$R_n$" means "the substituent R taken *n* times".

### *Stereochemical Formulas*

Often, the best structural formulas are ones that clearly convey the three-dimensional arrangement of at least some of the constituent atoms. This is particularly true of structures intended to illustrate an argument in which *stereochemistry* plays a role. Several standard conventions have evolved for indicating spatial relationships within the confines of a 2-dimensional drawing. For example, Fig. 8-3 shows two alternative ways of providing configurational and conformational information about substituted ethanes (*sawhorse* diagrams and *Newman projections*), as well as the usual ways of illustrating the *chair* and *boat* conformations of cyclohexane units. Other conventions for showing spatial relationships of atoms in molecules or molecular aggregates include *perspective* drawings of ring systems (cf. Figures 8-2b, c, and d) and the use of *wedges* and *dotted lines* for distinguishing between singly-bonded atoms located *above* and *below* the plane of the paper, respectively (Fig. 8-2b).

The most dramatic way of depicting spatial relationships among atoms (apart from physical models) is with pairs of *stereodrawings*, since these can be viewed as true 3-dimensional images with the aid of a *stereoscope*—or even

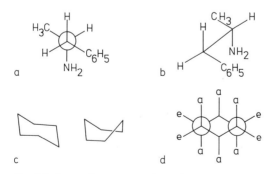

*Fig. 8-3.* Perspective representations of chemical structures: a) Newman and b) sawhorse projections of amphetamine, c) chair and boat forms of cyclohexane, d) Newman projection of the chair form of cyclohexane, showing axial and equatorial substituents. With the exception of one corner in the boat form of cyclohexane, these structures were created entirely with the *Chemistry Stencil 3D for Drawing Stereo Formulae and Stereo Diagrams* (Fig. 8-1f).

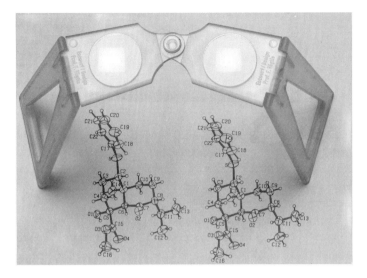

*Fig. 8-4.* Viewer for examining stereo images, together with a pair of stereo structures.

without, after sufficient practice. Fig. 8-4 shows a pair of stereodrawings and an example of a commercial stereoscope.* When such drawings first became common they were almost always produced by computer, but the *Chemistry Stencil 3D for Drawing Stereo Formulae and Stereo Diagrams* shown in Fig. 8-1f also makes it possible to prepare such drawings by hand.

## 8.2.2   Chemical Equations

*Chemical equations*,[†] like their mathematical counterparts, sometimes appear as elements of sentences:

---

\* The model shown is available from VCH Verlagsgesellschaft mbH, Weinheim, and VCH Publishers, Inc., New York.

[†] The "equality" implied in the phrase *chemical equation* (and formerly expressed symbolically by means of the equality sign, though the symbol is increasingly being replaced by an arrow) is quite different from that understood by a mathematician. "Chemical equality" refers to two constellations of a given set of atoms, and asserts that the one arrangement can be converted into the other. Thus, materials consistent with the first set of symbols (the *educts*) are transformed into new materials, the *products*. It has been suggested that less confusion would be engendered if the term "chemical equation" were replaced by an alternative such as "reaction symbol".

"The complete ionic separation of EA is best
represented as

$$EA \rightleftharpoons E^{\delta+}A^{\delta-} \rightleftharpoons E^+ || A^- \rightleftharpoons E^+ + A^-$$

where the process is initiated by induced
polarization of ...".

In such a case the placement of the equation itself is inflexible, and the preceding text line is usually incomplete. Generally speaking, however, it is better to frame an argument in such a way that equations serve to amplify or illustrate the text rather than becoming a part of it [eq. (14) is a case in point]. This has the advantage that the equations themselves can then be placed at convenient nearby breaks between paragraphs.* Whereever applicable, advantage should also be taken of the possibility of assembling several chemical equations into a single *block*.

$$IO_3^- + 3\ HSO_3^- \longrightarrow I^- + 3\ HSO_4^- \tag{14}$$

$$5\ I^- + IO_3^- + 6\ H^+ \longrightarrow 3\ I_2 + 3\ H_2O \tag{15}$$

Equations stand out most clearly if they are indented 3 to 5 spaces from the left margin (displayed, cf. Sec. 8.1.1). In typescripts, equations should be separated from text lines preceding and following them by two-and-one-half spaces. Several equations appearing in sequence should be double-spaced if possible; i.e., unless extensions in the vertical direction necessitate wider spacing. The resulting block of equations is made to resemble a *text block*. It is recommended that equations be numbered with arabic numerals (single or double) enclosed in parentheses and placed at the right margin. For the sake of appearance it is preferable to arrange for vertical alignment of all *chemical equality symbols* (*reaction arrows* or *equality signs*) appearing on a given page. Arrows are best constructed in typed text by rolling the paper down by one typewriter space (cf. Sec. 5.3.1) and using four or five successive strokes

---

* An equation should never be randomly placed between text lines *within* a paragraph. Apart from the fact that such an insertion is distracting, an occasion might arise later for the manuscript to be typeset (or simply retyped by someone else), in which case the person processing it would probably adhere rigidly to the original placement. The likely result would be a premature—and nonsensical—termination of the preceding line because the horizontal placement of specific words would have changed. The problem will be familiar to anyone who has had experience reformatting text on a word processor.

of the underline key. Arrows to be augmented with symbols or text need to be somewhat longer. *Arrow points* can be added by hand with pen and ink.

Should it prove necessary to divide a chemical equation into two lines, the part *following* the arrow should be placed on the second line; e.g.,

$$\left[ C_5H_5(CO)_2 Re=C{<}^{P(CH_3)_3}_{C_6H_5} \right] BCl_4 \; + \; P(CH_3)_3 \; \rightleftharpoons$$

$$\left[ C_5H_5(CO)_2 Re-C{<}^{P(CH_3)_3}_{C_6H_5}_{P(CH_3)_3} \right] BCl_4$$

Thus, "reaction symbols" (cf. the first footnote in this section) are treated somewhat differently from "true" (mathematical and physical) equations. Arrows in divided chemical equations are written only *once*, and always at the *end* of a line, while in other equations the symbol chosen as a break point is repeated at the beginning of the new line (cf. Sec. 8.1.1).

# 9     *Figures*

## 9.1     *Introduction*

### 9.1.1     *Types and General Characteristics of Figures*

A typical scientific manuscript relies primarily on *words* for the transmission of ideas, but complex verbal arguments can often be strengthened, or at least clarified, by selective use of a more *pictorial* mode of communication; that is, by the introduction of appropriate *figures*. The role of figures and certain formalities regarding their treatment have already been touched upon briefly in Sec. 3.4.3 within the context of journal article preparation. In this chapter we focus primarily on technical matters, especially the craftsmanship involved in the *preparation* of figures.*

Every element of a document—each heading, paragraph, list, table, or figure—has a unique assignment, and it is the writer's task to ensure that each is successful. As we have repeatedly stressed, this often translates into a quest for a maximum degree of *clarity* and *simplicity*. In the case of a figure it is important to use as few lines and as little wording as possible. It is a "picture" one is seeking, a graphic expression of a single crucial idea that cannot be conveyed efficiently in words.

Each figure must be an integral part of the overall manuscript, but no figure should be tied too closely to the text.[†] As we noted earlier, readers often look at figures first, hoping to gain some sense of what the surrounding text contains. Consequently, figures should be as *attractive* and *self-explanatory* as possible. They must be supportive of the text, but at the same time interesting in themselves. Thoughtful design and careful execution are necessary if a figure is to succeed on both counts.

---

* Additional information is available in MacGregor (1979). Figures intended to serve as the basis for *slides* or *overhead transparencies* call for special consideration, a subject treated in detail in Appendix A.

[†] A statement such as "for further details see text" should be used in a figure caption only in exceptional cases.

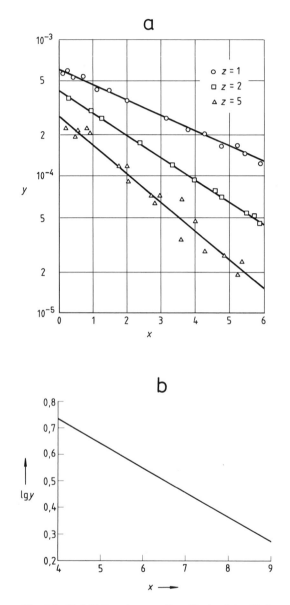

*Fig. 9-1.* Straight lines in a semilogarithmic plot.—(a) Least squares fit to a set of data points; axes used to establish an outline, and axis markings extended to produce a coordinate grid; (b) example of a straight line plotted within a simple pair of axes; no grid, *x*-axis linear, *y*-axis logarithmic.

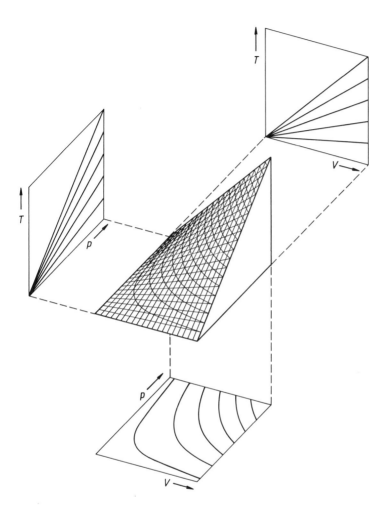

Fig. 9-2. Three-dimensional representation of the ideal gas law $pV = RT$ showing the three possible projections in the $V, p, p,T,$ and $V,T$ planes; the interaction of the three variables is illustrated by a surface in space.
(*An example of a figure based on a functional relationship.*)

Figures—sometimes referred to by professionals as *artwork*, or simply "art" —can be of many types: schematic diagrams, flow charts, nomograms, cross-sectional drawings, occasionally even photographs. However, in scientific manuscripts the most common type of figure is the *line graph*, a pictorial representation of *relationships* between quantities (cf. Figures 9-1 and 9-2). Because of their importance, line graphs will occupy most of our attention (esp. in Sec. 9.2.3).

Regardless of the form they take, all figures share certain important attributes. Therefore our discussion begins with a series of observations regarding figures in general. The following recommendations are meant to apply to figures for all types of documents, both typewritten and typeset.

● Figures within a manuscript are always *numbered* in the order of their appearance.*

In long manuscripts containing numerous figures it is customary to employ a "double-number" system that also identifies the chapter in which the figure is located; e.g.,

Fig. 3-11 (the eleventh figure in the third chapter)

● Every figure must be *cited* explicitly, by number, at least once within the text.

Figure citations need not be elaborate; they are normally simple phrases appended to or contained within appropriate explanatory sentences; e.g.,

... as is shown in Fig. 7-1.
... evident from the path of the middle curve in Fig. 3-4.
... assumes a sigmoid shape (Fig. 1-2).

---

* Authors occasionally wish to incorporate a different type of artwork: small illustrations that have neither numerical designations nor captions. These are referred to technically as *cuts*, and their illustrative function is such that they must appear precisely at a given location within the text. Unfortunately, this requirement can be the source of major problems, especially in the case of a manuscript that is to be typeset. Proper insertion of a cut is sometimes nearly impossible, particularly if the appropriate location falls near the bottom of a page. For this reason, publishers usually insist that items intended as unnumbered illustrations be extremely small, and even then they are permitted only in exceptional cases. Unnumbered drawings are submitted as part of the artwork portion of a manuscript. They should be identified either by the number of the text page with which they are associated ("art for MS 355"), or by means of auxiliary numbers enclosed in double parentheses. Chemical formulas to be inserted within text may be treated in the same way.

● A figure is always accompanied by a brief text known as a *legend* or *caption*.

Occasionally a distinction is made between the two terms. Strictly speaking, a "caption" is like a title: it identifies a figure's *content*. A "legend", on the other

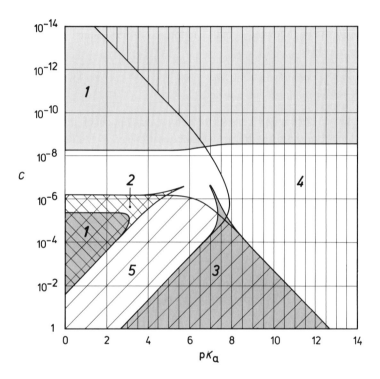

*Fig. 9-3.* Regions of applicability of five different approximations to the solution of $\Delta pH_i = |pH^* - pH_i| < 10^{-2}$ ($i = 1, ..., 5$); $C$ and $K_a$ are numerical values of concentration and acidity constant, respectively.

*1* $pH_1 = -\lg(C + 10^{-7})$

*2* $pH_2 = -\lg C$

*3* $pH_3 = (pK_a - \lg C)/2$

*4* $pH_4 = -\lg(K_aC + K_W)^{1/2}$

*5* $pH_5 = \lg 2 - \lg[-K_a + (K_a^2 + 4K_aC)^{1/2}]$

(*Quantitative graph; figure caption illustrates block arrangement of an explanation; the figure itself shows how patterns can be used to distinguish various regions.*)

hand, clarifies *details* associated with the figure. For example, a legend might be used to explain the significance of reference letters (a, b, c ...) in a figure, or to identify non-standard abbreviations, such as "V" for "valve" and "P" for "pump". Legend information can either be collected and placed in a separate block under the caption proper, or it can directly follow the caption text, set off only by a dash. Examples of these two standard ways of presenting legends are provided in Fig. 9-1 (continuous text) and Fig. 9-3 (block arrangement). Several identifications following a single caption should be separated from one another by semicolons.

• Each figure caption *begins* with the word "Figure" (or the abbreviation "Fig.") followed by a number and a period.

The remainder of the caption consists of a brief description of the figure's content. Note that the word "figure" and its abbreviation are always capitalized when used in conjunction with a number.

• All figure captions should be *placed* uniformly on the pages.

The usual choice is either centered or flush with the left margin (or alternating between the left and right margins). The same convention should be observed for figure caption placement as for chapter and section headings and the headings to tables.

• Figures themselves should be kept free of explanatory wording, apart from very brief identifications. Only in the case of a textbook is this rule somewhat relaxed (cf. Sec. 9.1.3).

• If one wishes to reproduce a published figure, *permission* must first be obtained.

The *source* of such a figure should always be disclosed in the accompanying caption, along with the fact that authorization was granted (cf. Sections 9.1.3 and 11.3.1, as well as the information on copyrights in Appendix C). If data reported by others have simply been *incorporated* in a newly-designed figure, or if a published figure has been modified, it is sufficient to *acknowledge* the source in the caption; e.g.,

> ... (after Smith, 1980).
> ... (adapted from Miller [4]).
> ... (cf. [5]).

Here, the usage "after" signals that a published figure has been slightly altered, while "adapted from" implies virtually a new illustration. As suggested by the above examples, literature citations occurring in figure captions are treated like those elsewhere in the manuscript, and they are numbered accordingly (cf. Sec. 11.3.2).

## 9.1.2    Figures for Typescripts

The rules in this section apply specifically to figures for *typewritten* documents.

● Figures for typescripts should be *inserted* directly into the text.

Ideally, every figure should be placed in close proximity to the first reference to it. However, if an argument requires the grouping of a large number of figures, one should seriously consider the alternative of collecting them in an appendix at the end of the manuscript. This has the advantage of maintaining greater continuity in the text, and it avoids a potential source of distraction for the reader (although at the cost of some inconvenience to anyone wishing to inspect the material).

● Assuming the typescript itself is double-spaced (cf. Sec. 5.3.1), all *figure captions* should be "one-and-one-half-spaced" in order to distinguish them clearly from the body of the text.*

● Figures should be *prepared* on separate sheets.[†]

It is impractical to attempt to draw directly on typewritten pages. Instead, appropriate sized *copies* of original drawings should be pasted into spaces provided in the typescript.[‡] The *width* of these inserts must be restricted to the breadth of the typed area of a page (typically ca. 16 cm).

---

* In printed documents figure captions are almost always set in special small type (cf. Sec. 4.3.7, "Choice of Type"). Narrower line spacing is an attempt to simulate this effect in a typewritten manuscript.

[†] An exception is a manuscript prepared with a word processor and a "laser printer" (cf. Sec. 5.2.2). In this case, figures and text can often be integrated prior to printing. The extraordinarily high resolution of laser printers results in figures and text of professional quality (the present book is a case in point).

[‡] This assumes that one intends to distribute only *photocopies* of the final document, as recommended in Sec. 5.5. Otherwise, figures should be introduced as separate full pages.

- The size of any *lettering* in figures should correspond to that produced by the typewriter (or word processor) used for the manuscript. This normally means capital letters should have a height of ca. 3 mm.

- Figure quality can usually be improved by *photoreduction* of oversized originals (cf. Sec. 9.2.2).

The corresponding originals may be hand drawings (including graphs) prepared on full sheets of standard 8 1/2 in. x 11 in. or A4 paper, or they may be instrumental records, such as spectra (cf. Fig. 9-4). Photoreduction may be accomplished in any of several ways, but the easiest approach takes advantage of a suitably equipped xerographic copier. As noted above, lettering should ultimately have a height of about 3 mm. This means that originals must contain correspondingly larger letters. It is important to remember that reduction also decreases *line widths*. Final copy should never contain lines less than 0.15 mm wide (cf. "Drawing the Lines" in Sec. 9.2.3).

Sometimes entire typescripts are photoreduced prior to wider distribution. This is the case, for example, with camera-ready copy submitted to a "letter journal" (cf. Sec. 3.4.6). The same technique is also used with conference proceedings and even certain specialized monographs. What results is a document with linear dimensions about 70% as large as those of the original (i.e., reduction by ca. one-third). Capital letters become approximately 2 mm high, similar to those in printed text. Oversized figures could thus suffer a "double reduction", and potential problems with respect to legibility must be taken into account.

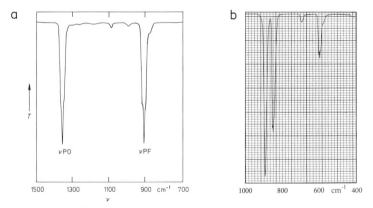

*Fig. 9-4.* Two different representations of portions of an infrared spectrum. (*Reproduced with the kind permission of H. Willner, Hannover.*)

## 9.1.3 Figures for Publication

Original figures that are to be part of a *typeset* document (i.e., a journal article or a book) must be treated somewhat differently. Figures for manuscripts of this kind are never mounted in the text manuscript, because text and figures proceed along separate paths through the publishing process (cf. Sec. 3.4.3). Figures are submitted together in what is sometimes called the *art manuscript*. Publishers usually require authors to supply several copies of their manuscripts, but one set of *original* figures will suffice provided it is accompanied by an appropriate number of xerographic copies.

The suggestions below should be studied before one begins to prepare figures for publication.

● Each figure should be prepared on a *separate sheet.*

● The preferred *size* for figure copy is 8 1/2 in. x 11 in. (or A4).

Smaller figures should be pasted or carefully taped onto standard paper, one figure per sheet. Small, loose originals are undesirable, because they are easily lost during processing. Exceptionally large figures (i.e., any larger than a full A4 sheet) should be avoided because of the risk of damage during mailing and processing. Moreover, the lettering associated with very large figures tends to become too small after photoreduction.

● All figures should have clear *margins* of 2-3 cm on each side to permit editorial markup at the publishing house.

● The preferred *media* for original drawings are black India ink and vellum (cf. Sec. 9.2.2).

Carefully inked original figures have long been a hallmark of high-quality scientific publications, and most publishers still regard them as superior.* Nevertheless, computer-generated figures are becoming increasingly common, particularly with the advent of sophisticated layout and design software tailored to personal computers. It is important to recognize, however, that successful use of computer techniques requires that one have access to a plotter or printer with excellent resolution characteristics.†

---

* High-contrast glossy photographs of one's figures may be submitted instead of the original artwork itself.

† The normal printer-plotters associated with word processors produce figures of decidedly inferior quality. At best these are adequate for use in an informal report.

● The publisher should be contacted before a decision is made to incorporate *patterned areas* into a figure.

Background patterns are often used for distinguishing various *regions* in a complex figure (cf. Sec. 9.2.3, "Patterns"). Publishers generally have no objection to the practice, but the proper *choice* of patterns may be dependent upon the printing method to be employed, as well as the extent to which a given figure will be reduced in size.

● An author wishing to include a previously published figure has the responsibility of acquiring from the original publisher a high-quality *copy* suitable for reproduction.*

An appropriate request can be added to the letter used to obtain permission for reproduction (cf. Appendix C.6 for an example of such a letter). The *caption* for a reproduced figure must always contain a statement such as

> ... taken from Fischer (1981); reproduced with permission of the XXX Press.

● As a safeguard against loss or errors, every submitted figure should be clearly *labeled* with at least the name of the author and the figure number.†

Full-size A4 drawings can be labeled on the front in pencil, preferably near one edge. Photographs should be marked lightly with pencil on the reverse side. Even a drawing or photograph that is to be pasted onto a larger sheet should be labeled, and the identification should be repeated on the corresponding carrier sheet. This extra precaution will prevent confusion should the two somehow become separated.

● Original figure copy should remain *flat*, and it should never be folded.

If figures are to be *mailed* they should first be placed between sheets of stiff cardboard and then carefully packaged, taking all practical measures to ensure that none of the items becomes soiled or wrinkled. Photographs must be

---

* High-quality originals are particularly important in the case of photographs. Attempts to reproduce photos directly from printed text lead to distinctly inferior results, for reasons discussed in Sec. 9.3.1. Acceptable results can be achieved, however, with photographs of published line drawings.

† It is also advisable to indicate explicitly which side of a figure is the top. This may be self-evident to the author—or any other scientist—but it may not be so obvious to a printer.

carefully protected against fingerprints, preferably by enclosing them in transparent cellophane envelopes.

● *Captions* should always be separate from the figures themselves.

All captions should be typed as a group—double-spaced!—and submitted as a supplementary part of the text manuscript.

● *Titles* are generally omitted from figures published in books or journals. The title function is assumed entirely by the figure caption. This is in contrast to the usual practice with figures prepared as slides or overhead transparencies to accompany a lecture (cf. Appendix A).

● Publishers discourage excessive use of *descriptive labels* in figures, preferring instead that individual lines, curves, or other components be identified only with letters. These are then elaborated upon in a legend following the caption.*

The reason for this advice is that text within figures cannot be typeset. Instead, it must either be inserted by hand (usually with the aid of a template) or superimposed photographically after the initial reproduction steps. Both approaches are time-consuming (i.e., expensive!), and if the work is not executed professionally the results are likely to be unacceptable.

In the past, much published artwork was the product of professional draftsmen employed by publishers. However, most journals have now abandoned this costly practice, choosing instead to reproduce illustrations exactly as they are submitted by the authors (after appropriate size reduction). Even the figures in many books are now prepared this way despite the fact that uniformity is sacrificed in any work with multiple authorship. Authors have thus inherited a major new responsibility: they must learn to prepare their own high-quality artwork.

Occasionally, publishers are willing to compromise: authors are asked to provide *unlettered* artwork, and to include on a *copy* of each original a pencil sketch of any necessary inscriptions. A draftsman at the publishing house then

---

* More latitude may be granted to the author of a textbook in recognition of the fact that explanations *within* a figure are especially helpful to the inexperienced reader. Another exceptional case is an item whose identification requires the use of symbols that would be difficult or impossible to typeset.

translates the sketches into high-quality lettering on the originals. The reason for this particular division of labor is that lettering is highly susceptible to the adverse effects of amateur workmanship (cf. Sec. 9.2.3, "Basic Rules"). Poor lettering can seriously detract from the appearance of an otherwise carefully prepared figure, and much can be gained by entrusting the task to a skilled craftsman.

Most science publications contain only black and white figures; *color* is a luxury restricted to those special situations in which the considerable expense and effort are clearly warranted. Even black and white figures can be expensive and troublesome if provision must be made for *gray tones*. Generally speaking this is not the case with *diagrams* or *graphs* (collectively: *line drawings*), a factor that explains much of their popularity. *Photographs*, on the other hand, almost always include shades of gray. This characteristic is reflected in the classification of photographic figures as *halftones*. The remainder of the chapter is devoted to suggestions for the preparation of these two fundamental types of figures, with primary emphasis (Sec. 9.2) on the more common line drawings.

## 9.2    Line Drawings

### 9.2.1    Planning

*Line drawings* (in contrast to halftones, Sec. 9.3) are figures consisting entirely of regions that are either wholly white or evenly black (or, for that matter, any other color).* India ink drawings are the most common type, as exemplified by most of the figures in the present chapter. Line drawings are nearly always preferred in scientific manuscripts because unlike halftones they can be photocopied rather successfully, and few technical problems arise when size reduction is required. Moreover, use of line drawings avoids the need for "screening" (cf. Sec. 9.3.1) prior to printing.

---

* This definition of a "line drawing" does not exclude diagrams with shaded regions, provided the shading results from dotting or cross-hatching (cf. Fig. 9-3).

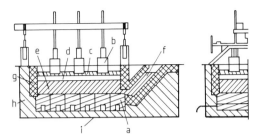

*Fig. 9-5.* Longitudinal and transverse sections of a 15 000-ampere aluminum refining cell.— (a) Anode (carbon blocks); (b) cathode (graphite electrodes); (c) refined aluminum; (d) electrolyte; (e) alloy serving as anode; (f) segregation sump; (g) magnesia brick; (h) refractory brick; (i) iron shell.

*(An example of a figure showing the design of a piece of equipment.)*

A properly conceived and executed line drawing can be an extremely effective device for conveying information, such as the fact that a trend is present (or absent) within a set of experimental data. Line drawings can also be used to depict functional relationships, clarify the design of a piece of equipment (cf. Fig. 9-5) or show the steps in a complex process (e.g., as a *flow chart*). Much of the power of a line drawing derives from its inherent simplicity, which permits rapid scanning and interpretation. Thus, the idea behind a complicated piece of apparatus, or even the way a device is to be used, is often far more obvious from a good *block diagram* than from text or photographs, although the latter may provide useful supplementary information. Likewise, a *graph* is often better than a table for revealing the existence and nature of a functional relationship. Graphs also make it possible to compare sets of data, to judge the significance of experimental error, and to estimate intermediate values or values lying outside the range of reported measurements. In short, a good figure represents a very efficient use of limited space.

Prior to the final preparation of any figures for a manuscript it is wise to *sketch* them all in very rough form. This not only provides an overview of the job ahead but also permits a preliminary comparison of the projected figures. Such a comparison is valuable, because it can point up imbalances in planned graphical treatment, or perhaps lead one to consider alternative presentation techniques.

Once sketches are available, the complete set should be carefully examined and subjected to the following tests:

- Are all the proposed figures both *useful* and *necessary*, or is it possible to combine or even delete some?
- Has the *space* within each been put to optimum use? Are all figures as compact as possible, consistent with the desired reading accuracy?
- Have all *scales* been properly chosen, or would the appearance or effectiveness of some figure be improved by changes (to logarithmic presentation, for example)?
- Are all the figures easily *understandable*, or are some too complicated? (Graphs containing more than five curves generally fall in the latter category.)
- Might a figure of a different *style* in some case be preferable (cf. Sec. 9.2.4)?

Only when the set of sketches has passed all the above tests should serious drawing begin.

## 9.2.2    Materials

In Sec. 9.1.2 it was recommended that line drawings be prepared on a relatively large scale and then *reduced* photomechanically to a size appropriate for the manuscript in which they are to appear. The advantage of this seemingly complicated process is two-fold: large drawings are more convenient to construct than small ones, and reduction makes the inevitable irregularities in hand-drawn originals less obvious, if not invisible (cf. Fig. 9-6). Reduction can be performed either photographically or by using a suitably equipped photocopying machine (xerographic copier). Photocopying must be done carefully, however, using a clean and well-adjusted machine; otherwise the resulting copies are likely to be very poor.

The standard method for preparing line drawings of professional quality has long been that of applying *India ink* to *vellum*. Heavy vellum (e.g., 115 g/m$^2$) is an ideal work surface, and it has the great advantage over paper that it readily tolerates corrections. Superfluous ink can be cleanly removed with a sharp knife or a razor blade, and the drawing surface remains virtually undamaged. Using vellum has the further advantage that one can begin by making a rough sketch of the intended figure using a soft pencil; pencil lines are easily and safely erased from the finished product.

Every part of such a drawing is prepared by hand with India ink, though generally with the aid of suitable *stencils* (*templates*). For example, one can purchase stencils for the creation of many types of lettering (upright, italic, Greek, etc.). A special set of stencils designed to meet the needs of chemists is

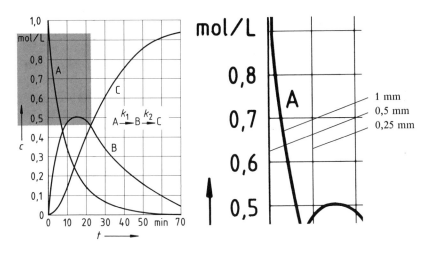

*Fig. 9-6.* Concentration vs. time curves for sequential first-order reactions.—$k_1 = 0.1$ min⁻¹; $k_2 = 0.05$ min⁻¹; $c_0(A) = 1$ mol/L.
*(An excerpt from the original drawing is shown at full scale to demonstrate the use of various line widths.)*

illustrated in Chapter 8 (Fig. 8-1). Various types of stencils, together with the appropriate pens, can be obtained from dealers in engineering and artists' supplies.

India ink lettering may also be applied with the aid of a Leroy™-type lettering set. In this case the guide is a wooden or plastic form into which an alphabet has been incised. A special "scriber" permits the desired characters to be reproduced on vellum, but at a point a few millimeters above the form itself (cf. Fig. 9-7).

Adhesive transfer letters (cf. Sec. 8.1.3) or other mechanically prepared inscriptions can also be affixed to an India ink drawing, but typewritten letters are never acceptable.

It has already been noted in passing that computer technology has recently opened exciting new avenues for the preparation of artwork. Some computer-based systems provide output of exceptionally high quality. However, it would take us beyond the scope of this chapter to discuss computer graphics in detail, especially since the subject is extremely complex and the field is evolving rapidly.*

---

* Relevant strategies and techniques are the subject of a forthcoming book by C. Bliefert and C. Villain (to be published in 1987 by VCH Verlagsgesellschaft).

*Fig. 9-7.* Schematic drawing of a Leroy™-type lettering set, consisting of a "three-point" scriber and a lettering guide.

Computers represent a particularly attractive tool when one wishes to plot curves that are easily generated by computer. Even if this is not the case, a computer can still be a useful drafting aid. For example, suppose one wishes to incorporate a complex spectrum into a figure. A computer can be used to generate and label a set of axes, and within these axes a high-quality original spectrum can be pasted. Since vellum is transparent, it is also possible to superimpose a hand-drawn curve on a computer-generated grid. In both cases the final figure copy would be a photographic reproduction of a composite original. Relatively unsophisticated computer drawings are also valuable: they may be used as starting points for the preparation of vellum drawings. Curves generated by a computer are easily traced, and a crude computer plot is an excellent guide to proper scaling, as well as to optimal placement of appropriate lettering.

Vellum sheets roughly the size of so-called "legal paper" (i.e., 8 1/2 in. x 14 in. or 21 cm x 38 cm) are ideal for the preparation of most figures. These can later be reduced to ca. 40% of their original size, transforming 5-mm lettering into characters 2 mm high, or the equivalent of printed text.* Margins of 2–3 cm should surround all graphs to permit labeling of axes (e.g., specification of quantities, numerical values, and units). Preparing all drawings to the same scale has two major advantages: only a single set of pens and stencils is required, and the reduction process is standardized.

In addition to special stencils and pens, there are several other *drawing aids* with which one should become familiar. For example, satisfactory preparation of lettering requires the availability of a firm *straightedge* on which to rest the stencils. A heavy metal ruler will suffice (preferably one backed with rubber), but a better choice is a drafting rule firmly affixed to the work surface. A

---

* *The ACS Style Guide* (Dodd, 1986) recommends reduction to two-thirds of original size. Following these guidelines means originals should be one and one-half times their printed size, so 3 mm lettering is appropriate.

selection of *triangles* and guides for drawing *curves* is also nearly indispensable, as noted in the following section. Moreover, anyone seriously interested in preparing high-quality drawings is strongly advised to invest in a small *drawing board*. Several excellent models are available commercially (e.g., the Koh-I-Noor Rapidograph Drawing Board). The best are provided with movable rulers or straightedges that travel freely in slots incised near the edges of the drawing surface. These rulers can be firmly locked at any point, thereby providing complete flexibility in the establishment of any desired horizontal or vertical line. A drafting board has the additional advantage that vellum can be firmly (but temporarily) taped to the surface, as can any figure copy that must be traced.

### 9.2.3    The Drawing Process

*Basic Rules*
The following guidelines are primarily intended to aid the novice working for the first time with pen and drawing ink.

● The *pens* chosen should always be of professional quality, and they must produce lines of the proper width (cf. the discussion under the subsequent heading, "Drawing the Lines").

Drawing pens are so designed that they must be held strictly perpendicular to the writing surface. This orientation may at first seem unnatural, but it is essential for achieving satisfactory results.

● The only acceptable writing fluid is high-quality *India ink*; ordinary ink flows unevenly, and it can easily clog a drawing pen.
● Pens must be scrupulously *cleaned* after each use, with careful attention to all manufacturer's recommendations.
● It is strongly advised that drawings be prepared on heavy *vellum* rather than paper, because razor blades or special erasers can then be used to remove unwanted ink marks.
● Before serious drawing is begun one should always *test the pen* on a spare scrap of vellum.
● Lines must be drawn *slowly*, with the pen always moving at a constant rate.
● One should avoid any temptation to *retrace* lines already drawn. Doing so results in unsightly line-width irregularities.
● Mutually *perpendicular lines* should always be constructed using a pair of

draftsman's triangles or some other device that ensures the production of a precise 90° angle.

● Smooth *curves* are best prepared with the aid of either a "French curve" (a template combining many different curve shapes) or a special "flexible ruler" that can be deformed at will to simulate any desired curved shape.

● The *distance* between any two adjacent lines (or letters) in a drawing should be at least as great as the width of the pen with which they are drawn.

Stencils, rulers, and similar drawing aids are extremely convenient, but their use increases the risk of smearing freshly drawn lines. For this reason, all such devices should be of the type with *raised edges*. Beginners in particular should try to avoid moving their drawing aids before the applied ink is thoroughly dry.

The preparation of *lettering* usually presents the greatest challenge in figure construction. Indeed, some publishers urge their authors not to attempt the task, but leave it to professionals (cf. Sec. 9.1.3). Irregularities in lettering are distressingly apparent, and inexperienced draftsmen find it very difficult to achieve proper spacing even when lettering stencils are used. The chances of success are improved by employing special stencils that contain explicit *space markings*. The importance of carefully balanced, "homogeneous" sets of letters is what leads us to recommend direct lettering stencils rather than Leroy™-type lettering devices; only with the former is it possible to see clearly in advance what spacing will result and how each new letter will appear relative to others already drawn.

The safest procedure is to prepare all hand-lettered text fragments on individual pieces of vellum. These can later be attached to the separately drawn figures, permitting any unsuccessful attempts to be replaced without re-drawing an entire figure. The fact that a "collage" is produced presents no problem since the edges of individual sections become invisible when a figure is photoreproduced. A further advantage of this procedure is that it permits one to experiment with the *placement* of lettered inscriptions.

No India ink drawing should ever be undertaken prior to the preparation of a complete preliminary sketch. Proper scaling and placement of the various elements can be determined only by a trial-and-error procedure. Provided one uses a soft pencil, the working sketch can be executed directly on the vellum intended for the final drawing. Pencil markings are easily and safely removed with a soft eraser. Alternatively, the sketch can be made on paper, which is then securely taped to the drawing board, covered with vellum, and traced.

## Providing a Framework for a Graph

Every line graph requires an appropriate structural *framework*, and some thought should be devoted to choice of style. At least three alternative formats should be considered:

- The simplest framework consists of nothing more than a pair of *coordinate axes* (cf. Fig. 9-1b).
- Axes can be used as the basis for an *outline* or border around a figure (cf. Fig. 9-4).
- A graph can be further elaborated by providing a *reference grid* or *network* within the outline. Usually such a grid consists of a set of mutually perpendicular lines drawn parallel to the axes and spaced to indicate regular and convenient subdivisions of the values associated with the plotted quantities (see, for example, Fig. 9-1a).

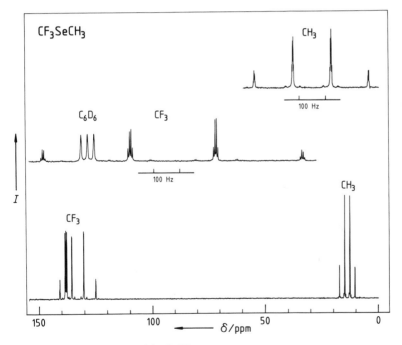

*Fig. 9-8.* $^{13}$C NMR spectrum of $CF_3SeCH_3$.
(*Sample spectrum; reproduced with the kind permission of W. Gombler, Emden.*)

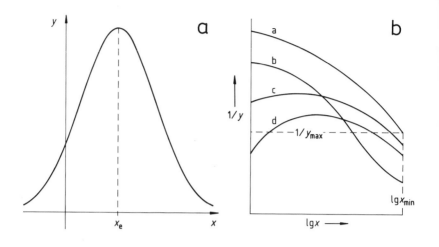

*Fig. 9-9.* Qualitative graphs.

All the above styles are common, but it is important to remember that journals (and even universities) sometimes have special editorial rules that limit an author's freedom of choice.

The horizontal axis (or *abscissa*) of a line graph is normally used for plotting the values of an independent variable, while the vertical axis (*ordinate*) is most often assigned to a dependent variable. Axes may or may not contain *scale markings*, although such markings are usually obligatory on a *quantitative* graph (see the discussion below under "Quantitative vs. Qualitative Graphs"). If a graph is enclosed in a full outline, scale divisions should be indicated by drawing short perpendicular lines *into* the graphing space, starting from the insides of the axes.* Such lines are drawn upward from the abscissa and to the right from the ordinate (see Figures 9-1b and 9-8).

Although axes are commonly subdivided on a *linear* basis, this is certainly not mandatory. *Logarithmic* division is often preferable, particularly when the quantities to be plotted vary over a wide range (i.e., several orders of magnitude). *Semilogarithmic* plots—graphs with one axis linear and the other logarithmic—represent another possibility (cf. Fig. 9-1). Indeed, almost any

---

* Unfortunately, this rule is frequently violated even in published works. Too often one encounters external scale marks.

mathematical function can be employed for axis division. Whatever the final choice, it should reflect the nature of the data and the purpose of the graph. A case in point is the *reciprocal plot*, in which one or both axes are subdivided in units of a function such as $1/y$ (cf. Fig. 9-9), often in order to transform a hyperbolic curve into a straight line.

### Drawing the Lines

Every figure contains different *kinds* of lines, and the differences should be reflected in the corresponding *line widths*. Careful control of line width emphasizes the most important parts of a figure and avoids a cluttered appearance. Preparation of a high-quality graph typically requires as many as four different line widths—usually 0.3 mm or 0.35 mm, 0.5 mm, 0.7 mm, and 1.0 mm.*

The following relationship can be taken as a useful rule of thumb for the correct choice of relative line widths (cf. Fig. 9-6):[†]

$$\text{grid} \;:\; \text{axes} \;:\; \text{curves}$$
$$1 \;:\; 1.4 \;:\; 2$$

Because the reader's attention is to be drawn primarily to the curves, these should be made most prominent (0.7 mm or 1.0 mm). Background grids or other supplementary lines included as an aid to the reader (such as *cross-hatching* or isolated *reference lines*) should always be more subtle. All graphs in this chapter were prepared to the specifications quoted above.

If more than one curve is to be included in a single graph, the corresponding lines should all be of the same width. Curves can be distinguished from one another either by the use of different data point symbols (cf. Fig. 9-1a) or by employing lines of different types; e.g.,

———————    —  —  —    — · — · —

Whenever two nearly parallel data lines intersect, one of the two should be interrupted so that the path of each remains clear.

---

* These values apply to original graphs that will subsequently be reduced (cf. Sec. 9.1.2.).
[†] Some published norms (e.g., Standard *DIN 461, 1973*) recommend the ratios 1:2:4. It is occasionally appropriate—or even necessary—to deviate from any set of guidelines to ensure that a particular figure accomplishes its intended purpose. Flexibility is particularly important with nomograms, spectra, or other graphs from which the reader may want to ascertain specific numerical values. The accuracy with which a curve can be read is frequently a sensitive function of line width.

If at all possible, any lettering associated with data curves should be placed horizontally. Such labels should be prepared with the pens used for the lines themselves in order to ensure that lines and labels have the same visual impact. One should plan figures so that it is unnecessary to introduce extra lines for connecting labels with the corresponding graphic elements. If such reference lines are required (as in Fig. 9-5) they should not be embellished with arrow points.

### Qualitative vs. Quantitative Graphs

Graphs are normally plotted so that values associated with the abscissa increase from left to right and those for the ordinate from bottom to top. *Arrows* can be added to show the direction of ascending values more explicitly. The graphs in Fig. 9-9, for example, would be ambiguous in the absence of arrows. Arrow points are sometimes added to the axes themselves (cf. Fig. 9-9a); it is preferable, however, to incorporate separate arrows into the axis labels, placed after the corresponding quantity symbols (Fig. 9-9b). Arrows are of course superfluous in a figure with scaled axes, because the reader can infer the same message from the numbers. Nevertheless, such arrows are sometimes included as a concession to the impatient reader (cf. Fig. 9-1b).

Fig. 9-9 shows two examples of *qualitative* (or semi-quantitative) graphs, figures intended to suggest only the general nature of the relationship between two quantities. Significant regions of such graphs are sometimes identified, but

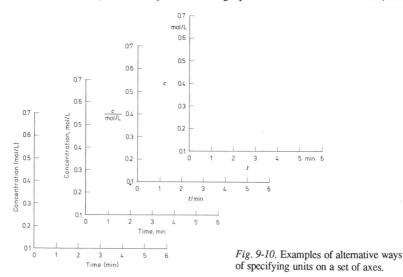

*Fig. 9-10.* Examples of alternative ways of specifying units on a set of axes.

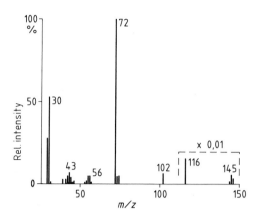

*Fig. 9-11.* Mass spectrum of an amino acid ethyl ester.
(*Sample spectrum; scale markings are indicated below the abscissa to prevent their being confused with mass signals.*)

no axis scales are supplied. Qualitative graphs are always assumed to be linear unless otherwise noted (as, for example, in Fig. 9-9b, where one axis is identified as logarithmic while the other refers to the reciprocal of a variable). Fig. 9-2 is a rather complex qualitative graph in which a relationship is expressed among three independent variables (*p*, *T*, and *V*) in a simulation of three-dimensional space.

Figures 9-1, 9-3, and 9-6 are *quantitative* graphs. Here the coordinate axes are subdivided numerically (*scaled*, as described previously under the subhead "Providing a Framework"). If desired, some of the subdivision markings can be extended to create a background grid. *Unit specifications* are required for both the ordinate and abscissa. These can be shown in any of several different ways:

● Units can be written near the right-hand end of the abscissa and the top of the ordinate, directly between the last two scale numbers. If space is limited, the next-to-last scale number can be omitted (cf. Fig. 9-6).
● The units can be made a part of the corresponding axis labels, separated from text or quantity symbols by either parentheses or commas (see Fig. 9-10).
● Unit specifications can be relegated to the legend following the figure caption (see Fig. 9-3).

Fig. 9-10 illustrates several of these alternative strategies. Regardless of the choice, care must be exercised to ensure that unit specifications are unambiguous. As discussed in Sec. 7.2.3, the preferred way of indicating units is to form quotients of the type "quantity/unit".

The *quantities* corresponding to plotted variables are identified under and to the left of scale markings on the abscissa and ordinate, respectively. In order to conserve space it is best to represent quantities by their standard *symbols*. Extensive wording should be avoided, especially along the ordinate. If a long ordinate specification is unavoidable the text should be arranged vertically and proceeding upwards (see Fig. 9-11 and the two innermost sets of coordinates in Fig. 9-10).

### Letters and Symbols

As previously noted, the *lettering* required on a graph (in the form of *figure inscriptions* or *labels*) can be created with the aid of pens and stencils, but adhesive transfer letters (cf. Sec. 8.1.3) should be considered as an alternative. In either case, care must be exercised to prevent the results from looking unprofessional. The choice of letter size must reflect planned scale reductions. *Subindices* and *exponents* should be constructed from characters somewhat smaller than the rest of the lettering. A lettering stencil like the one shown in Fig. 8-1d is especially convenient because it incorporates a separate set of small numbers.

Convention dictates capitalization of the first letter (but *only* the first letter) of figure inscriptions; e.g. "Valve", "Vacuum pump", "All other elements". (A similar rule applies to words used as headings or entries in tables; cf. Sec. 10.2.1.) It is important to remember that symbols for *variables* should be lettered in *italics*, just as italics are used for variable symbols appearing within text (cf. Sec. 7.1.3). Strict adherence to this rule may not be crucial for a figure destined to be part of a typewritten document, since even in the body of the text normal lettering will probably be used for variables. Nevertheless, there is still merit in observing the italicization convention whenever figures are prepared, since this makes it possible later to publish without change a figure originally prepared for a typescript.

Chemical symbols and structural formulas are common in figures, and there is a wide range of special and highly sophisticated stencils available for their preparation (cf. Sec. 8.2.1). Stencils are also available for creating the standard symbols used in other disciplines, including those for components of electrical or electronic systems, vacuum systems, chemical engineering arrays, and computer flow charts.

Publishers generally recommend that only the following symbols be used for denoting data points in a line graph:

○ □ △ ▽ ◇ + ×
● ■ ▲ ▼ ◆
◐ ◖ ▲ ▼ ◗

In general, open data point symbols are preferred. No matter what styles are chosen, no symbol should be larger than a small letter such as "a" or "o". The only exception to this rule is a data point symbol designed to provide an indication of the reliability of the plotted data. In no case should data points be smaller than the width of the line that joins them.

## Patterns

Occasionally it is necessary to designate or emphasize one or more *regions* within a graph or other figure. One way to accomplish this is to use a fine pen for drawing narrowly spaced diagonal lines over the region of interest, but the

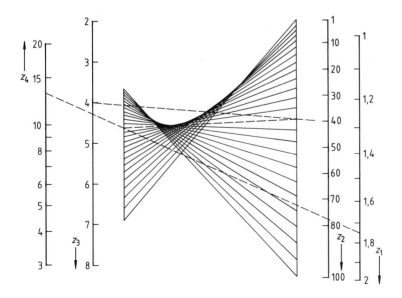

*Fig. 9-12.* Nomograph for the relationship $4z_4 = 3z_1(2z_2 + 5z_3)^{1/2}$.—Example of interpretation: for $z_1 = 1.75$, $z_2 = 40$, and $z_3 = 4$, $z_4 \approx 13.1$.

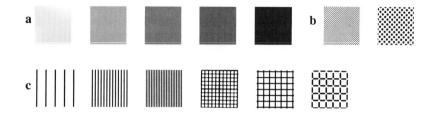

Fig. 9-13. (a) Gray scale values of black and white tone patterns; (b) alternative tone patterns (60 and 20 lines per inch) with identical gray scale values; (c) examples of various line and dot patterns.

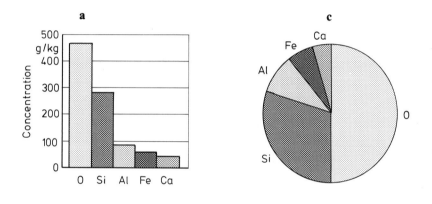

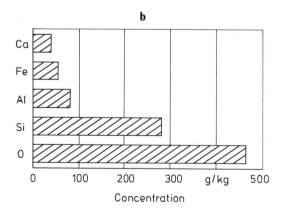

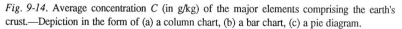

Fig. 9-14. Average concentration C (in g/kg) of the major elements comprising the earth's crust.—Depiction in the form of (a) a column chart, (b) a bar chart, (c) a pie diagram.

job is a tiresome one, and the results are not likely to be satisfactory. A better solution is to overlay each distinctive region with a carefully cut piece of transparent adhesive *transfer pattern*. Sheets of such material are available from engineering supply houses in a wide variety of types. Generating the corresponding figure by computer is another tempting alternative, since patterns can often be incorporated automatically.

Proper choice of cross-hatch density and orientation, and the use of a combination of dotted and lined patterns (cf. Fig. 9-13), makes it possible to differentiate several overlapping regions within the same diagram (as in Fig. 9-3). However, two contrasting areas will be readily distinguishable only if their gray values differ by at least 30% (cf. Fig. 9-13a).

Publishers generally recommend caution with respect to adding patterns to figures in manuscripts that will be typeset. At the very least one should contact the production staff at the publishing house in order to ascertain what patterns are appropriate. The extent of planned size reduction is the crucial factor: it is important that individual dots in a pattern maintain their integrity after reduction (i.e., dots must not run together). Patterns that reduce well are characterized by a relatively small number of lines per inch for a given gray value (cf., for example, Fig. 9-13b). An alternative approach is to submit figures without patterns, instead indicating either on a photocopy or on a transparent overlay (e.g., a separate piece of vellum) what treatment is desired.

## 9.2.4   Column Charts, Bar Charts, and Pie Charts

Column charts, bar charts, and pie charts are all variants on the principle of representing the relative *magnitudes* of quantities by corresponding *areas* on a surface. Use of this technique is less common in scientific manuscripts than in writings from other disciplines, primarily because most scientific data lend themselves more effectively to treatment as part of a *continuum* of values, as implied by a line graph. Nevertheless, occasions do arise in which discrete values are to be compared, in which case such alternative presentations should be considered.

*Column charts* and *bar charts* both employ *rectangular* surfaces for illustrating magnitudes. In both cases, the length of the rectangle is varied in proportion to the magnitude of the displayed quantity. The two types of chart differ only with respect to the directional placement of the rectangles: column charts have a vertical orientation, bar charts are horizontal (cf. Figures 9-14a and 9-14b, respectively). The following suggestions apply to the preparation of both types of charts:

- All bars or columns should be of equal width.
- All elements should be equally spaced relative to one another.
- Spaces between bars (or columns) can be omitted if the number of elements is large.
- Members of a *set* of bars or columns should normally be distinguished from one another by the use of patterns (cf. the preceding setion).

A *pie chart* (such as Fig. 9-14c) is a particularly dramatic way of illustrating the relative magnitudes of portions of some complete entity (i.e., a whole that is to be seen as the sum of a set of parts). The whole is represented as a *circle*, and values associated with the individual members of the set are symbolized by *segments* of the circle. A well-constructed pie chart can be very effective, although it does not lend itself to exact *comparisons* as well as a bar or column chart. The principal considerations to bear in mind when constructing a pie chart include the following:

- Pie charts should be reserved for situations requiring a maximum of six segments.
- No segment should comprise less than 5% of the whole (equivalent to an arc of 18°).
- If possible, segments should be arranged in order of decreasing size.
- It is sometimes useful to introduce patterns for distinguishing segments from one another, but the practice may cause a "busy" look.

Additional advice on the preparation of column charts, bar charts, and pie charts is offered in the book by MacGregor (1979).

# 9.3 *Halftones and Color Illustrations*

## 9.3.1 *Photographs as Figures*

*Halftones*, like line drawings, are "black and white" illustrations, but they differ from line drawings in that they also contain at least some regions that appear to be gray. Another and perhaps more revealing expression is *continuous-tone* illustration.* Most figures within this category originate as *photographs*, often pictures of apparatus or photomicrographs. Halftone figures are expensive to reproduce; their use in chemistry or physics manuscripts is rare—and is not encouraged. In the earth and life sciences, however, halftones are nearly indispensable.

The principal justification for introducing *photographs* into any manuscript is a desire to provide a precise record of some event or object, a record that faithfully reproduces the most important characteristics of the original. Assuming halftones are absolutely necessary, and if only a few copies of a document are to be prepared, the simplest procedure is to attach glossy prints of original photographs to each copy of the typescript. If the number of copies makes this approach impractical, one is obliged to engage a suitably equipped printing firm to prepare what are known as *screen prints*. The modern screening process is designed to be used with *offset printing*, and it involves photographing the original illustration through a transparent film inscribed with a fine, regular latticework or mesh of opaque horizontal and vertical lines. The finer the mesh—150 lines per inch, for example—the better will be the final illustration.

The result of screening is that the image comprising the illustration is transformed into an orderly matrix of minuscule dots. In high-quality publications the dots are so fine that close inspection is required to verify that the printed illustration actually differs from the original. The difference is real, however, and it is crucial from the standpoint of printing technology. Every printed dot in a screened offset illustration is uniformly black. Different regions of the picture vary only in the number and size of the dots present. Provided the dots are sufficiently numerous and small, the reader's eye will fail to

---

* Occasionally a distinction is made between the two expressions, with "halftone" used only to describe a published photograph.

perceive the resulting discontinuities; instead, those regions containing relatively more (and larger) dots will be interpreted as possessing a darker shade of gray.*

Any screening required in conjunction with the publication of a manuscript is normally attended to by the publisher, though a fee may be assessed for the service. Authors need only provide standard glossy photographs of sufficiently high quality (see Sec. 3.4.3). One limitation is particularly important to remember, however: a "photograph of a (printed) photograph" is unsuitable for publication. Any published photograph will already have been subjected to screening. Further reproduction would require a second screening, resulting in considerable loss of detail and perhaps in the creation of unexpected distortion or even optical illusions. For these reasons, successive reproduction of halftones is strongly discouraged.

The printing of a *color figure* is an even more complex process. The associated costs are so high that authors are normally urged to refrain from use of color entirely unless they are willing to paste original color photos or drawings into each copy of their text. Publication of a color illustration requires that the original first be "color separated" in an operation that leads to four separate printing plates, one for each of four ink colors. As a rule, each of the four separate images must then be screened. Only in the course of a multiple printing operation are the original colors re-created as a result of superimposing images based on the three primary colors and black.

## 9.3.2    Choice and Preparation of Photographic Copy

If photographic illustrations are to be included in a printed document the submitted prints must be glossy and rich in contrast. *Contrast* is particularly important, since publication always results in some "leveling" of tonal variation. The *resolution* in submitted originals should be so fine that even tiny details are clearly visible.

---

* An alternative technique, known as the gravure process, adds an additional "dimension" to screening, in that tiny "wells" of varying depth are produced on the printing plate. The depth of any given well is dependent on the gray shade of the corresponding region of the original photograph. The deeper the well, the more ink it will hold. Hence, shallow wells produce lighter dots, deeper wells result in dark tones. An analogous tonal option is not available with the common offset process, which is only capable of distinguishing between black and white: the presence or absence of ink.

Much of the success of a photographic figure depends on the care with which the subject is first chosen and then captured on film. The possibility that photographic documentation may someday be incorporated in a manuscript should be taken into account at the time the original exposure is made. If at all possible one should consider engaging the services of a professional photographer.

Observance of the following suggestions will increase the likelihood that a given photograph will prove suitable for publication:

● Only what is absolutely necessary should appear in the picture. In particular, the *background* must not be allowed to become a distracting element. Observing this rule may require changing the proposed viewing angle, depth of focus, or illumination, or perhaps masking parts of the negative before it is printed.

● It is generally wise to include in each picture a *ruler* or some other standard by which the size of the portrayed objects can be judged.

● The *principal subject* (normally the center of the picture) must be clearly recognizable. It is often helpful to choose a background tone that emphasizes the contrast between background and subject.

● Careful *focusing* is essential; even the edges of a finished print should appear sharp.

● If possible, all members of a proposed *series* of photographs should be prepared with the same type of film and developing process, as well as the same lens setting, similar background and lighting, and the same viewing angle and distance. Identical conditions lead to a desirable uniformity of appearance, including comparable gray tone distribution and content size.

Submitted prints should either be the same size as the desired figure or somewhat larger. Illustrations that are to appear side by side should, if possible, be of equal size, or at least of equal height. When reduction is necessary, the proposed scale should be carefully noted. *Photomicrographs* or other very detailed photographs that are too large for direct reproduction should be limited as much as possible by "cropping" (i.e., by deleting unimportant regions). This will minimize the loss in resolution that inevitably accompanies photomechanical size reduction. Desired cropping can be indicated using an *overlay* securely taped to the back of the print (but visible from the front when viewed against a light). Simply cutting the print to size is not a recommended practice. The risk of damaging the surface is too great, and it is also quite likely that the cut would be imperfect, requiring further trimming by the printer with possible loss of some important detail. Moreover, the publishing process usually requires that a small border area be present on every photograph submitted.

All photographs should be marked with the author's name, the figure number, a proposed scale of reduction, and the correct bottom-top orientation. Writing is permitted only on the *reverse side,* and it must be done with a soft pencil and light pressure to avoid damage. Tracing paper *overlays* are convenient for indicating the location of desired arrows, labels, or scales, though care must be exercised with this technique to ensure that final positioning will be exact.

Special attention is required to protect photographs from damage. Any flaws present in an original will surely be visible after printing, and repair is virtually impossible. In particular, all use of paper clips should be avoided because these tend to leave permanent marks.

# 10    Tables

## 10.1    The Nature and Proper Function
of a Table

A *table* permits the concise and orderly presentation of a set of *related* facts or observations. In scientific manuscripts tables are most frequently used for summarizing either experimental conditions or experimental results.* Much of the popularity of tables can be attributed to the perception that they represent the most efficient means of communicating large amounts of precise information, particularly numerical data. Moreover, authors assume tables are easier to prepare than the *figures* discussed in Chapter 9. In fact, a properly constructed table is far more than just a space-saving device, and the preparation of a good table can be rather complicated. A narrative is often considerably enhanced by the presence of tables, but only if the author has taken full advantage of their potential.

The first step in creating a table is *selection* of the appropriate information or data. The assembled material must then be *organized* in a logical way, after which attention can be directed to the best means of *displaying* it. If sufficient care is exercised at all stages the resulting table will clearly reveal *patterns* present within the data—and reveal them more distinctly in some cases than would a figure.

A typical table can be regarded as a *two-dimensional data array*, in which the two dimensions are somewhat analogous to the *x*- and *y*-axes of a line graph. Unlike a line graph, however, the "axes" of a table need not have any quantifiable meaning. This makes it possible to use a table for presenting "two-dimensional information" in which only one of the two "dimensions" is numerical; for example, the per capita steel production capacities of a group of nations. A table showing whether the responses of various groups of patients

---

* Tables are also used (perhaps too frequently!) in slides designed to accompany lectures. Those intended for this purpose are subject to special considerations, as discussed in Appendix A.

to a particular type of therapy were positive (+) or negative (–) would be completely devoid of numbered quantities. In neither of these instances would it be sensible to present the data as a line graph, and even a column or bar chart (cf. Sec. 9.2.4) would probably be a poor choice for the second example. Tables can also be more useful than graphs for displaying results that vary over several orders of magnitude, or data influenced by several independent variables.

*Table 10-1.* Yield of **3** from the reaction of **1** with **2** in different solvents at various temperatures.

| Solvent | Temperature | Yield |
|---------|-------------|-------|
| $CCl_4$ | 40 °C | 79% |
| $CCl_4$ | 75 °C | 79% |
| $C_2H_5OH$ | 20 °C | 24% |
| $C_2H_5OH$ | 75 °C | 24% |

The question of the *appropriateness* of a given table should always be resolved before time is spent on its *design*. For example, a little reflection may make it apparent that a figure would actually be a wiser choice, or the data in question may be so meager that they can be included to better advantage in the text. Table 10-1 is a case in point—its content can perfectly well be summarized in words:

> Compound **3** was prepared in 79% yield by reaction of **1** with **2** using carbon tetrachloride as solvent. Results were identical at both 40 °C and 75 °C. Changing the solvent to ethanol lowered the yield to 24% independent of the reaction temperature (tests conducted at 20 °C and 75 °C).

Nothing is gained by presenting the information in tabular form; the reader acquires no new insights, and no space is saved. The only real consequences of substituting a table for text are that the reader is forced to take note of the corresponding *table reference*

> "... the results are shown in Table 10-1, where ..."

then look for the table, and finally (and without assistance) interpret the data. Thus, a superfluous table represents wasted effort for the reader—and the author who must prepare it.

A good table, on the other hand, is a valuable asset. The following observations and suggestions are intended to help authors create tables that will add to the clarity and impact of their manuscripts.

## 10.2    The Preparation of a Table

### 10.2.1    Titles, Headings, and Footnotes

A table is primarily a collection of information, but just as in the case of a figure, every table requires a suitable framework. The key structural elements of a table are a *title, dividing lines, column headings*, and often *footnotes*. Each warrants separate examination before we confront the organization and presentation of the actual *data* one wishes to display.

A *table title* is necessary so the reader can quickly identify the subject. The text of the title is always preceded by the word "Table" and a sequentially assigned *number*. Tables in books should be assigned "double numbers" that incorporate a chapter identification (see the analogous case of "figure numbers" in Sec. 9.1.1).

The *title text* should be brief, but it should also convey as much information as possible. For example, one might allude in a title to the significance of the reported data, or to the way the results were obtained. "Solubility Measurements on $CaF_2$" is certainly concise, but the average reader is probably better served by

> *Table 3-4.* Solubility ($s$, in g/L) of $CaF_2$ in water at various temperatures ($t$, in °C). Data used in determining the temperature-dependence of the corresponding solubility product. Each reported value is a mean derived from seven independent measurements.

Table titles, like figure captions (cf. Sec. 9.1.1), should all be *placed* and *punctuated* uniformly. Usually they are either set against the margin or centered on the page. The placement should be the same as that adopted for figure captions and section titles. Periods should be placed after the table number and at the end of the title text. Sometimes one sees table titles in which all important words are capitalized. Presumably this practice is an attempt to stress the analogy between "table titles" and the titles (headings) of elements such as sections. However, the *function* of a table title more nearly resembles that of a figure caption, and in the latter case only first words are capitalized. We have chosen to refrain from superfluous capitalization in our table titles.

The title is placed above the table itself, from which it is always separated by a single *dividing line*. Next appears a set of *column headings* (also known as *box headings* or *boxheads*) comprising what is known as the *overhead key line*, equivalent to an "*x*-axis". Only a limited amount of space is available for headings, so they must be formulated as concisely as possible. Nevertheless, brevity must not be carried to the point of ambiguity. Headings are intended to explain the meaning of the data appearing within the columns, and that meaning must be absolutely clear. It is also advantageous to include *units* within column headings, since doing so results in a more compact presentation and eliminates much unsightly repetition (compare, for example, the tables in Fig. 10-1).

Often the key "line" is divided into more than one actual line of text. For example, a first line might identify the general subject of a column, while a line below could provide such details as the relevant units of measure (cf. Tables 10-3a and 10-3b in Fig. 10-2), or identify factors that distinguish two or more columns sharing a common first-line text (cf. Table 10-4a in Fig. 10-3). If a single first-line text applies to two or more adjacent columns simultaneously it should be separated from the succeeding line by what is called a *straddle* (or *spanner*) *rule*, a type of "underline" indicating the set of columns to which a given portion of the key line applies (cf. Table 10-4a).

All entries in the overhead key line are written with the first letter of the first word capitalized, as are most non-numerical data entries in the body of a table. First-letter capitalization (previously specified for inscriptions in figures, cf. Sec. 9.2.3, "Letters and Symbols") serves to emphasize a set of words and designate it as a single entity.

The bottom of a table should be marked with a single horizontal line, under which any *table footnotes* should appear. For example, a footnote might be used to indicate relevant experimental details, or to inform the reader of restrictions that apply to some of the reported data. Table footnotes can also provide such peripheral information as the number of experiments performed or the statistical significance of the data. An external source can also be cited in a table footnote, but it is preferably not the location of a *first* citation of a given source (cf. Sec. 11.3.2). The *citation symbol* used to call attention to a table footnote is generally a lower case letter (a, b, c, ...) rather than a number. This

Table 10-2a. Time-dependent
decomposition of **1** in
aqueous solution (pH = 6.8).

| t | c |
|---|---|
| 1.2 s | 0.998 mol/L |
| 45 s | 0.927 mol/L |
| 1.2 min | 0.886 mol/L |
| 8 min | 0.449 mol/L |
| 0.4 h | 91 mmol/L |
| 1.5 h | 0.15 mmol/L |

Table 10-2b. Time-dependent
decomposition of **1** in
aqueous solution (pH = 6.8).

| t/min | c/(mmol L$^{-1}$) |
|---|---|
| 0.02 | 998 |
| 0.75 | 927 |
| 1.2 | 886 |
| 8 | 449 |
| 24 | 91 |
| 90 | 0.15 |

Table 10-2c. Time-dependent
decomposition of **1** in
aqueous solution (pH = 6.8).

| 100t/min | c/(mmol/L) |
|---|---|
| 2 | 998 |
| 75 | 927 |
| 120 | 886 |
| 800 | 449 |
| 2400 | 91 |
| 9000 | 0.15 |

*Fig. 10-1.* Examples of typewritten tables.

avoids confusion with reference or note markings in the text, as well as with numeric table entries. If a footnote applies to a table as a whole, the corresponding citation symbol should appear in the table title. Similarly, a note that applies to an entire column is cited within the key line. The table in Appendix J provides several examples of the use of table footnotes.

The structural components we have considered above ensure the fulfillment of one important criterion for a good table: just like a figure (Sec. 9.1.1; cf. also Sec. 3.4.3), every table should be *self-contained.* A table must be capable of successfully communicating a clear message even in the absence of its host document. This injunction places special demands on the title and key line, and it means that explanations must be present (preferably in the form of a table footnote) for any unusual abbreviations or symbols.

Table 10-3a. The effect of
temperature on the formation
of **1** according to equation (12).

| Temperature °C | Yield % |
|---|---|
| 10 | 5 |
| 20 | 12 |
| 30 | 25 |
| 40 | 51 |
| 50 | 76 |
| 60 | 84 |
| 70 | 83 |
| 80 | 79 |
| 90 | 62 |
| 100 | 32 |

Table 10-3b. The effect of temperature on the formation
of **1** according to equation (12).

| Temperature °C | Yield % | Temperature °C | Yield % |
|---|---|---|---|
| 10 | 5 | 60 | 84 |
| 20 | 12 | 70 | 83 |
| 30 | 25 | 80 | 79 |
| 40 | 51 | 90 | 62 |
| 50 | 76 | 100 | 32 |

Table 10-3c. The effect of temperature on the
formation of **1** according to equation (12).

| $t$, °C | 10 | 20 | 30 | 40 | 50 | 60 | 70 | 80 | 90 | 100 |
|---|---|---|---|---|---|---|---|---|---|---|
| Yield, % | 5 | 12 | 25 | 51 | 76 | 84 | 83 | 79 | 62 | 32 |

*Fig. 10-2.* Examples of typewritten tables.

<u>Table 10-4a</u>. Product distribution upon reaction of 1 mol of **12** with alcohols in various solvents.

| Alcohol | $t/°C$ | Solvent | Product Yield | | |
|---------|--------|---------|-----|-----|-----|
| | | | **2** | **3** | **4** |
| methanol | 20 | $CCl_4$ | 0 mol | 0.285 mol | 0.49 mol |
| ethanol | 20 | $CCl_4$ | 0 mol | 0.120 mol | 0.48 mol |
| ethanol | 30 | $CCl_4$ | 0 mol | 0.137 mol | 0.49 mol |
| 2-propanol | 20 | $CCl_4$ | 0 mol | 0 mol | 0.08 mol |
| octanol | 20 | $CHCl_3,CCl_4$ | 0 mol | 0 mol | 0.94 mol |
| decanol | 20 | $CHCl_3,CCl_4$ | 0 mol | 0 mol | 0.92 mol |

<u>Table 10-4b</u>. Product distribution upon reaction of 1 mol of **12** with alcohols (20 °C in $CCl_4$); concurrent formation of **2** could not be detected.

| Alcohol | $n(3)/mmol$ | $n(4)/mmol$ |
|---------|-----------|-----------|
| methanol | 285 | 490 |
| ethanol | 120 | 480 |
| ethanol[a] | 137 | 490 |
| 2-propanol | 0 | 80 |
| octanol[b] | 0 | 940 |
| decanol[b] | 0 | 920 |

[a] Reaction run at 30 °C.
[b] Identical results obtained in $CHCl_3$.

<u>Table 10-5</u>. Selected physical properties of elemental fluorine.

| | |
|---------|---------|
| Boiling point | 85.0 K |
| Critical pressure | 55.71 bar |
| Critical volume | $1.74 \times 10^{-3}$ $m^3/kg$ |
| Density at 77.8 K | 1562 $kg/m^3$ |

*Fig. 10-3.* Examples of typewritten tables.

In the process of planning a set of tables (or figures) one should guard against the almost universal tendency to summarize too much information at once. A table must never be looked upon as a mere repository for all the numbers one feels obliged to report. Rather, it should be viewed as another tool for developing an argument. The best table is one whose entries convey a single, clear message. Anything not absolutely essential to the communication of that message should either be moved elsewhere—to the text, the table title, or a table footnote (cf. Table 10-4b in Fig. 10-3)—or deleted entirely. If too many "messages" are present they tend to obscure one another. Consideration should always be given to the possibility of dividing a complex table into two or more smaller ones.

## 10.2.2    Presenting the Data

### The "Dimensions" of a Table

A good table normally consists of at least three *columns*.* In the same way that an overhead key line resembles an *x*-axis, the lefthand column of a table, known as the *reading column* or *stub*, can be regarded as a type of "*y*-axis". All data in the remaining columns depend upon these two "axes" for their interpretation. A stub entry provides a definition for one *row* in a table. This "definition" may be a verbal explanation (cf. Table 10-5 in Fig. 10-3), but it could also be an experiment number or a compound identification. Sometimes a stub entry resembles an additional piece of numerical data (e.g., a time or a concentration).

The utility of a stub can often be enhanced by introducing a certain amount of *internal structure*. For example, the stub might be divided into *main entries* and *sub-entries* (cf. the Table of Appendix K). The result is an additional level of grouping of the reported data, which are transformed into a three-dimensional array. Repetition of the procedure leads to even more dimensions. Stub sub-entries should always be indented to ensure that the intended structure will be apparent. As a further sign of distinction, main stub entries in printed tables are frequently set in italics or some other typeface. Blank lines can also be introduced to alert the reader to the fact that a table contains several segments.

---

* One-column and most two-column information sets are usually better treated as *lists* rather than tables (cf. Sec. 10.3).

All the columns to the right of the stub contain *data*. Several alternative modes of data presentation should be explored before a final choice is made, since the best arrangement is rarely obvious. For example, one should examine the effect of varying the sequence of columns, or of altering the units of reported quantities. Even alternative ways of formulating the information should be considered. For example, it is almost always possible to process raw data from an experiment in a multitude of ways, and some treatments will be more informative than others. In particular, one should attempt to present information so that the most significant points are explicit rather than implicit. Finally, "blocking" of the data (either by use of stub sub-entries or introduction of appropriate blank lines) should be examined to see if this might enhance the impact of key relationships (cf. the table comprising Appendix K).

### Units, Data Alignment

Most scientific tables consist primarily of numbers, usually *quantitative experimental data*. It is essential that the *units* associated with reported numbers be so stated that their meaning is absolutely straightforward and devoid of ambiguity. The removal of ambiguity is not always easy, partly because what may seem perfectly clear to an author may not be equally clear to the reader. Consider, for example, a column with the heading "mol/L x $10^{-3}$". Such formulations are common, but they are singularly unfortunate. Some readers will interpret the above statement as an instruction to *multiply* the numbers in the corresponding column by $10^3$, while others will see it as a sign that each of the reported numbers should be understood to have the "suffix" $10^{-3}$, which would require *division* of the entries by 1000! In contrast, there is no ambiguity whatsoever about the meaning of the numbers in the second column of Table 10-2b (Fig. 10-1); the heading "$c/(\text{mmol } L^{-1})$" permits only one interpretation (cf. Sec. 7.2.3). The entry "998" on the line beginning with "0.02 min" must mean that the concentration observed after 0.02 min was $c = 998$ mmol $L^{-1}$. The unit specifications in Table 10-2a (Fig. 10-1) are equally clear, although the table itself is somewhat less appealing.

Considerable thought should be devoted to the optimal *choice* of units. Two separate criteria should be applied: convenience for the reader and overall visual effect. Units appearing in a table should always be consistent with current international standards; that is, *only* SI units are appropriate (cf. Sec. 7.3.1). An author who fears that readers will find SI units unfamiliar or inconvenient can easily meet their needs by supplying conversion factors in a table footnote. This way the reported data will become most accessible to the growing number of scientists familiar with the new standards only.

Often decisions are called for with respect to the *scaling* of data, particularly if the numbers destined for a given column vary over several orders of magnitude. In general, it is better to show scale factors by means of the internationally accepted decimal prefixes (cf. Sec. 7.3.3) rather than through extensive use of "10-factors" (see below, however). One technique that sometimes eliminates awkward units and scale factors is the presentation of *normalized* data. "Unitless" numbers are used to express *comparisons* between various results in terms of an appropriate *standard* or *control experiment*. However, this approach requires that the corresponding control data also be reported, complete with proper unit specifications.

Tables of all types are normally intended to encourage the reader to make comparisons among the various numbers appearing in a given column. To this end, figures within a column should always be arranged so that their decimal points are aligned;* e.g.,

| | | | | | |
|---|---|---|---|---|---|
| 3.25 | | 3.25 | | 3.25 |
| 10.7 | not | 10.7 | or | 10.7 |
| 214.834 | | 214.834 | | 214.834 |

Trends within the numbers thus become more readily apparent. Unfortunately, the result is sometimes a rather unsightly table. The problem of appearance can usually be solved by changing units in the middle of a column (cf. Table 10-2a in Fig. 10-1), but this solution is counterproductive because the interrelationships one wishes to emphasize are then obscured. This is one occasion when judicious use of "scientific notation" (i.e., $4.3 \times 10^6$ instead of 4 300 000) may be warranted. It may also help to introduce a multiplicative factor in the key line unit specification (cf. the first column of Table 10-2c in Fig. 10-1).

If the data comprising a particular column are not intended to be comparable (e.g., because successive entries are only loosely related, or because different lines display entirely different entities), then *left justification* of the data is preferred, as in the first columns of Table 10-2a (Fig. 10-1) and Table 10-5 (Fig. 10-3). Consistent left justification is also preferable with non-numerical data.

---

* The beginning of the corresponding key line entry is usually aligned with the character farthest to the left within the corresponding column, although centering of the key line information is another possibility.

## 10.2.3   Table Format

Tables should never be embellished with superfluous *dividing lines* or "picture-frames". These do little to improve appearance, and any lines, especially vertical ones, make typesetting more difficult. As noted above, three horizontal lines generally suffice to define a table: two are used to set the key line apart from the table title and the data, respectively, and a third is placed at the bottom, immediately above any explanatory notes. The need for vertical lines is eliminated by careful arrangement of column entries and proper column spacing.

The optimum *format* for a given table depends on the nature and number of entries on each line.* One should always avoid long, narrow tables (i.e., those with only two or three columns, but a length greater than one-half page). Short, wide tables are also inappropriate, especially if they must be set *broadside* (i.e., sideways, so that the lines follow the long dimension of the paper).

The simplest remedy for a table that is too long and narrow is doubling it back upon itself: that is, the table is broken into two halves, and these are placed side by side. The column heads from the first half are then duplicated over the second half (cf. Tables 10-3a and 10-3b in Fig. 10-2). Alternatively, the functions of columns and rows can be reversed, as illustrated in the arrangement of Table 10-3c in Fig. 10-2. By means of a similar inversion, an overly wide table can be converted into one that is long and narrow, and this in turn can be split into parallel segments, as described above.

Only if measures such as these are impractical should a table be arranged broadside, and even then it is important that the corresponding pages be fully utilized (as with the table in Appendix J). Headings for broadside tables should always be placed on the left side of the page. Broadside tables are a nuisance to the reader, but sometimes they are unavoidable.

Some tables are so long that that they extend over several pages (a typical example is that in Appendix J). A table like this requires a set of column headings at the top of each page.

Before final decisions are reached concerning table shape and spacing it is important that a few of the proposed data lines be carefully typed on scratch paper and the results examined. The sample should be chosen so that it contains the widest of the data entries. Only when such a "mock-up" has been

---

* Editorial guidelines may also be a factor (cf. Sec. 10.2.4).

prepared will the author be in a position to anticipate and evaluate accurately the probable appearance of the completed table. (The author with a word processor may find table construction to be far less onerous; cf. Sec. 5.2.2, "Basic Advantages").

## 10.2.4    Miscellaneous Matters

Every table must be linked to the text itself by a specific *reference*. References within a narrative typically take such forms as

... are presented in Table 8.
... the *y* parameters of all compounds so far investigated
(see the middle column of Table 5-7) clearly reflect ...

Note that the first letter of "Table" is always capitalized when used in conjunction with a number.

For the reader's convenience, a table should be placed as close as possible to the point at which the first reference to it appears. Nevertheless, if several full-page tables must appear together it is better to create an appropriate appendix at the end of the manuscript.

As with other "imported" elements (including figures, cf. Sections 9.1.1 and 9.1.3), a table derived from an *external source* must always be so identified, and the origin must be explicitly acknowledged (either formally or informally, depending on whether or not the data have in some way been modified; cf. Sec. 9.1.1). It is the author's responsibility to secure permission from the copyright holder before reproducing a previously published table. In practice, one seldom finds the need to request such permission. Tables, unlike figures, can easily be revised and presented in a slightly modified form, permitting the author to claim legitimately that a "new" table has been created.

We conclude with a few suggestions regarding the *typing* of tables, and a look at the process of preparing tables for publication.

● *Typewritten* tables should be either single- or one-and-one-half-spaced. The proper choice between these two alternatives is largely a matter of visual effect, and is dependent on overall size as well as on the scope and nature of the data (cf. Tables 10-4b and 10-5 in Fig. 10-3).

● The titles for typewritten tables should always be "one-and-one-half-spaced" in order to distinguish them clearly from the surrounding double-spaced narrative (an attempt to imitate the effect of the smaller type used for headings in typeset tables).

- In a table title, the word "table" with its accompanying number is usually underlined, corresponding to the use of italics (or sometimes boldface) in most typeset works.
- Tables that are to be *typeset* (i.e., those accompanying manuscripts for publication) should be double-spaced in their entirety.

It is best to prepare all tables on separate sheets of paper. If a document is to remain in typewritten form, each completed table is later trimmed to an appropriate size and pasted into a blank space within the text, preferably either at the top or the bottom of a page. For manuscripts that are to be *typeset*, however, all table material should be collected together, separate from the text manuscript.* In this case it is *essential* that each table occupy its own page. Authors should be aware that tables prepared for insertion in a typeset document are usually subject to special space restrictions designed to ensure that the material will fit properly in the available space. Such limitations can cause serious problems, particularly in the case of journal articles. Editors' guidelines are often very specific, and they must be scrupulously observed.

Tables are a subject of discussion in most general treatises on writing (e.g., *The Chicago Manual of Style*, 1982), and we suggest that the writer seriously interested in the subject consult one or more such references for additional guidance.

# 10.3    Lists

A *list* can be regarded as a rudimentary table consisting of only one or two columns. The list of thesis components in Sec. 2.2.1 is an example of a one-column list; another is the materials summary frequently introducing the experimental section of a report. Symbols used in a discussion are often identified by means of a two-column list (cf. Fig. 7-1); the first column is devoted to the symbols themselves, while the second provides verbal counterparts.

Lists are usually *displayed* (cf. Sec. 8.1.1) and arranged vertically (i.e., with each entry on a separate line). All entries in a displayed list should be justified

---

* This rule has been relaxed somewhat in recent years thanks to the development of more flexible typesetting equipment. Most tables can now be treated as text rather than as elements subject to special handling.

on the left, and the list as a whole should be indented somewhat from the left margin. Two-column lists are often set in the center of the page. No special designations are required to distinguish one list entry from another (e.g., they need not be numbered), but if a single entry occupies more than one line, the continuation line(s) should be uniformly indented somewhat farther than the list as a whole. Unlike tables, lists seldom bear titles, they are usually not numbered, and only rarely do they require headings or footnotes.

Sometimes it is advantageous to treat whole sentences or even paragraphs as items in a list, even though the length of the components rules out a format like that described above. In such cases it is customary to provide *item markers* in the form of large black dots, called *bullets* (●). We have made frequent use of this device in the present book. Lists so designated need not be indented. Indenting is a way of attracting the reader's attention, and the dots themselves fulfill this function; indentation would be a waste of valuable printing space. Moreover, the use of two signaling devices at once causes pages to acquire a distressingly "busy" look.

A list in which the items are preceded by *numbers* is known technically as an *enumeration*, e.g.,

> ... will be treated after the pattern
> 1. physical properties
> 2. synthesis
> 3. chemical properties
> 4. applications

Individual items within an enumeration are distinguished by Arabic numerals,* and the numbers themselves are followed by periods. Punctuation at the ends of the items can be omitted if the enumeration is displayed (as above), and the items need not be capitalized. One can also forego the display approach and retain instead a standard paragraph format; e.g.,

> ... will be treated in the sequence (1) physical properties,
> (2) synthesis, (3) chemical properties, and (4) applications.

In this case the items comprising the enumeration are separated from one another by arabic numerals enclosed in parentheses, and standard sentence punctuation is employed throughout.

---

* The once-common use of (lower case) Roman numerals [i.e., (i), (ii), (iii), (iv), ...] is now considered old-fashioned.

# 11 Collecting and Citing the Literature

## 11.1 Introduction

The fact that scientists are singularly dependent on the scientific literature has been an underlying theme throughout much of our book. As its principal assignments, this chapter will (1) show how literature citations can best be introduced into manuscripts, and (2) provide advice on the preparation of reference lists. The second facet leads inevitably into a disconcerting tangle of often conflicting recommendations regarding the proper treatment of various kinds of sources. In an attempt to establish some order in this chaos we develop here a set of comprehensive but nonetheless relatively straightforward guidelines for constructing references. The basis for our suggestions is the subject of Sec. 11.4, and Appendix J illustrates the results in the form of an extensive collection of sample references organized by type. Our suggestions may occasionally require minor adaptations, but they should nevertheless make the preparation of almost any reference list a relatively easy task. It must be emphasized, however, that many publishers—and some universities—have their own referencing rules, and these must of course take precedence whenever conflicts arise.

However, before we embark on a pursuit of our main objectives we propose a brief side excursion. Almost daily, as a part of either casual reading or a more structured literature search, every active scholar encounters books, articles, or other pieces of information that appear to warrant closer examination. For the chemist these may be intriguing new synthetic or analytical methods, unexpected results with potentially interesting ramifications, or theoretical explanations that invite later, more careful study. The need soon arises to find a way to collect such random discoveries—i.e., to preserve and organize them—so they can be readily retrieved at the proper time, perhaps even cited someday in one's own writings.

The techniques involved in *collecting* a part of the literature are rather more closely related to the techniques of *citing* the literature than may first be obvious. It is thus appropriate that our chapter begin with a brief look at literature logistics. In particular we shall consider possible ways to establish and manage a personal and evolving "index" to relevant portions of the literature.

## 11.2    Organizing a Personal Collection of Sources

### 11.2.1    The Master Card File

Today, anyone contemplating the preparation of a personal document catalogue should give serious thought to basing it in a computer. The advantages are discussed at some length in Sec. 11.2.3, and many have found them compelling. Nevertheless, some readers may be in no position to follow this advice; we therefore begin with a summary of proven strategies for developing a document catalogue on file cards. Fortunately, much that must be said is also applicable to a computerized document file.

Probably the most frequently encountered cataloging system, and certainly the one with the longest tradition, begins with an *author card file*. Creating and later expanding such a file is a straightforward matter. The first step is preparation of a master file card for each *document*\* to be catalogued. At the top of the card one writes the appropriate authors' (or editors') names,† and under this one adds other information required for establishing the identity of the document (its *bibliographic data*). In other words, one reference (or document)—one card.

The cards then need to be arranged in a specific way. Most prefer what is called the "name-date system", details of which are provided in Sec. 11.3.3. It is sufficient here to note that the system produces a file in which all document

---

\* The word "document" here includes journal articles, books, reports, patents, theses— virtually any recorded piece of information.

† Patents are classified according to their *inventors'* names (cf. Sec. 11.4.4). Anonymous documents can be filed under the name of their *issuer*: a corporation, an institution, a society, or an agency.

cards are arranged alphabetically by first author, with multiple works by the same author ordered chronologically. This is also the approach specified by a number of journals (e.g., *Biochemistry*) for the *reference lists* appearing at the ends of articles, and we have used it for the literature section at the end of this book. Since the primary ordering in such a filing system takes account only of *first* authors, some type of cross-referencing is necessary if one wishes to be able to search the file by additional authors as well (cf. the first footnote in Sec. 11.2.2, but also Sec. 11.2.3).

There are, of course, other ways of arranging a master file. One tempting approach is a system based on the principal *subject* discussed in each document, perhaps utilizing the Universal Decimal Classification (UDC) described briefly in Sec. 11.2.2. Generally speaking, however, a subject-based master file is less satisfactory than one based on authors because of inevitable ambiguities regarding the "proper" subject assignment for a given document.

The optimal *format* for one's cards is a matter of personal preference. Most researchers find that standard 3 in. x 5 in. (or A7) cards serve the purpose nicely, though some choose instead the larger 4 in. x 6 in. (or A6) size. Initially, a researcher's card collection is likely to fit conveniently in one or two file boxes, but eventually it may grow to the point of occupying several desk drawers, or even an entire filing cabinet. This points up one of the advantages of small cards: they consume less storage space.

Proper choice of the bibliographic data to be included in the *card headings* is crucial. Thought should also be given to the best *style* for the headings. Consistency in style is important because it makes a set of cards easy to scan. The bibliographic style recommended in Appendix J has the particular advantage that it produces card headings similar to the references employed in journals. Master cards so organized instantly yield the information required later when documentation must be added to a newly written manuscript.

The most important prerequisite for a system of card headings is that it be sufficiently *complete* and *detailed* to meet any foreseeable need. Care in this regard is well worth stressing because it is both inconvenient and irritating to discover that one's "complete" bibliographic notes are insufficient for purposes of publication or, worse, that they fail to permit retrieval of an important document. The sample document citations in Appendix J are reliable guides because they include all the elements considered essential by professionals for proper referencing.

A minimal *bibliographic entry* for a document consists of:

- the full names of all authors,
- the complete title of the document, and
- an exact bibliographic description of its source.

It is the last point that causes the most problems, as will be seen in Sec. 11.4.

The body of a master reference card may be used for summarizing the contents of the corresponding document and for making note of any particularly interesting facts from it (or about it). Some researchers routinely prepare their own document summaries; others find it more convenient to incorporate and then annotate photocopies of published abstracts.

Our experience has shown that there is value in also assigning running catalogue numbers (*entry numbers* or *card file registry numbers*) to all cards as they are filed.* These chronological numbers have little intrinsic significance, but they do provide a convenient way of cross-referencing associated material filed elsewhere (e.g., reprints, photocopies, extended excerpts, or hand-written notes). Numbers also facilitate internal cross-referencing and indexing of the collection. Card file registry numbers can even be used during the early stages of manuscript preparation as temporary substitutes for proper citation numbers (cf. Sec. 11.3.2).

However, use of registry numbers does necessitate some form of "double-entry bookkeeping". The cards themselves will be filed systematically (i.e., alphabetically by authors or subjects), so a separate numerical list is needed for locating a given document solely from its entry number—unless the "card file" is maintained on a computer (cf. Sec. 11.2.3).

## 11.2.2   Locating Pertinent Entries

An author card file is obviously very useful whenever one wants to identify and recover publications on the basis of their authorship, but a little reflection shows that this type of search capability is unlikely to be sufficient. The natural sciences are fact-oriented, not author-oriented.[†] Consequently, some type of *subject index* is also an indispensable part of the traditional author-based bibliographic file.

---

* A numbering (paginating) stamp is useful for the purpose.

[†] Actually, an author file has shortcomings even when a particular author's work *is* being sought. The most prominent scientist or the leader of a research group is not necessarily the first author of all of "his" (or her) publications (cf. Sec. 3.2.4). Thus, a cross-referencing system is required even to locate all publications from a single laboratory.

Creating and maintaining an index can be a problem. In principle, one could start by developing a *thesaurus*, a list of the keywords most closely associated with one's interests. One would then need to associate each newly found document with appropriate keywords from the list. It seems unlikely, however, that such an *a priori* topic list would prove adequate over time, partly because a researcher's interests change. Alternatively, one might simply designate document keywords one at a time, as the need for them arises. This approach certainly provides greater flexibility, but the resulting index would not be very systematic, and it is extremely difficult to maintain an "evolving" index.

No matter how the index comes into being it should also be a card file, this time one comprised of *keyword cards*. However, keyword cards differ in a fundamental way from master cards: they require frequent updating. Additional entry numbers must be added under the appropriate keywords every time a new document is entered in the master file.

Maintaining a subject index is a very laborious task. Considerable time and thought are required to select an adequate set of keywords for each document (10 would not be unreasonable), and attending to the resulting index entries is likely to be even more time-consuming. Furthermore, the use of such an index can also be cumbersome. Suppose, for example, that the average keyword is applicable to every twentieth document. This means there will be 50 entries under each keyword by the time the document collection itself reaches 1000 items (a modest number!). It is a demoralizing prospect to anticipate having to check 50 master cards every time the subject index is consulted for a particular document. This is one of the reasons why a great many well-intentioned attempts to set up personalized bibliographic files end in disarray, with the once cherished cards left to gather cobwebs.

A possible escape from the indexing problem was noted briefly in the preceding section: one could base the whole collection on a subject-oriented filing system instead of relying on the author-oriented name-date system. However, if such a system is to be effective it must permit one to associate the content of each document with the narrowest possible classification in a very elaborate scheme. The most extensive and detailed model available is the *Universal Decimal Classification* (UDC), developed by the Fédération Internationale de Documentation (FID)* working in cooperation with national standards organizations. The UDC was designed for use by bibliographers and

---

* PO Box 30115, Den Haag, The Netherlands.

documentation specialists, and it is subject to constant review and expansion. Some authors find the UDC quite appropriate for organizing their private literature collections. The bibliographic card related to a particular document is simply labeled with the most appropriate UDC number (or its associated term) and then filed numerically. There is no shortage of potential classifications: tens of thousands of UDC numbers are allocated to the sciences alone.

Any author relying on this system will of course require access to a copy of the UDC handbook, although one can also secure relevant *segments* of the unabridged version, covering only the few subject areas most likely to be of primary concern. Another possibility would be to use the *abridged version* of the handbook, which lacks some of the most subtle subclassifications. It might then be necessary to invent one's own numbers to cover missing subjects, but there is certainly no harm in doing so since the object, after all, is to adapt the system for personal use.

An overview of all published issues and editions of the UDC, in various languages, may be obtained (free of charge) either from the Classification Department of the FID or through one of the national standards organizations. Mathematics and the natural sciences are covered in Division 5 of the UDC. Successful management of a personal literature collection according to the UDC system requires a certain investment in both money and time, and for most people it entails the acquisition of a set of new skills. Unfortunately, it is not a complete solution to the problem of cross-referencing, because most documents defy classification under a single subject heading.

## 11.2.3    Taking Advantage of Computers

One of the early attempts to solve the problems posed by document indexes involved storing bibliographic data on *punchcards*. Holes were punched in specific places on each master card to signify important characteristics of the associated document, and the cards were then accessed with the aid of some type of mechanical sorter. However, with the advent of inexpensive personal computers, punchcards, like the slide rule, have been relegated to the realm of nostalgia. Today's solution, a far better one, is the "electronic file" (bibliofile, *electronic library*).

To create such a file one begins by establishing a set of categories for characterizing documents. No categories are required for information that is part of the bibliographic data (e.g., year, journal, etc.) since this information can be searched in any case. However, one might want to include a category for document type (book, journal article, patent, etc.), or one for UDC numbers, perhaps also a category for tailor-made keywords or even one for the country in which the work was carried out.* The nature of the most appropriate category set will vary depending on what types of document searches are envisioned. As each document is introduced into the computerized catalog—complete with the usual bibliographic information— explicit entries (*descriptors*, also known as *properties* or *index terms*) are made under each of the defined search categories.

A considerable amount of "typing" may be necessary at first, especially if one is attempting to computerize a catalog previously kept on file cards,† but once the file has been built it is extremely simple to use—and it occupies virtually no space. In a matter of seconds a computer can locate within the file every document characterized by any given property. Sometimes a search will produce too many documents (called "finds" or "hits"). If so, the search can be repeated, this time with the restriction that two or more relevant properties must be present simultaneously. Complex searches of this type are almost impossible to perform manually, but they present no challenge at all for a computer. The additional time required for a two-parameter search is negligible, and the number of finds is greatly reduced because the search is narrowed. Even the number of documents on file is rarely a limiting factor with respect to a computer search.

Storage and subsequent retrieval of bibliographic information in the manner described can be accomplished with the aid of any of a host of commercially available computer programs. Some prefer software designed specifically for the purpose, but general database or even "warehouse management" software packages can also prove quite satisfactory. Increasing numbers of scientists are relying on computerized document files, and most are enthusiastic about the benefits. There is admittedly a certain "activation energy" to be overcome before one reaches the point of having a functioning system, but the

---

* Almost anything can serve as the basis for a category, and categories can be added to the file retrospectively as a need for them becomes apparent.

† There is one attractive way of reducing the typing load, and at the same time minimizing the likelihood that errors will creep into the file: references can be selectively assembled from a remote computerized database and then "downloaded" directly into one's own computer with the aid of a modem.

advantages that accrue quickly make up for the effort expended. It seems only a question of time before most scholars will adopt this strategy.*

Extensive computer-based literature files are also available for public use. Some are incredibly comprehensive, and they can prove invaluable when one is confronted with a special project such as preparing a review article or writing a monograph. Indeed, computers are rapidly becoming the preferred resource even for routine literature searches, although their popularity is still limited somewhat by the costs entailed.† Nevertheless, most scientists affiliated with large corporations or major research institutions already have access to computerized literature databases, often with the support of in-house information specialists trained to perform efficient and comprehensive searches (although as systems become more and more user-friendly a trend toward primary-user searching is again evident). The extraordinary possibilities created by such tools are fascinating, but beyond the scope of the present chapter (cf., however, the brief discussion of literature searching in Appendix D).

---

* Computer programs are even available for *rearranging* bibliographic data from such a file to correspond to any desired reference style (e.g., in preparing manuscripts suitable for submission to any of several journals). These are designed for use on standard personal computers, and in conjunction with various types of word processing software. An example is *Reference Manager*, a software package available from Research Information Systems, Inc., 1991 Village Park Way, Encinitas CA 92024, USA.

† Large scientific databases are maintained by Chemical Abstracts Service (CAS), the Institute for Scientific Information (ISI, whose *SciSearch* files provide access to *Science Citation Index*), and others. These databases can in turn be accessed either directly or through third-party systems such as DIALOG (operated by the Lockheed Corporation). Brief introductions to the structure and virtues of the CAS databases are to be found in the brochures *Chemical Abstracts Service—An International Resource* and *CAS Chemical Abstracts Service Statistical Summary 1907-1984*, available from Chemical Abstracts Service, Columbus, Ohio 43210, USA. A more comprehensive description of the present state of the art of information processing, as well as its implications for chemistry, is provided in Ash, et al. (1985); cf. also Schulz (1985).

# 11.3    Citation Techniques

## 11.3.1    The Purpose and Nature of a Citation

Virtually every scientific report relies to some degree on results, measurements, observations, or theories already described in the published literature. Whenever an author borrows a fact or an idea from an external source, the transaction must be acknowledged and the exact origin of the information revealed. The precise relevance of the imported material to the new document must also be made evident,* a factor the significance of which will become clearer as we proceed.

The formal acknowledgment within a document that an external source has been employed is called a *citation*. The citation serves as the link between the text in which it occurs and a formal *bibliographic reference* (or simply *reference*). The latter is a concise statement enabling the interested reader to find the borrowed information or idea in its original context, the *source publication*.

Standard *ISO/DIS 690-1979* (cf. Sec. 11.4.1) defines a citation as "a brief form of reference inserted parenthetically within the running text." Another authority, Standard *ANSI Z39.29-1977*, substitutes for "citation" the term *in-text reference*. In scientific writing these brief references may take either of two forms:[†]

1. Numeric symbols (*numeric system*)[‡]
2. Author-date notations (*author-date system* or *name-date system*).

The two systems are described in the following subsections (Sections 11.3.2 and 11.3.3).

---

* Michaelson (1982) in his well-written treatise *How to Write and Publish Engineering Papers and Reports* devotes an entire chapter ("How to Cite References Properly in Text") to this matter.

[†] The first is the most common representative of what is called the *note-bibliography system* of citation, because a note in the text (in this case an Arabic numeral) serves as a guide to a source described elsewhere, usually in a separate bibliographic section. The name-date system, also called the *Harvard system*, is a typical example of the alternative scheme of *parenthetical documentation*, where at least some essential data pertaining to the source are disclosed in the text itself, set off in parentheses.

[‡] Non-numeric symbols would also serve the purpose, but numbers are almost always employed for citations in scientific documents.

Every citation (in-text reference) must provide an unambiguous link between the marked text element and a complete bibliographic reference, perhaps pointing to a specific site (*location*) within the source publication (cf. "Depth of Referencing" in Sec. 11.4.1). Occasionally the corresponding reference identifications are typed or typeset as *footnotes* at the bottom of the pages on which citations appear. More frequently they are collected in a separate list called "References".* This is usually placed at the end of the document, although books sometimes have reference lists after each chapter. Publishers prefer consolidated lists because they are more convenient to typeset than footnotes and because they are less costly to produce. Readers, on the other hand, are better served by footnotes, since these permit easier and more immediate access to what may be highly relevant information.

It is not unusual for an author to mention explicitly in the body of the text the names of at least some of those people whose work is cited; indeed, if the name-date citation convention is followed then names *must* be given. Conventionally, only *last* names are used except in cases where ambiguity might result, or where attention is to be called to the role of a well-known individual (e.g., Charles Darwin, Emil Fischer). When referring to a piece of work published by two or three authors, all names should be included, but publications with more than three authors are traditionally ascribed in the body of the text only to the first author, whose name is augmented by the phrase "et al." (an abbreviation for the Latin *et alii*, "and others"; note that "et" is *not* an abbreviation and is therefore not followed by a period).† The once-common alternative formulation "and co-workers" is probably best avoided in an era in which implied distinctions based on rank are considered poor taste.‡ Besides,

---

* Sometimes "notes" of the more usual kind—substantive notes: supplementary elaborations or explanations of the text (cf. Sec. 3.4.5)—are integrated together with the references, although this is possible only if the numeric system of citations is used. This is the practice, for example, in *J. Am. Chem. Soc.*, where a numbered note—an actual footnote in this case—may even contain both explanatory material and references. Thesis writers also sometimes adopt the integrated approach, although it is not always permitted under university or departmental rules. Integrated notes appearing as a separate section of a document should be headed "References and Notes", never simply "References". The more orthodox practice is that in which explanatory (substantive) notes are separated from references, and are designated differently (e.g., using asterisks and similar symbols to call attention to the former, as in this book).

† Some journals (for example, all those published by the American Chemical Society) permit the use of "et al." even if there are *only* three authors (Dodd, 1986; p 107).

‡ Strictly speaking, the term "co-workers" refers to associates who are "full partners", and thus co-equal with the named individual. However, it is usually not so interpreted, hence the suggestion that it be avoided.

it is not uncommon for the senior member of a research team to be named last among a set of authors, or for authors' names simply to be arranged alphabetically. A citation that degrades the party responsible for a project to the apparent status of a "co-worker" among junior colleagues (or graduate students) might, even today, be regarded as inappropriate.

If one wishes to incorporate *literally* (i.e., word for word) passages written by others, these must be identified as *quotations* by setting them in quotation marks, e.g.,

> ... The assertion by Miller et al. [7] that "... this reaction occurs only in the presence of traces of heavy metals" is contradicted by the present results.

Publications one has not personally seen should never be acknowledged directly. If a source is known only "second-hand" then it is the "source of the source" that should be revealed. Original sources should of course be identified (and consulted!) whenever possible, but the subject of a citation should always be the document to which the author actually had access. One compromise is permissible: citation of an unseen original source followed by a semicolon and a description of the appropriate *secondary* reference, e.g.,

> ... ; cited in [3].
> ... ; quoted by Braun (1979).
> ... ; *Chem. Abstr. 1966*, **64**, 4953a.

The above admonition with respect to actual sources is partly dictated by a concern for honesty, but it also results from another important consideration: the reader deserves to be warned that certain relevant information has been obtained indirectly, and that the author is therefore in no position to vouch for its accuracy. A *CA* reference like that in the third example above has the advantage that it provides the reader with a place to look for more information regarding a source likely to be difficult to obtain.

An author may alert readers to useful sources other than those cited directly in the text by including a section entitled "Other Literature", "Further Reading", or "General Bibliography". Any list of this type must be kept separate from the list entitled "References". Usually it would be placed before the reference list. Elementary textbooks often contain *only* general references of this kind, which makes them the sole exception to the rule that all scientific writing is to be thoroughly documented.

## 11.3.2    Numerical Citation Systems

In Sec. 11.3.1, two methods were introduced for calling the reader's attention to the fact that information has been drawn from the literature. Here we examine in detail the first of these, which employs numeric symbols—*citation numbers* (or, less precisely, *reference numbers*).

Citation numbers may be introduced into a text in any of three ways: in brackets,* in parentheses, or as superscripts, e.g.:[†]

| | | |
|---|---|---|
| ... text [7], ... | ... text [7,8], ... | ... text [4,7-10], ... |
| ... text (7), ... | ... text (7,8), ... | ... text (4,7-10), ... |
| ... text,[7] ... | ... text,[7,8] ... | ... text,[4,7-10] ... |

Whatever form is chosen, the number should be placed immediately after the text element most closely related to the external source. Usually this will be either a technical term or a brief definitive phrase, as in the examples below. It is only on rare occasions that optimum placement is at the end of an entire sentence (cf. Sec. 11.3.4).

A more recent investigation[7] has shown ...
... from spin-decoupling experiments[7] and ...

Use of the numeric system of references does not preclude also mentioning an author's name; e.g.,

The results reported by Smith[7] are ...

Numbers in brackets or parentheses have one significant advantage over superscripts: they can be augmented by page numbers or other *locators* in order to specify more precisely where the passage of interest occurs within a source publication; e.g.,

... in a recent review [4, p 765] ...
...in the literature /e.g., (2), Sec. 3.4/...

---

* Slashes (e.g., ... /27/ ...) are often considered an acceptable substitute in typewritten documents if square brackets are unavailable.

[†] Note the placement of the numbers relative to sentence punctuation, here illustrated by a comma. A superscript is written *after* any punctuation mark. The only exception is the dash; dashes always follow superscripts. Notice also that superscripts are never preceded by blank spaces, while numbers in parentheses or brackets are.

Alternatively, supplementary guidance of this type could be added to the bibliographic information in the reference list, but the in-text approach is decidedly superior for sources that need to be cited repeatedly.*

Numbering of citations is normally sequential throughout a manuscript: the first reference cited receives the number "1" and the document referred to last is given the highest number. This procedure sounds simple enough to implement, but it can cause major problems during the development of a manuscript, because citation numbering is likely to change frequently as early drafts are revised. Some authors find it more convenient to use document file numbers (cf. Sec. 11.2.1) as citation numbers through all stages of manuscript preparation up to the point of preparing the final version. Only then are the temporary numbers replaced by proper citation numbers, a process requiring relatively little effort. Another possibility is the temporary use of name-date citations (cf. Sec. 11.3.4).

One additional complication can accompany the assignment of sequential reference numbers if a manuscript will later be typeset. Figure captions or tables may contain citations, and the author has no way of knowing in advance precisely where these elements will appear in the printed document. Probably the best solution is to treat such citations as if they were located at the point of the *first reference* to the table or figure in question.

Reference lists in lengthy works (e.g., books) are often compiled by *chapter*, permitting the first citation in each chapter to bear the number "1". The most obvious advantage of this practice is that it keeps citation numbers from becoming large and unwieldy. There is an additional benefit as well: a few last-minute changes can be made in a set of literature references without the need to change every subsequent citation number in the book.†

Occasionally reference numbers are assigned in an entirely different way. First, all referenced works (as yet unnumbered) are arranged alphabetically by first author, as in the name-date system (Sec. 11.3.3). Then the alphabetical list is numbered sequentially, after which the resulting numbers are distributed

---

* Locators can also be provided in abbreviated reference list entries; e.g., [12] Ref. 6, p 100. The relevant information would then be found at the specified site in the source whose bibliographic data are listed under entry 6.

† A reference added late in the publishing process is sometimes accommodated by creating a citation like "9a" between citations "9" and "10", but this solution is both awkward and unsightly. Occasionally letters are introduced for an entirely different (and justified) reason: to permit a distinction among several documents (e.g., 9a, 9b, and 9c) that have enough in common to warrant a joint citation (as "9"), but also differ enough so that at least one will be cited separately.

as citation numbers in their proper places throughout the text. Obviously the citations themselves will not occur in numerical order. This system has its advocates, but it seems to us an unwise choice. It is true that in a reference list so arranged it is easy to locate any work by a particular author, but no method exists for finding the place in the text where such a paper has been cited. Only if the numbers arise sequentially is it possible to work backward from a reference list to the first citation of a given source.

### 11.3.3    The Name-Date System

Many scientists and editors, particularly in medicine and the biosciences, prefer another approach to citations—the *name-date system* (or *author-date system*). This is the system that has been employed in the present book. Reference numbers are totally avoided. In their place one introduces the *name* of the first author of the corresponding source publication (or the names of the first and second authors if there are only two*), together with the publication *year* of the document.

The following examples illustrate how name-date citations can be incorporated unobtrusively into running text:

> Young (1981) and also Peterson (1983) report that ...
> Recent in vitro studies (Smith and Johnson, 1986) have shown ...
> In fact, however, the actual value must be somewhat higher
> (Meier et al., 1978).

The use in the last example of "et al." is an indication that the cited paper has three or more authors (cf. Sec. 11.3.1).

The corresponding list of cited references appears as usual at the end of the manuscript (or chapter), but this time it is arranged in a sequence partly *alphabetical* and partly *chronological*. To construct such a list one begins by grouping all cited references alphabetically by *first author*. If a given author is represented by more than one work, *subsets* are then formed, in order, as follows: (1) works by the one author alone; (2) works in which the author has been joined by one co-author; and (3) works of the author and two or more co-authors. All references falling in subset (2) are then grouped alphabetically according to the name of the second author. Finally, all items within each set or

---

* Some publications permit up to three names in citations.

subset are arranged chronologically, oldest references first. If a given subset contains multiple items from the same year these are differentiated by appending *letters* a,b,c,... to the year, doing so in the order in which the documents have been *cited*. These letters are then also made a part of the citations themselves; e.g.,

> On the basis of kinetic (Beckman, 1975a) and spectroscopic (Beckman, 1975b) evidence ...

It may seem odd that further alphabetization has not been suggested within subset (3). After all, third authors' names are just as easy to alphabetize as those of second authors. Nevertheless, it is rarely done. The purpose of organizing a reference list is to simplify searching, and a paper with numerous authors is probably most easily located within a list by scanning the dates. Alphabetization by a fourth author, and perhaps even a third, would in any case be almost pointless, because the corresponding in-text references would have the "et al." format.*

What follows is a hypothetical collection of authors and publication dates arranged according to the standard *sequencing rules* for name-date references:

> King, F.J. (1985)
> King, H. (1969)
> Kong, P.S. (1984)
> Kong, P.S. (1986)
> Kong, P.S., Pong, R. (1985)
> Kong, P.S., Song, U. (1983)
> Kong, P.S., Song, U., Pong, R. (1980)
> Kong, P.S., Pong, R., Wang, W.E. (1986a)
> Kong, P.S., Chu, V., Wang, W.E. (1986b)

Note that the work of Kong with Song and Pong is listed before that with Pong and Wang due to the former's earlier publication date. The final two entries appear to be arranged arbitrarily, but one assumes they are actually listed in the order of their in-text citation. No other systematic rule applies in this case; the number of co-authors precludes alphabetization, and both papers were published in the same year.

---

* As previously noted (Sec. 11.3.1), "et al." is *always* substituted in citations for the names of all but the first author whenever more than three authors have collaborated on a paper, and often in the case of *only* three.

*Alphabetization* itself is usually straightforward, but even here complications can arise. This is perhaps best illustrated by some of the rules applied to family names consisting of more than one word. For example, a *prefix* ("particle") within a name, such as

de, von, von der, zu

is normally treated for alphabetization purposes as *not* being a part of the corresponding surname (unless its owner chooses to write it fused into the surname, in which case it has lost its identity as a prefix). Instead it is placed after the author's first name or initials, e.g.:

Kékulé, August von

Names beginning with "Mc" have traditionally been treated as if this were an abbreviation for "Mac", just as "St." is understood to be an abbreviation for "Saint"; neither case is looked upon as involving a prefix, so alphabetization is based on the entire (implied) first word.

*Modified letters* also warrant attention. For example, a name containing a letter with an "umlaut" ( ¨ ) should actually be alphabetized as if the modified letter were followed by an "e", since this is the meaning (in German) of the umlaut. Thus, "Bäder" would fall between "Badger" and "Bartlett," since "ä" is equivalent to "ae". Many English language publications disregard this rule, however, and simply ignore umlauts when lists are alphabetized.*

A different type of citation and alphabetization problem might appear to arise when the names most closely associated with a given document are those of editors, corporations, or organizations rather than authors in the everyday sense of the term. The solution is to treat these names as if they belonged to authors and let them serve as the basis for citation and alphabetization. Occasionally a situation arises in which no name at all can be found to characterize a document (e.g. an anonymous newspaper article). In this case

---

* Actually, the problem of alphabetization is even more complex. For example, one should expect to find a distinction between the "prefixes" in the Dutch name "De Boer" and the French name "de Guibert" particularly since one is capitalized and the other is not. Similarly, there is clearly a difference in the role of "van" as it appears in the Dutch name "van der Waals" compared to the American name "Van Slyke." Furthermore, if "Mc" is to be alphabetized as "Mac", should not similar treatment be accorded the "du" in "Du Plessis"? The French word "du" is, after all, a contracted form of "de le", where the "le" is a masculine definite article like the feminine "la" in the name "de La Fontaine". We choose to leave the ultimate solution of such dilemmas to professional bibliographers.

the name-date system becomes very awkward, although the term "Anon."—for *anonymous*—has sometimes been invoked in the past.

It should be noted that the sequence of bibliographic elements for entries in a name-date citation reference list differs slightly from what we recommend as a "standard form" in Sec. 11.4 and Appendix J. The year of publication must always be moved forward from what would otherwise be its normal position so that it falls directly* behind the name(s) of the author(s), editor(s), or issuing body. All other information is left undisturbed. The References section of the present book serves as an illustration of such a list.

## 11.3.4    Which System to Choose?

There is far from universal agreement as to which citation system is best. From the author's standpoint it is often an academic question anyway, because an editor will usually dictate the practice to be followed. Advocates can certainly be found for both systems, and every author is likely to have occasion to use both.

From the reader's perspective the name-date system might seem to be the most advantageous. Unlike an arbitrary number, a citation comprised of a name and a date conveys a certain amount of valuable information: whether the cited work is recent, for example, and who is responsible for it.[†] Many authors (and publishers) also find the name-date system more convenient because it avoids the introduction of a set of numbers highly susceptible to errors.

The name-date system does have disadvantages, however. It is more costly in terms of space utilization than the numerical system, and the presence of lengthy citations, especially if they are numerous, can prove very distracting to the reader trying to fathom a complex argument. One way to minimize disruption by name-date citations is to group them at the ends of sentences; e.g.,

> ... shown by IR, UV, and NMR spectroscopic studies
> (Arniaud, 1978; Miller, 1982; Smith, 1980; Rossini, 1980).

---

* Cf., however, the next-to-last footnote in Sec. 11.4.3.
[†] This statement requires some qualification. What one actually learns is the identity of one or two authors. The identity of a group leader whose name appears last among several authors is obviously not disclosed in a citation.

Unfortunately, this solution is antithetical to the goal of establishing a precise link between a particular assertion and its source. In the above example, for instance, it is unclear which of the three sources should be consulted for UV information. Using the numeric system, proper and precise placement of citations presents no problem at all:

... shown by IR,[8] UV,[9] and NMR[10] spectroscopic studies.

Perhaps the most one can say is that while the name-date system has important advantages, these are increasingly challenged by disadvantages as the number of citations in a document increases.

No matter which system is followed in a final document, it is convenient to use name-date citations in early drafts. Doing so certainly facilitates the author's own editing and revising; if sequential citation numbers are assigned at the beginning they become subject to revision every time the manuscript is rearranged, expanded, or condensed. Especially with a long document such revisions are extremely tedious, and mistakes almost invariably creep in.*

## *11.4    Reference Format*

### *11.4.1    Introduction*

**Basic Requirements**

A proper *bibliographic reference* is a formal description of a *literature source*. It consists of a set of prescribed *elements* presented in a more or less standardized *sequence*, and it contains all the information necessary for the *identification* and *retrieval* of a particular *document*. It is perhaps unfortunate that *reference formats*—a term used to describe the specific arrangement of elements within a reference—are not subject to an even greater degree of standardization. This is

---

* Another solution to the problem, already mentioned in Sections 11.2.1 and 11.3.2, is citation by file card registry numbers in the early stages of manuscript development.

one area in which editors and publishers have proudly maintained their independence, leaving authors with the responsibility of conforming to diverse prescriptions.*

Every reference format is an attempt to achieve an optimal balance of the same basic requirements. Thus, most would agree that the ideal reference should have the following characteristics:

- It should be *easy to comprehend.*

This requirement alone encourages the development of standard practices. The hurried reader has little interest in practicing puzzle-solving skills while screening a list of references. It should be clear in advance where within a reference each particular type of information is to be sought.

- A reference should be *unambiguous.*

There should be no doubt, for example, whether evidence to which a reference "refers" can actually be examined in a published volume of conference proceedings, or if the author's word must be accepted regarding an unpublished lecture he or she happened to hear. Likewise, a reference should clearly distinguish between a researcher actually responsible for a particular piece of information and an editor who only arranged for its publication.

- A reference should be *complete.*

It must contain information sufficient for a reader to gain direct access to the specific document in question. This means that one must make clear, for instance, whether a book to which reference is made is the second or the third edition; whether the item in question was published in Cambridge (UK) or Cambridge (Massachusetts); or whether the journal cited is *Scientia (Milano)* or some other journal called "Scientia". Completeness is thus closely related to lack of ambiguity.

- Finally, a reference should be as *short* as possible.

Brevity has the virtue of conserving printing space, but it also results in a substantial saving in time—not only for typists, but also for readers as they try to scan a list of references or interpret its contents. One step in this direction

---

* One interesting consequence, already noted above (in a footnote to Sec. 11.2.3), is the development of commercial computer software designed to convert an author's stored references into precisely the format required for any given journal.

has been international agreement that scientific journals will be designated not by their full names but rather by *abbreviated* (or *short*) *titles* extracted from the former by application of standard rules (cf. Appendix G). However, brevity and completeness are to some extent antithetical demands, and this is one of the reasons why differences of opinion exist regarding the "best" way to structure a reference.

One factor complicating the discussion of reference formats is the diverse nature of the *documents* that may need to be cited. Provision must be made for reports, theses, journal articles, and books, but it is also necessary to take into account lectures delivered at meetings, corporate and government publications, patents, translations of original works, and "documents" stored on microfilm or magnetic tape. Moreover, the peculiar needs and practices associated with scientific work in various *disciplines* impose their own referencing demands. These differ considerably, for example, from problems encountered in the humanities, although the latter serve as the source of most of the commonly taught referencing guidelines. Precedents drawn from other fields such as medicine, the social sciences, or jurisprudence are also of limited value to the scientist, and for similar reasons.

The following sections (sections 11.4.2 through 11.4.5) are designed specifically for the writer in the natural sciences. The recommendations provided are based on a careful evaluation of numerous style guides and official standards,* as well as on editorial policies developed by a number of scientific societies and publishers. The diversity one finds in contemporary referencing practice is staggering. Indeed, one of the principal outcomes of our study has been a clearer awareness of the extent of inconsistency from one publication to another with respect to both reference formats and citation techniques. The situation is somewhat alarming, though some may take comfort in the freedom it implies for the author privileged to prepare an independent document subject to no particular set of rules.

Freedom always has its price, however. An author who has meticulously prepared a list of references for an article, carefully following the guidelines of journal A, may find that the job has to be completely redone after the article has

---

* The most elaborate of these is Standard *DIN 1505, Teil 2, 1984, Titelangaben von Dokumenten—Zitierregeln.* In fact, it is so sophisticated that few science authors (or editors) will be prepared to study or apply its recommendations. It also proposes certain reference formats that contain a considerable amount of German vocabulary, so its international use is impeded. DIN (Deutsches Institut für Normung) probably has not served the scientific community well in this particular case.

finally been accepted for publication in journal B. Flexibility never extends beyond the bulwark of rules erected by a journal editor anxious to maintain consistency within his or her journal. The overall situation is far from satisfactory despite sporadic efforts to introduce reform. The scientific community, including science publishers, would do well to continue to press for standardization. The present chapter should be viewed as another attempt to highlight the problem and to plead for change, but at the same time to provide as much systematic guidance for the present as possible.

Within the narrower realm of chemical literature some significant steps toward uniformity have in fact been taken. One of the more important was the American Chemical Society's distribution of the *Handbook for Authors* (ACS, 1978) and its successor edited by Dodd (1986). We have relied heavily on ACS guidelines in the course of preparing our own recommendations, although at a number of points we have felt the need to propose extensions, elaborations, or, in the interest of consistency, minor changes.

Much of the diversity in referencing practice is a matter of detail: preferred punctuation, spacing, typeface, etc. While consistency within a single document or publication is extremely important, the particular choice made from among a set of stylistic possibilities is less crucial. For this reason we largely withhold comment on such minor matters. Our attitude is conveniently reinforced by a pair of statements found in a current draft of an international recommendation, Standard *ISO/DIS 690-1979,** with the title *Documentation —Bibliographic References—Content, Form, and Structure*: "A consistent system of punctuation should be used for all references included in a publication."; and "Variation in typeface or the use of underscoring may be used to further emphasize the distinction between elements ...".

## *Document Classification*

Bibliographers classify all documents into one of two fundamental categories: *independent*, and *dependent* (or *subordinate*). The first category includes reports, theses, books, conference proceedings, patents, standards, government bulletins, and company brochures. In contrast to these are contributions to journals, chapters in edited multi-author reference works,

---

* "DIS" is an acronym for "Draft International Standard", and it serves as an alert that at the time of publication of the present book (spring, 1987) this standard had not yet been adopted by the ISO Council. Its forerunner is *ISO 690-1975, Documentation—Bibliographic references—Essential and supplementary elements*.

individual lectures incorporated within published conference proceedings, and portions (e.g., chapters) of books, all of which fall in the second category, dependent or subordinate works. This distinction has already been noted in the opening remarks of Chapter 4 (Sec. 4.1.1).

A subordinate contribution is always contained within an independent publication (an article in an issue of a journal, for example), resulting in what might be called a "host-guest relationship" between the two. *Host document* has actually become a technical term with a formal definition (Standard *ISO/DIS 690-1979*): "A host document is a document with separately identifiable component parts which are not physically or bibliographically independent."

This definition points out what is perhaps the single most important characteristic of any bibliographic reference in its relation to a particular document.

• The document must be *identifiable*.

Rigorous establishment of an identification is the purpose of a reference, and it is this purpose that underlies the claims made earlier that a reference must be comprehensible, unambiguous, and complete. Armed only with the information in a proper reference, any reader should be able to acquire the corresponding document through a library, or order it through a book dealer (provided it is not out of print). Often it would even be possible to call up the document (or at least an abstract of it) from a computerized database.

Unfortunately, altogether too many reference lists are defective. Publishers have long since become accustomed to receiving requests from individuals, book dealers, or librarians—inquiries and orders presumably derived from defective published references—which are so obscure they cannot be accommodated. Even if the publication sought is one that originated with the publisher who has been contacted the order may still prove impossible to fill: orders are frequently received for items not listed in the publisher's *catalog* because they are not independent publications—chapters in books, for example. To put the matter another way, virtually the only publications that can be ordered directly are ones assigned their own ISBN or ISSN (cf. Appendix E), a criterion that is, to some extent, a restatement of the definition of an independent publication.* The same limitation applies if a publication is to be sought within a library collection.

---

* Independent publications—what Standard *ISO/DIS 690-1979* calls "physically or bibliographically independent units"—can be viewed as analogous to "molecules" within the

It should therefore be obvious why references must clearly distinguish between dependent and independent publications, and at the same time guide the reader from the former to the latter (from "guest" to "host"). The reference guidelines that follow were constructed with these ends in mind. In particular, we strongly recommend that *titles* of independent publications be printed in *italics* (the equivalent of underlining in a typescript). Titles of subordinate publications, if they are listed at all, are to be printed in normal upright letters, but set in quotation marks.* Other distinctions are more subtle and tend to involve punctuation or the relative placement of various elements within an overall reference.

### Depth of Referencing

Although we have stressed that references need to be complete and unambiguous, this does not necessarily mean that "more is better". On the contrary; brevity was also cited as a virtue. We may express the ideal in this way:

● References should be made proportionate.

That is to say, the *depth* of a reference should be in reasonable proportion to the purpose at hand. If an author wants to convey the message that handbook XY is a good general source of basic information, or that it provides details on a wide range of subjects, then reference should be to the handbook as a whole —even if it is a multi-volume work. On the other hand, if only a single volume of a treatise is relevant in a particular context, then it is certainly better also to specify that volume by number or title. The same philosophy can be extended to apply to cases that would require reference to an individual chapter in a book, to a range of pages, to a single page—even to a particular table on a given page.

In the humanities it is standard practice for an author to refer to a specific page in someone else's work, and even to quote directly a sentence from that

---

publication and documentation universe. Just as molecules can be broken down into atoms and even smaller elementary particles, so independent publications can usually be disassembled into a smaller or larger number of subordinate, or dependent, publications: books into chapters, series into volumes, journal issues into articles. Standard *ANSI Z39.29-1977* in this context speaks of a hierarchy of "bibliographic levels".

* This convention was proposed in Standard *DIN 1505, Teil 2, 1984*, and has been incorporated into several other standards. In the 1978 ACS *Handbook for Authors* it was stipulated that book titles should be printed in normal upright type and enclosed in quotation marks; fortunately, matters have changed, and ACS now recommends the newer style (Dodd, 1986).

page. Scientific writers are more often content simply to provide generalized references. While this fulfills the minimum requirement of giving credit where it is due, it often does a great disservice to the interested reader who actually wants to inspect the source and learn additional details. It is difficult enough to try to find a particular set of experimental conditions within a 12-page journal article. It can be extremely irritating, however, to be forced to search an entire review—often a hundred pages or more—in order to locate one specific statement. In other words, authors should accept the responsibility for ensuring that references serve the *reader's* needs, balancing the potential benefits of increased specificity against the corresponding costs—including extra time required for reference preparation and proofreading, and added printing expenses.

A reference that is highly specific must contain what bibliographers call *locators*—guides to particular sites within a document. Standard ANSI *Z39.29-1977* uses an even more refined term in this context, the *work-unit fraction*: a work-unit fraction is "a designator that indicates which item on a page, band on a sound recording disc, ... table in a book, ... etc. is being referenced".

Particular pages on which information appears can be noted in either of two ways: in conjunction with the citation (the in-text reference, cf. Sec. 11.3.2), or in the corresponding reference. If the latter course is chosen, the desired page number, preceded by "p", can be enclosed in parentheses immediately after specification of the initial page number (or page range) of the overall source.* Alternatively, a location designator may be placed at the very end of the reference, in which case it is preceded by a semicolon.

Appendix J contains examples of a wide variety of references. The styles employed are those we regard as most satisfactory, taking into consideration both the range of current practice and the recommendations of others. In the remaining sections of this chapter we examine in detail references to each of the major types of commonly cited works, providing a critical consideration of alternative views where controversy exists or where practices seem to be changing. Illustrative examples also appear in these sections, some of which will be recognized as inconsistent with the styles we have ultimately recommended in Appendix J.

---

* Note that "p 55" refers to the location of a specific statement within a document, whereas "120 pp" at the end of a review reference means that the review comprises 120 pages (*extent of work*). Another measure commonly used to imply the extent of a review article, by the way, is the number of references cited therein, e.g., "147 references".

## *11.4.2    References to Books*

### *General*

*Books* are independent publications which, to use ANSI terminology, reflect either the *monographic* or the *collective* levels of information. A proper reference to a book always contains the full, unabridged title of the work, preferably set in italics (cf. the previous section). Other essential elements of a complete and unambiguous book reference, listed in their usual order of appearance, are

● the name(s) of the author(s) or editor(s),
● the location and name of the publisher, and
● the date of publication

where the latter two constitute the *publication data.* Thus, a typical book reference might read:

> Smith, J.D., Crawford, F., *Hormonal Mediators in Bees.*
> Greenvillage: The New Science Press, 1985.

Additional examples can be found in Appendix J. Note that the above reference has been closed with a period, a style we recommend for references of all types.

There is actually more to be discussed here than might be expected. We shall therefore treat each of the various elements of a book reference separately according to the pattern established above, although even the optimal sequence is a matter of debate.

### *Names of Authors*

The *authors* in our example have been listed with their *last names* (surnames, family names) preceding the *initials* of their first names (given names). This order is usually considered preferable (and is recommended by the American Chemical Society*), although the reverse order is also sometimes found. Last names appear first because most authors are known principally by their last names. First names do serve as additional identifiers, but they are normally less important for immediate name recognition—except in the case of a name

---

* The recent work by Dodd (1986) summarizes current American Chemical Society practice. Hereafter, the abbreviation "ACS" refers to this source (unless otherwise specified).

like Smith.* The last-name-first rule *must* be followed in the case of a reference list structured and arranged according to the name-date system (cf. Sec. 11.3.3); here, as in an index, an alphabetical arrangement by last names is employed, and the structure of the list must stand out as clearly as possible. Were initials to be placed first they would tend to obscure the alphabetization. Convenience and consistency in referencing are powerful arguments in favor of routine use of a single convention with respect to names, whether citations follow the name-date system or the more common numerical system (cf. Sec. 11.3.2). Hence, we recommend *always* placing last names first in references.

The foregoing observations regarding the ordering of first and last names seem straightforward enough, but ambiguous cases sometimes arise. For example, authors of non-European heritage (e.g., the Chinese) often write what we would regard as their principal names (or family names) first, and this order should obviously *not* be reversed in a reference list. The *Chemical Abstracts Author Index* is a convenient guide if one is uncertain how a particular name should be listed. The same source can also provide help in the transcription of names originally written in an alphabet other than the Latin.

It should be noted that in our sample book reference no conjunction was placed between "Smith, J.D." and "Crawford, F." This omission was intentional. Use of a conjunction would require that changes be made in the reference if it were to be translated into another language ("and" becoming, for example, "und" or "et"), and change invites errors. Moreover, a conjunction occupies considerably more printing space than the equally suitable comma.

Recently, some editors have begun to follow the rather different ANSI recommendations, which call for use of semicolons instead of commas—and not just between authors' names, but any time several elements from what ANSI defines as a *group* (authorship group, title group, imprint group, etc.) appear together. These rules also specify that different groups are to be separated from one another by periods (Standard *ANSI Z39.29-1977*). We are not convinced of the wisdom of these suggestions. The resulting proliferation of semicolons lends an odd appearance to the text, and both semicolons and periods have other assignments more in keeping with their usual functions (cf.

---

* While first names are normally abbreviated by initials, they should be written out in full if to do otherwise would result in the loss of essential information; e.g., Smith, Harold, vs. Smith, Henry.

the examples in Appendix J).* It should be noted, however, that we have deliberately refrained throughout this chapter from being dogmatic with respect to use and placement of punctuation, because choice among various options is largely arbitrary. Indeed, modern computer-generated bibliographies are often entirely devoid of punctuation marks, an expedient that simplifies both processing and typesetting.

Sometimes in printed works authors names appearing within the text are distinguished by the use of a special type face, *small capitals* (cf. Sec. 4.3.7, "Choice of Type"). In some ways this is a useful practice; it certainly does call attention to the fact that a name is being used. In a reference list, however, the benefits are rather limited. Names abound in such a list, and the frequent repetition of a peculiar style element engenders monotony. Moreover, yet another typeface provides one more opportunity for the creation of errors. In this book we have elected to dispense with the use of small capitals entirely.

All the foregoing comments regarding the treatment of authors in book references are equally applicable to authors of all kinds of documents. We turn now to aspects of "authorship" that are more specific to books.

It is not unusual to encounter works issued under the name of an *editor* rather than its authors. Such volumes are referenced in basically the same way as those by discrete authors. The name of the editor(s) begins the reference, but this is followed by the parenthetical abbreviation "(Ed.)" or its plural "(Eds.)". A book not written in English will have its editor(s) designated by some foreign language equivalent, such as "(Hrsg.)" (an abbreviation for the German *Herausgeber*), but the English "(Ed.)" should always be employed in an English-language reference list. The same principle applies to other abbreviations common in references, such as "ed." for edition,[†] "Vol." for volume, and "p" or "pp" for page or pages; i.e., the English abbreviations should be used in an English reference list regardless of the language of the original work. *Supplementary information*, such as "2., durchges. Aufl." (German for "second, revised edition"), can properly be quoted in the original language of the document, however, since it can be regarded as part of the title.

---

* ACS recommendations seem to represent a compromise with ANSI rules—an awkward one in our judgment. Semicolons are employed by ACS for separating members within a group as well as for separating different groups from one another.

† Some publications insist that "edition" be abbreviated with the first letter uppercase (i.e., "Ed."), and the period following it is occasionally omitted ("Ed"). Similarly, "editor" is occasionally abbreviated "ed." It is somewhat surprising that Standard ANSI *Z39.29-1977* makes no distinction between the two; "ed." is used for both "editor" and "edition".

There would appear to be a problem raised by those works—such as encyclopedias or handbooks—that make no obvious mention of either authors or editors. Usually the absence of names means that a work has resulted from the joint efforts of many. Examples include the *CRC Handbook of Chemistry and Physics* and *Ullmann's Encyclopedia of Industrial Chemistry*. A reference to a work of this type simply begins with the title, and the title dictates placement of the reference within an alphabetized reference list.

### Title

The *title* of a document has been rather loosely defined in Standard *ISO/DIS 690-1979* as a "word or phrase, usually appearing on the document, by which it is convenient to refer to it, which may be used to identify it, and which often (though not invariably) distinguishes it from any other document."* Whenever possible, the title employed in a book reference should be precisely that which appears on the main title page (title leaf) of the book itself. Authors who take this suggestion seriously will endeavor to inspect a copy of every book they cite. This advice may seem onerous, but it is strictly consistent with the high ethical standards expected of a scientific writer.

Often the title of a book consists of a main title and a *subtitle*, the latter "completing the title proper of a document appearing on the title page" (Standard *ISO/DIS 690-1979*). Both main and subtitle should be quoted in a reference, with the two separated by a dash or a colon. Titles should always be given in their original languages, but a title appearing in an alphabet other than the Latin may be replaced by a *transliterated* version, or a transliteration may be added (in parentheses) to the original. If for some reason a title is translated (or even if it is only transliterated), proper referencing practice requires that it be followed immediately by a statement such as "In Russian".

The title page of a book sometimes contains other information required in a reference, such as an edition number or specification, a volume number and series title if the book is part of a set, or the names of those who held important *subordinate responsibilities* in the creation of the work (e.g., a translator). Several examples of complex book references containing such information are to be found in the next subsection and in Appendix J.

---

* Interestingly, Standard *ANSI Z39.29-1977* essentially ignores the challenge of defining the bibliographic meaning of the word "title", choosing instead only to *use* it to create a series of subcategories such as "Analytic Title", "Parallel Title", and "Monographic Title".

Titles of books are usually written with the first letter of each significant word *capitalized* (cf. Sec. 4.3.7, "Miscellaneous"). This includes the *first* and *last* words, regardless of their "importance". Standard *ANSI Z39.29-1977* also permits the alternative of capitalizing only the first letter of the first word of a title. The latter mode is always followed in the case of journal article titles (Sec. 11.4.3); we prefer that the convention of more complete capitalization be followed for book titles, thereby introducing a subtle means of differentiation.* The fact that book titles should be set in *italics* has already been discussed in Sec. 11.4.1 under the heading "Document Classification"; the rationale appears there as well.

### Publication Data

The publication data required in a book reference consist of three major elements: the *place* of publication, the *name* of the publisher, and the *year* of publication. These normally appear in the sequence given, a practice consistent with both ISO and ANSI recommendations.†

The *name of the city* in which the book was published should be recorded exactly as shown in the source; i.e., it should not be "translated" into English. Further specification in terms of a state or country (perhaps in abbreviated form) is often desirable, particularly where this is required to distinguish the city of publication from others of the same name (e.g., London, England, and London, Ontario) or in the case of a city that is not widely known. If the source itself indicates more than one place of publication it is considered sufficient to name only the most prominent; little is accomplished by reciting all the locations in which a large publishing conglomerate has subsidiaries. There is no cause for alarm, incidentally, if a reference list is found to contain various titles from a single publisher associated with different cities; an apparent

---

* Titles of documents in languages other than English are to be cited as they would be in the original language (i.e., just as they appear in the document itself). The capitalization rule given here applies to English, and it is not necessarily appropriate for publications in other languages.

† Current ACS practice is to name the publisher before noting the place of publication, and this actually seems more logical to us. After all, it is the identity of the publisher that is of importance to someone wishing to order a book, not where the publisher's offices happen to have been located at the time of publication. Publishers not infrequently move from one city (or even country!) to another, and some maintain offices in several locations simultaneously even though they issue but one catalog of all their titles. Nevertheless, the ISO and ANSI standards carry official sanction, and they should probably be honored.

discrepancy of this type is likely to be a logical consequence of one of the factors just discussed.

According to Standard *ISO/DIS 690-1979*, the *publisher's name* "shall be given either as it appears in the source or in a shortened or abbreviated form, provided there is no ambiguity introduced by abbreviating." Corporate notations such as "Ltd.", "Inc.", "GmbH", etc., can be omitted entirely. The aim is simply to ensure clarity with respect to the publisher's identity in order that a librarian or book dealer can easily secure a complete mailing address and successfully order the book.*

Occasionally, a book is *co-published* as a joint venture by two (or even more) publishers. In such cases it is correct to use either the name of the most prominent publisher or that of the original copyright holder. Another possibility is to list the publisher holding distribution rights in the market of primary interest to the recording author. If two or more names seem to be given equal prominence it is sufficient to list only the first.

The *year of publication* is quoted in Arabic numerals exactly as shown in the source. Reference to a book that has gone through two or more *editions* should always be to one particular edition (usually identified by number),[†] and the year listed should correspond to that edition. Publication of multipart works typically extends over more than a single year, and reference to such a set should note the appropriate inclusive dates (e.g., 1983-85). If publication has not yet been completed, the initial year is given, followed by a hyphen and a space.

The sample reference below illustrates the treatment of a rather complex case. The example is based on the rules outlined in Standard *ISO/DIS 690-1979*:

> Lominadze, D.G., *Cyclotron waves in plasma.* Translated by
> A.N. Dellis; edited by Hamberger, S.M. 1st ed. Oxford:
> Pergamon Press, 1981. *International series in natural philosophy.*
> Translation of: *Ciklotronnye voluy v plazma.*
> ISBN 0-08-021680-3.

---

* In most countries there are national catalogs that contain such addresses (e.g., *Books in Print* in the United States, *British Books in Print* in the United Kingdom, and *Verzeichnis lieferbarer Bücher* in Germany). The *CASSI Cumulative 1984* contains a listing (with addresses) of all publishers responsible for serials in chemistry and related fields (cf. Sec. 11.4.3). Publisher's addresses can also be found in the databases of computerized book ordering services.

† Terms such as "second" or "third" (i.e., ordinal numbers) should be expressed as their Arabic numeral equivalents; e.g., 2nd, 3rd, 18th.

Note that the book's ISBN (cf. Appendix E.2) has been included at the end as a way of eliminating all ambiguity. In this case only the first word of the title has been capitalized simply to reinforce the message that both capitalization conventions are permitted (cf. "Title", the subsection preceding this one).

### Chapters within Books

Standard *ISO/DIS 690-1979* defines a *chapter* as a "numbered or titled division of a written document which is generally self-sufficient but stands in relation to the divisions that precede or follow it." Chapters, like articles or other brief contributions, are subordinate (guest) publications, and often in scientific monographs they have their own authors. It is these authors who assume first place in a reference to such a contribution. Typical examples of volumes composed of chapters by various authors include edited multi-author works and conference proceedings. A corresponding reference might read as follows:

> Jonathan, A.B., in: Saul, C.D. (Ed.), *Krypton Chemistry.*
> Athens: Academia, 1986; Vol. I, Chap. 10.

As in the above example, it is not necessary to quote the full title of a subordinate publication; often the name(s) of the author(s) and perhaps a chapter number will suffice. However, if the title is made a part of the reference it should be set in quotation marks.

Standard *ISO/DIS 690-1979* grants considerable freedom in structuring references to subordinate publications: "The physical relationship between the guest and the host may be expressed by means of a word such as 'in', or it may be made implicit by means of punctuation, typeface, etc."

### Series and Multi-volume Works

Sometimes occasion will arise to make reference to literature on what ANSI calls the *collective level*—an entity comprised of more than one physically discrete unit. A distinction is usually made between a *series*, by which is meant a set of monographs with an unlimited number of volumes, and a *multi-volume work*, a set whose number of volumes is limited because each represents a defined fraction of the whole, one component within some clear and finite scheme.

A sample reference for one volume in a multi-volume work might read:

> Kuiper, G.P., Middlehurst, B.M. (Eds.), *Planets and Satellites*. Chicago: University of Chicago Press, 1961 (Kuiper, G.P., Ed., *The Solar System*, Vol. 2).

If one wished to refer to a particular chapter in *Planets and Satellites*, the reference would be further elaborated; e.g.,

> Minnaert, M., "Photometry of the Moon". In: Kuiper, G.P., Middlehurst, B.M. (Eds.), *Planets and Satellites*. Chicago: University of Chicago Press, 1961 (Kuiper, G.P., Ed., *The Solar System*, Vol. 2).

The preceding example has been taken, with minor modification, from Standard *ANSI Z39.29-1977*. It nicely illustrates the ANSI-recommended strategy for preparing a complex reference: information at the *lowest* possible level appropriate to the situation at hand is cited *first*, followed by higher level material arranged in an ascending sequence corresponding to the hierarchic order of the levels.* Thus, the chapter is defined first, and primary credit is thereby awarded to the actual author of the information cited. This is followed by identification of the volume, and then of the collective work. Structuring within and among the levels is accomplished by careful choice and placement of punctuation, type style, and adjuncts such as "In:" (or "in:"; cf. the example under the previous heading "Chapters in Books").

## 11.4.3    References to Journals and Other Serials

Contributions to *serials*, including *journal articles*, resemble book chapters in that they are dependent (subordinate) publications. A reference to a serial contribution may or may not report the title of the article. Indeed, it is customary to omit the titles of journal articles both to save printing space and to simplify the preparation of references. This is unfortunate, since the information loss is considerable.

The title of the serial itself must appear, of course, but the requirement for conservation of space also enters the picture here. Convention dictates that serial titles be reported in an abbreviated form known as the *serial short title* [Standard *ISO 4-1984(E)*]. Standardized abbreviations have been adopted, and

---

* The only exception is a work-unit fraction (such as "p 100") which, because it is devoid of separate authorship, is usually specified at the very end of a reference.

these are exceedingly well documented. A selection of accepted abbreviations for some of the words most commonly encountered in journal titles is presented in Table G-1 of Appendix G.

Chemists are fortunate in having available a particularly comprehensive and regularly updated guide to the vast periodical literature of their discipline: the *Chemical Abstracts Service Source Index* (*CASSI*), a publication of Chemical Abstracts Service. The scope of *CASSI* is remarkably broad, extending well beyond the strictly chemical sciences to include large segments of the biological, engineering, and physical sciences. Moreover, *CASSI* coverage is not limited to journals, but includes annual series of the "advances" type (e.g., *Advances in Inorganic Chemistry*), various government bulletins, publications emanating from more or less regular scientific gatherings, and a number of other sources, some of which would not normally be considered serials.*

Sources in *CASSI* are listed alphabetically by title. Each title is followed by relevant publication data and a selective list of libraries subscribing to the publication. Those portions of the title that are to be used to form the serial's *abbreviated title* (or *short title*) are printed in boldface type. Thus, *CASSI* becomes a convenient place to find such abbreviations.†

The abbreviated title is the only descriptor used in a reference to a serial. Additional descriptors, such as the publisher's name and location and the identity of the editor, are considered superfluous since, in principle, this information can be obtained from a work like *CASSI*. Other possible sources include *Ulrich's International Periodicals Directory*, published by Bowker (New York). Abbreviated titles could also be used to reference certain edited monographic sets and multi-volume works, because many of these appear in *CASSI*. This practice is rare, however.‡

---

* The *CASSI 1907-1984 Cumulative* also contains an index of all publishers of chemistry-related serials. This is a handy collection of information not otherwise easily accessible.

† In fact, alphabetization in *CASSI* makes use of only those parts of a title that are printed boldface. As a consequence, one must be able to anticipate how a periodical title might be abbreviated in order to look it up. Official short titles are, or at least should be, printed on the serials themselves; hence, these may occasionally be more convenient sources than *CASSI*.

‡ Using a *CASSI* abbreviated title for such a source (*Biotechnology*, for example) obscures the fact that the reference in question is actually comprised of only a limited number of volumes published over a relatively short period of time. An unsuspecting reader seeing such a title would be likely to assume the source was a perodical (serial) in the usual sense of the word, and would automatically seek it in the wrong places. Also, the absence of publication data would make the task of ordering the publication more difficult.

Official short titles of some of the major journals in chemistry and related areas are listed in Table G-2 of Appendix G. Note that the individual abbreviations and words comprising a short title are always separated from each other by spaces.

A reference to a journal article may take the following form:

Falstaff, M., *J. Phys. Chem. 80* (1976) 1739-42.

The italicized portion is the official abbreviation of the journal title. Notice that the volume number is also set in italics, since it functions as part of the title.* The year of publication of the cited volume, 1976 in this case, is appended in parentheses, presumably a subtle way of acknowledging that the information is redundant; a given volume number is virtually always associated with a single year. However, this particular redundancy is tolerated, primarily because the average reader is unlikely to know the relationship between the year and volume number of a journal, but may well be concerned to know how current a given reference is. Some journals are denoted only by year rather than by volume number, in which case the year must be italicized (e.g., *J. Chem. Soc. 1924*).

The numerals following the year identify the first and last pages of the cited article.† Some serials are unusual in that each issue is paginated separately (i.e., each begins with "page 1"). References to publications of this type must include specification of the issue number in addition to the volume number. Issue numbers are usually set in parentheses (but not italicized) and placed immediately after the volume number.

Unfortunately, the style and sequence of information recommended above —volume number, year of publication (in parentheses), page number—has not been universally adopted. ACS journals, for example, employ a reference format in which the year of publication precedes the volume number and is set

---

* It was once common to single out the volume number by setting it in boldface rather than italics, and this practice is still occasionally followed.

† Until recently, few references included the number of the last page, but the practice is becoming more widely accepted. The information is especially useful to the reader interested in knowing the length of the work before deciding whether to read it in a library, photocopy it, or write to the author for a reprint.

in boldface.* The following example is taken from Dodd (1986):

> Stinson, S.C., *Chem. Eng. News* **1985**, *63*(25), 26.

The ACS sequence has been accepted as an alternative by Standard *ANSI Z39.29-1977*, which also permits further abbreviation of journal article references in either of two ways:

> Libr. J. 99: 2573-2578; 1974.
> Libr. J. 1974; 99: 2573-2578.

For many decades it was customary to list the year of publication *last* in a serial reference, as is the case with books. The change to placement *between* the volume number and the page number (as in our first example) was incorporated by the Deutsches Institut für Normung into its Standard *DIN 1505, Teil 2* issued in January, 1984.

References in publications that use the name-date system must be arranged somewhat differently because of the key role played by the year of publication. Thus, the year is moved forward until it immediately follows the name(s) of the author(s); i.e.:

> Song SP, PD Walton 1975 Inheritance of leaflet size and specific
> leaf weight in alfalfa. Crop Sci 15:649-52.

Notice the unusual spacing and punctuation (as well as the lack of italicization of the journal short title) in this example drawn from a recent issue of *Plant Physiology*.†

Whenever possible, *all* names of listed authors should be cited, in the order in which they appear in the original publication. Some published standards (including ANSI) do permit the use of "et al.", but suppression of an author's name in a reference amounts to withholding credit where it is due, clearly an impolite gesture.

---

* The year of publication is a useful piece of information, and its placement so near the middle of a reference tends to make it somewhat hidden. ACS counters this danger with use of boldface type for the year; we regard the ACS sequence as unsatisfactory for numerical references in the absence of boldface. When references are arranged by the name-date system the ACS recommendations are not unreasonable (again assuming the availability of boldface). Indeed, their "bifunctionality" in this sense is an argument in their favor.

† Actually, this particular journal is one of the few examples of a publication that employs the hybrid alphabetical-numerical citation system described at the end of Sec. 11.3.2; however, the arrangement of its references is the same as it would be in the name-date system.

As already mentioned at the beginning of Sec. 11.4.3, references to serial contributions are sometimes expanded to include *titles* of articles. If so, these should be set in normal type but placed in quotation marks (cf. Sec. 11.4.1, "Document Classification"). Only first words (and, of course, names) are capitalized in the case of English-language titles.

Even more inclusive reference formats are occasionally encountered (e.g., ones containing authors' affiliations), as are ones in which the components are arranged differently (e.g., title-first references). Standard *ANSI Z39.29-1977* in particular permits a surprising amount of latitude, extending to approaches that many would regard as rather bizarre. Appendix J contains a wide variety of examples of journal references constructed according to more conventional rules. Those shown have been chosen to cover not only the situations described above but also certain specialized and more complicated cases.

## 11.4.4    Patents

A *patent* is analogous in some ways to a copyright (cf. Appendix C): it is a legal guarantee of an inventor's right to control the way an invention will be used. "Invention" is a technical term in patent law encompassing not only mechanical, electrical, or electronic devices, for example, but also newly-discovered chemical processes, or applications of chemical compounds. In order to obtain a patent an inventor must disclose in writing the details of the invention, or at least give the appearance of doing so, and it is for this reason that patent documents are of potential interest to chemists and others. Indeed, patents are often the only published sources of important scientific information.*

In bibliographic terms, *patent documents* are considered a part of the serial literature, and they should be referenced accordingly.† This generalization applies not only to the final documents upon which legal patent rights rest, but also to *patent applications* and similar documents.

---

* It should be pointed out, however, that information derived from patents must be regarded as less reliable than that from the standard scientific literature. Patents are often less than candid in revealing full details, and many apparent "procedures" described in patents are in fact hypothetical; e.g., they may be *plausible extensions* (for which patent protection is sought) of what is known, rather than factual reports of what has actually been observed.

† A cursory search of the literature nevertheless reveals a bewildering variety of patent reference formats, presumably because the serial nature of patents is not well understood.

The specification of a patent document *type* together with a *country* is equivalent to giving the title of a journal. When these elements appear in a reference list they should therefore be set in italics. By convention, the country is named first, so a typical patent reference might contain the "title"

*Germany (Fed. Rep.) Offenlegungsschrift.*

However, the combination of country and document type is only one of the *essential* elements in a patent reference. Two others are the *patent document number* and the *patent-issue date*. The shortest permissible form of a patent reference prepared according to ANSI specifications is illustrated by the following example:

*U.S. Patent* 3 654 317. 1972 April 4.

Sometimes the exact date of issue is omitted in favor of simply stating the year.

While these four pieces of information—country, type, number, and date— fulfill the minimum requirements for an unambiguous patent reference, there are several additional pieces of information that a reader might find useful. These include the name(s) of the *author(s)*—here actually the *inventor(s)*—of the document, the identity of the *assignee* (usually a company or institution to which the right to exploit the patent has been delegated), and the descriptive *title* of the document itself. Standard *ANSI Z39.29-1977* classifies these elements as *recommended* components of a patent reference.

Still other elements could be included, but they are truly *optional*.* One example is the *International Patent Classification* (IPC) code. This cryptic combination of numbers and letters (e.g., "Cl A01N9/36") immediately tells an expert to which field of technology a particular patent belongs. *Chemical Abstracts* usually includes IPC codes in its patent summaries. More important to the general reader than an IPC code is a *Chemical Abstracts* reference, since this provides ready access to more information about patent content, e.g.,[†]

Niimoto, S. *Japanese Patent 56 167 650*, 1981,
Nippon Kayaku Co. Ltd.; *Chem. Abstr.* **1982**, *96*, 162 305.

---

* The essential/recommended/optional classification of bibliographic elements can be applied to other types of references as well. The distinction has been introduced here only because this is a case in which its significance is particularly clear.

[†] Note that *Chemical Abstracts* references contain *abstract numbers* rather than page numbers.

This example is based on the format recommended by ACS in the 1978 edition of its *Handbook for Authors.\** *Chemical Abstracts* employs a somewhat shorter notation for patents, one in which both the country of issue and the patent type are abbreviated. The most commonly encountered *CA* abbreviations are presented as the middle column of Table 11-1. A more complete list appears in the "Introduction" section to each volume of *Chemical Abstracts* (e.g., *Chem. Abstr.* **1984**, *101*, xii).†

Use of the abbreviations listed in Table 11-1 as *patent short titles* would be analogous to naming journals by their *CASSI* short titles (cf. Sec. 11.4.3). If, as we have implicitly recommended, *CASSI* is accepted as the authoritative guide to titles in the serial literature, then *CA* rules should also be followed for patents, because they are also part of the serial literature.

*Chemical Abstracts* summaries typically contain two-letter ISO country codes (drawn from Standard *ISO 3166-1974*) appended to the patent short titles; e.g.,

> *U.S.* US 4 393 718

This style involves unnecessary redundancy, but we have nonetheless included country codes as the last column of Table 11-1.

The patent literature is severely complicated by the fact that patent requests normally pass through various public stages of evolution prior to final approval (e.g., "unexamined applications", "examined applications", etc.). Several nearly identical "publications" result. Furthermore, it is common for a given invention to be the subject of patent proceedings in numerous countries, leading ultimately to "parallel patents". For these reasons, what are called "patent families" arise, sets of documents that all deal with the same subject in virtually identical ways. Usually it is sufficient to refer to only one member of such a family, preferably the "parent" document. An interested reader could easily locate other members of the same family by consulting the *Chemical Abstracts Patent Index*. However, situations can arise in which more than one document must be cited, such as in an argument based on a comparison of parallel patents or a discussion of the evolution of a given patent.

---

\* More recently it has been suggested (Dodd, 1986) that adjectives referring to countries be abbreviated (e.g., "Br." for British). However, no general rule for country abbreviations—such as use of uniform country codes—seems to have been adopted by ACS.

† *Chemical Abstracts* currently covers patent documents issued in 27 countries. A more specialized patent documentation service is the *Chemical Patent Index*, distributed by Derwent Publications Ltd., London (cf. Appendix D.3).

*Table 11-1.* Abbreviations used by *Chemical Abstracts* in patent abstract headings, along with corresponding ISO country codes (cf. Standard *ISO 3166-1974*).

| Country and type | Abbreviation<br>(*CASSI* short title) | Country code |
|---|---|---|
| Australia (examined application) | *Pat. Specif. (Aust.)* | AU |
| Austria | *Austrian* | AT |
| Belgium | *Belg.* | BE |
| Canada | *Can.* | CA |
| Czechoslovakia | *Czech.* | CS |
| Denmark | *Dan.* | DK |
| European Patent Organization | *Eur. Pat. Appl.* | EP |
| France (patent) | *Fr.* | FR |
| German Democratic Republic | *Ger. (East)* | DD |
| Germany, Federal Republic of<br>(unexamined application) | *Ger. Offen.* | DE |
| Germany, Federal Republic of<br>(examined application and patent) | *Ger.* | DE |
| India | *Indian* | IN |
| Israel | *Israeli* | IL |
| Japan (unexamined application) | *Jpn. Kokai Tokkyo Koho* | JP |
| Japan (examined application) | *Jpn. Tokkyo Koho* | JP |
| Netherlands | *Neth. Appl.* | NL |
| Norway | *Norw.* | NO |
| Poland | *Pol.* | PL |
| Spain | *Span.* | ES |
| Sweden | *Swed.* | SE |
| Switzerland (patent) | *Patentschrift (Switz.)* | CH |
| Union of Soviet Socialist Republics | *U.S.S.R.* | SU |
| United Kingdom (examined application) | *Brit.* | GB |
| United States of America<br>(unexamined application) | *U.S. Pat. Appl.* | US |
| United States of America<br>(patent) | *U.S.* | US |

## *11.4.5    Miscellaneous Sources*

Citations are occasionally necessary for sources other than books, journals, and patents. One should nevertheless refrain from citing inconvenient sources if comparable information is available in the more traditional print media.

*Lectures*—e.g., a "paper read at the XY Conference"—are a case in point, and they also exemplify the category sometimes called "gray literature". This term refers to material that in a sense is or has been public, but has not been widely distributed and is unavailable in most libraries. It also includes works of which only a small number of copies have been produced.

Lectures, in particular, are often cited in the scientific literature, because the first announcements of many major research findings occur at conferences attended primarily by specialists. Sometimes such oral presentations are subsequently made available to the public as part of a volume of *conference proceedings*. If so, the individual contributions can be treated exactly like chapters in a multi-author book (cf. Sec. 11.4.2). However, many conferences are handled more informally. Typically, only a set of "abstracts of papers" is printed, for distribution primarily to those who attend the conference and perhaps to others who specifically request them, or who are on the sponsoring organization's mailing list.

"Publications" of this kind are amenable to citation, but the corresponding reference must be exceptionally thorough. For example, it should include precise information about the date and location of the meeting, as well as unambiguous identification of the sponsoring organization. A calendar year is often an integral part of the name of a conference, in which case the year usually constitutes part of the title of the corresponding "proceedings" volume as well. Nevertheless, the actual issue year of the abstracts must also be included in a reference, because the two dates may not coincide. Sample references to conferences and conference proceedings are included in Appendix J.

It is also permissible to cite material that has been presented at a meeting but never released in any printed form. A strictly orthodox interpretation of the term "reference" would certainly hold that the corresponding acknowledgment belongs not in a reference list but rather in a footnote. This would be consistent with the treatment of other unpublished information, such as "personal communications". However, in recent years editors have become progressively less legalistic about the definition of a "publication". As a result, it has become common practice to treat references to lectures presented at conferences and even small symposia just like references to the regular literature, even if the information never actually appears in print.

The author wishing to cite a lecture should feel a special obligation to provide as much accurate and pertinent information as possible. The following example illustrates how such a reference might be constructed:

> Hamprecht, B., lecture presented at the meeting of the Deutsche Forschungsgemeinschaft—Schwerpunktsprogramm Biochemie des Nervensystems. Max-Planck-Institut für experimentelle Medizin, Göttingen, 13-14 December 1984.

*Theses* quite properly class as independent publications, even though they may not be readily accessible to most readers. The awarding university assumes the role of "publisher" in any reference to a thesis. Whenever possible, thesis references should include University Microfilms International registration numbers (cf. Sec. 2.3.3) since this facilitates access to copies of the work. Examples of thesis references are given in Appendix J.

*Reports* qualify as public documents for referencing purposes only if they have been submitted to a recognized agency, organization, or sponsoring body, in which case they are treated as monographs. Reference to a report should contain any applicable identifiers, such as *acquisition numbers* or *report numbers,* appended to the titles of the documents. Copies of many official reports issued in the United States are available from the National Technical Information Service (NTIS)* in Springfield, Virginia. Reports in the NTIS collection bear catalog identifications (numbers preceded by the letters "AD" or "PB"), and these should also be incorporated in references whenever possible.

The agency to which a report was originally submitted must always be clearly identified. The following sample reference to a report in a language other than English illustrates some of the potential complexities:

> Duelen, G., *Mathematische Grundlagen für die Steuerung von Industrierobotern.* Fraunhofer-Institut für Produktionsanlagen und Konstruktionstechnik. Karlsruhe (Germany, FRG): Kernforschungszentrum Karlsruhe, 1982 (KFK-PFT-E6).

If the research reported here had been carried out under contract to a

---

\* NTIS operates under the sponsorship of the United States Department of Commerce, Washington, D.C.

government ministry or some other major institution, even more information might have been required. Only the author who has taken the trouble to obtain such a report will be in a position to decide what data are "most likely to be used to obtain the report" (quoted from Standard *ANSI Z39.29-1977*). Whatever their nature, however, these data should be made part of the corresponding reference.

A document classed as a report is much easier to deal with if it happens to belong to a series to which a "CODEN" has been assigned. CODENs are a creation of the American Society for Testing and Materials (ASTM), and they are widely used by *Chemical Abstracts* and others as shorthand equivalents for periodical (serial) titles (see also Appendix E.1). A CODEN usually consists of six capital letters, often chosen so that it is possible to guess the full title, and it provides unique identification of a particular serial. Any CODEN that applies to a cited report should be incorporated into the report reference. Information should still be provided regarding the report's origin, however, particularly the name, city, and country of the issuing agency. In the interest of better serving the reader it is even a good idea to provide an exact mailing address. Again, examples are to be found in Appendix J.

Advice similar to that given for reports applies to references to *company brochures*, *trade catalogs*, and similar documents. Every effort should be made to provide sufficient information to permit the reader who so desires to acquire a copy of the original document. In this case the name of a company takes the place of an author or editor. The date of issue, place of publication (i.e., location of the company), and document or catalog number are all important, and these should be specified with as much precision as possible.

It is an intriguing fact that the standards developed for bibliographic references are largely silent on the question of the way standards *themselves* should be referenced. The 1984 issue of Standard *DIN 1505, Teil 2* is an exception; it at least touches on the matter, recommending that all such references begin with the word "Norm". Apart from substituting the equivalent English word "Standard" we have based our own recommendations on this source, which would itself be fully referenced as:

> Standard *DIN 1505, Teil 2, 1984. Titelangaben von Dokumenten—Zitierregeln.*

Other examples are provided in Appendix J. Placing the word "Standard" in front of the reference has the effect of grouping all such references in an alphabetical list (cf. the References section in this book).

Following the introductory word ("Standard") specifying the *type* of document is a shorthand notation comparable to an abbreviated journal title (*DIN* in this example). This notation, usually a combination of three or more capital letters (an acronym, cf. Appendix H), is a brief *identifier* of the issuing authority, in this case the Deutsches Institut für Normung (Federal Republic of Germany, FRG), and it is also a trademark. Other similar and frequently encountered acronyms of this type include:

ANSI  American National Standards Institute (USA)
BS    British Standard (UK)
ISO   International Organization for Standardization

Since these acronyms are comparable to journal short titles they should be printed in italics. The same holds true with respect to the *standard number* (see below), which loosely resembles the volume number of a journal. While it is reasonable to hope that the shorthand notations listed above will be generally understood, others—such as the acronyms for less familiar technical societies or institutes—certainly will not. In such cases it is probably best to spell out completely the names of the organizations issuing the standards or recommendations.

The remaining essential elements in a reference to a standard are the *number* of the document and the *date* or *year* of its publication, along with the *edition number* (except in the case of a first edition). The complete *title* of the standard, again printed in italics (since a particular norm is clearly an independent document), is a "recommended" element. Further information, such as the location of the issuing authority and the name and location of the publisher (if any), is optional.

Day (1983), in his "References" section, adopts a format for references to standards that includes all of these additional data, e.g.:

**American National Standards Institute, Inc.** 1969.
American national standard for the abbreviation of titles of
periodicals. ANSI Z39.5-1969. American National Standards
Institute, Inc., New York.

The reference is structured like a book citation, in contrast to our approach which more nearly resembles a journal reference. Thus, the full name of the issuing authority appears first, effectively serving as the editor of the standard. Day chooses to set this name in boldface, consistent with his treatment of authors and editors in general. This is strictly a matter of preference; small capitals or normal type would be equally appropriate. The name of the issuing authority, along with the place of publication, is repeated at the end of the reference—this time to indicate the publisher (here identical to the "editor"). The entire reference is arranged in the manner required by the name-date system.

A source that could not be considered "published" in any of the ways discussed above almost certainly should not be included in a reference list. Instead it should be acknowledged in a *footnote*, e.g.:

> \* Smith, J., personal communication to the author.
>
> † Author, A.B., unpublished results.
>
> ‡ Author, A.B., Coworker, C., submitted for publication.
>
> § Author, A.B., Tables of atomic function. Tables supplementing the author's paper in *IBM J. Res. Dev.* **1965**, *9*, 2 are available from the author on request.

As mentioned previously, some would hold that a footnote is also the only proper place to refer to an unpublished lecture.

There is only one major exception to the generalization that reference lists are reserved for *publications*: a manuscript that has already been submitted *and accepted* for publication can be referenced in the usual way, complete with all currently available information, even though the published document has not yet appeared. The fact that a source is of this type must be made explicit, but otherwise it is treated as if it *were* published:

> Galen, H.C., et al., *J. Med. Chem.*, in press.
>
> Knott, B., *J. Irreproducible Results*, accepted for publication.

*Appendixes*

# Appendix A
# Oral Presentations:
# Organization and Visual Materials


## A.1 Spoken and Written Presentation of Scientific Results


Although the principal purpose of this book is to provide practical advice on the *written* presentation of scientific information, we were reluctant to neglect completely the related topic of *oral communication*.* Today, opportunities for formal oral presentations are more plentiful than ever, particularly talks dealing with research results. Even students are expected to present their work in seminars,[†] and senior scientists regularly address their peers at *scientific meetings*.[‡] Indeed, it is generally accepted that at least occasional active participation in such meetings is an obligatory part of a scientist's "continuing education", and employers tend to be quite sympathetic to requests for the required "time off".[§] One alternative mode of participation eliminates the need for a formal talk: the "poster session", discussed in Sec. A.5.

---

* Oral presentations are also the subject of chapters in Booth (1985), and by L. Venable in Dodd (1986).

† The topic of student seminars is addressed in a concise but effective article by K.J. Laidler (*J. Chem. Ed.* **1971**, *48*, 671-674).

‡ With respect to this subject we highly recommend a recent book by Neuhoff (1986); it offers both informative and amusing "further reading".

§ Less certain, however, is the availability of *financial support* to attend such a meeting— unless one is delivering a paper. This circumstance has the unfortunate consequence that some oral presentations arise more out of a desire to obtain a travel cost reimbursement than from an inner conviction that others deserve to be apprised of new and important information. However, a lecturer's audience should not be forced into an awareness of such questionable motives. Listeners have a right to expect more than a dull and unrewarding lecture, poor because it was poorly prepared.

Effective oral presentation demands an approach entirely different from that appropriate for a written report. The differences are partly due to the fact that the spoken word—unlike the written word—is designed only for *immediate* consumption. After its last echo, the spoken word can no longer be called forth for review and further study.

Nevertheless, written and oral communication do have certain characteristics in common, and these are worth noting. Every lecture—regardless of whether it occupies 10 minutes or a full hour—should resemble a good written report in the following ways:

- It should have a logical *structure*, including a clear division into *sections*.
- Its subject must be set within a meaningful *context*.
- Its *content* should be restricted to what is relevant.
- All thoughts must be conveyed *intelligibly* and with *precision*.
- The *development* should proceed from one point to the next without unreasonable gaps.
- The *level* of presentation should be tailored to the expected audience.

However, it is the *differences* between spoken and written communication that have dictated inclusion of this appendix. These differences are perhaps best illustrated by the following set of practical rules, applicable to nearly any lecture:

- A lecturer should begin with a carefully crafted *introduction* in order to ensure that (captive!) listeners will eagerly anticipate what is to follow.
- The *structure* of the lecture should be revealed at the beginning, including the nature and role of its individual segments. This will enable the listener to envision what is to come, and also maintain a sense of the progress of the talk.
- After the scene is carefully set, the *methods* applied and the *results* achieved should be sketched as concisely (but clearly) as possible.
- The speaker must ensure that *discoveries* remain fresh in the listener's consciousness when the time comes to discuss their significance.
- Lengthy excursions into *experimental details* should be avoided unless they are absolutely essential for understanding the nature of the results. Only in a talk whose principal subject is techniques is it sensible to present a thorough exposition of individual experiments.
- It is pointless to include vast amounts of *data* in a lecture, because the facts will quickly be forgotten. A few well-chosen examples are far more illuminating than an exhaustive catalog of details.

- Primary stress should be laid on *generalizations* the listener should remember.
- Frequent *repetition* of key points is essential; so is a clear explanation of underlying *concepts*.

The final injunction in particular is a sharp contrast to the advice given for a written report, where redundance is frowned upon and the reader is the one expected to take the initiative if background review is required.

Because of the strictly *sequential* nature of orally transmitted ideas, results and discussion should be intermingled. A formal separation of the two is only appropriate in a context that permits recipients to move freely back and forth between the corresponding sections. Listeners, unlike readers, have no way of re-examining what came before, so "sections" of the familiar kind lose their effectiveness.

# A.2    Preparing a Lecture

Everyone brings his or her own particular set of rhetorical gifts and past experiences to the preparation of a lecture. Consequently, each speaker prepares a lecture in a unique way, in most cases conditioned by the nature of the assignment. This severely limits the extent to which we can provide meaningful generalizations, although there is merit in stressing (as in the foregoing paragraphs) the special nature of the lecture as an *information-transfer medium*. What follows is a set of observations that every would-be lecturer should consider. We hope these observations will stimulate some to undertake a bit of critical self-analysis of their speaking habits; a heightened awareness of personal weaknesses can be a powerful incentive to working toward their correction.

If one plans to use either notes or a complete text during the delivery they should be prepared in an especially legible form. Nervousness can cause an otherwise perfectly legible text to become meaningless hieroglyphics, particularly if (as often happens) the lighting is poor. All text should be at least double-spaced, and points that represent the introduction of a new idea, the beginning of a new section, or an especially important conclusion should be highlighted to make them readily visible (for example, with color). Many speakers also insist that their manuscripts be prepared with extra-large type. It is important to ensure that all manuscript pages, as well as any slides or transparencies, are clearly and consecutively numbered.

Only the most practiced and experienced lecturer can afford to forego a "trial-run" of an oral presentation. At the very least such a *practice session* serves the purpose of establishing the true *length* of a proposed talk—important since the novice lecturer usually attempts to present at least twice as much material as the time available would permit. Practicing will also indicate what sections are likely to prove most troublesome, and it may show that some parts require thoughtful reconsideration.

A practice session can be a totally private affair, but it is often better to arrange it as a small seminar. Alternatively, one can practice a talk before one or two critical and experienced colleagues. Modern technology has made another option available: recording one's talk on *tape*—or even *videotape*. Playing back such a tape enables the lecturer to experience—at leisure and perhaps repeatedly—just what the audience will hear and see. Previously unrecognized problems often become extraordinarily obvious in the process, and they can then be systematically analyzed and corrected.

For many, the worst moments in a lecture occur at the very beginning, a time characterized by extreme nervousness and excitement. For this reason it is wise to pay particular attention to the preparation of the *opening* of any lecture, perhaps even *memorizing* the first few sentences. The same holds true for the beginnings of any *sections* with which the speaker feels ill at ease. Memorizing must be done carefully, however. One of the keys to an effective lecture is conveying an impression of self-confidence and spontaneity. Listeners must remain completely unaware of the fact that parts of a talk are being delivered from memory.

Intense preparation of the kind suggested above is time-consuming, and although it requires a high level of commitment, the effort will be rewarded. Moreover, if one conscientiously prepares several lectures in succession, it will soon become apparent that each new presentation demands less effort than its predecessor.

# A.3    Slides and Transparencies

At an early stage in the preparation of a lecture one should give careful thought to the possible use of various *tools* to supplement the spoken word—e.g., blackboard, pointer, slides, overhead transparencies, or even demonstrations. Our remarks here are restricted to the subject of *slides* and *transparencies*.

Most readers will recall at least one lecture that began with "The first slide please!" and then erupted into a spectacular display of 65 additional slides in the course of 15 minutes. However, it is unlikely that much was retained of the *content* of such a "talk". Overuse of slides (generally poor ones) is a very common error, and is almost certain to result in an inattentive, frustrated audience. A useful rule of thumb is that each slide should be permitted to remain on the screen for *at least* three minutes. This means that no more than five slides or transparencies should be employed in a 15-minute lecture. If 45 minutes are available—usually the maximum sensible length for a presentation—the limit is about fifteen. To put the matter more succinctly, a rate exceeding one slaud* is usually fatal; 0.3 sl is an acceptable mean.

There is hardly any activity having to do with information transfer in which more errors are made than in the preparation of slides[†] to accompany scientific lectures. Too often slides are difficult to read, awkwardly constructed, and overloaded with data. The following suggestions should help the reader surprise and delight an audience with exceptionally *good* slides.

Diagrams and figures originally prepared for *publications* are generally unsuitable for slides. Such materials will presumably have been designed to convey a maximum amount of information in a minimum amount of space, an assignment inappropriate for a slide. It is for this reason that slides prepared from journal article copy so frequently require embarrassing introductions such as "Please ignore all but the last set of numbers in the table" or "In this case we are only concerned with the first of the five curves shown"; they would be far better omitted entirely if they cannot be replaced by materials appropriate for the occasion.

---

* The "slaud" (symbol: sl) is a unit whose introduction we are taking this opportunity to suggest. Its name is taken in analogy to a standard unit of measure in the computer world, the "baud"; one slaud corresponds to one slide per minute: $1 \text{ sl} = 1 \text{ min}^{-1}$.

[†] Most of the following comments apply equally to *transparencies*, discussed in detail at the end of this section.

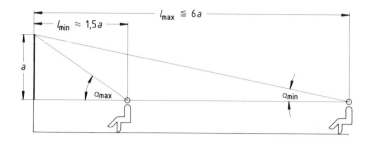

*Fig. A-1.* Maximal and minimal angles of vision $\alpha$ at a viewing distance $l$ from the screen; $a$ is the image height of a slide projected so that its longest dimension is vertical.

An effective slide focuses audience attention on what is being said. Its purpose is to provide a visual supplement to an essentially oral exposition. The slide embodies and continuously reinforces a single idea—a single message—and it does so clearly and succinctly. The *content* of the slide must be so organized that observers will quickly grasp the essentials—even if their attention strays from the speaker's words—and retain the corresponding message at least until the end of the lecture.

The *graphical* and *typographical* elements of a slide should be designed for easy scanning. A projected image will meet this requirement only if certain guidelines are observed, the most fundamental of which are based on geometric considerations. Fig. A-1 depicts the arrangement of a typical lecture room. Assuming the facility is properly designed, the distance of the audience from the screen will correspond to between 1.5 and 6 times the maximum height $a$ of any projected image. Thus, a medium-sized lecture hall should have a screen (or projection wall) whose size is about 2 m x 2 m, and the distance between the screen and a viewer should range from a minimum of ca. 3 m to a maximum of ca. 12 m. From these dimensions it follows that $\alpha$, the *viewing angle* occupied by the image, varies from about 10° to about 35° depending upon where one is sitting. The resolving power of the eye—its ability to distinguish between two adjacent parallel lines—is roughly 1' (1°/60). This means that a single character on the screen will be clearly legible from the back row of a medium-size lecture room only if the height of the character is at least 3% of the overall height of the image (and, therefore, of the original from which the slide was prepared). This in turn limits to about six or eight the number of *lines of type* that can be usefully incorporated into a slide (assuming reasonable margins and the equivalent of "double-spacing"). In other words, it is unlikely that the amount of available information will prove to be the limiting factor in the preparation of good slide copy.

Today, lectures are usually accompanied by so-called "35 mm" slides, whose actual dimensions are 24 mm x 36 mm. Originals for such slides should be prepared using paper with the same relative dimensions; i.e., 2 : 3. A particularly convenient paper size is 14 cm x 21 cm (roughly A5). Adoption of this standard permits the use of stencils based on 5 mm, 6 mm, or 7 mm letters. If one prefers to use a typewriter instead, the appropriate paper size is about 6 cm x 9 cm (ca. A7).

It has been recommended (cf. Standard *DIN 108, Blatt 2, 1975*) that all slide materials be subjected to a simple preliminary test for readability and clarity. The proposed test is based on the assumption that the maximum distance of the viewer from the screen will not exceed six times the screen height (cf. Fig. A-1). Thus, the original from which any slide is prepared should be clearly readable from a distance corresponding approximately to eight times its longest dimension.* If copy is prepared on A5 paper (as suggested above) it must be clearly legible at a distance of ca. 1.5...1.75 m. The DIN guidelines also underscore the theoretical conclusion we derived previously: they assert that a slide should never contain more than eight lines of text.

It should be noted that the above considerations all involve *limiting* conditions, whereas limits rarely correspond to *optimum* values. Viewers should never be subjected to "packed" slides unless it is absolutely necessary that they have simultaneous access to large amounts of information. In general it is better to divide an "8-line slide" into two or three simpler presentations.

*Tables* are particularly difficult to transform into effective slides. The tables prepared for written reports are almost always too extensive to be useful. The problem may be one of *length*—too many lines of data—but the *breadth* is equally critical. In no case should one attempt to utilize as a slide a table containing more than eight columns. Three- or four-column format is preferable, even with a horizontal slide.

Slides can be a particularly effective means of presenting chemical formulas, and they also lend themselves to various types of figures—provided the considerations discussed above are suitably taken into account. The corresponding originals should always be prepared as professionally as

---

* An alternative test is to see if the slide itself is legible at a distance of 25...30 cm. Of course, the safest test is to examine the actual *projected* image from an appropriate distance.

possible, utilizing all the techniques discussed in Chapters 8 and 9 for the preparation of high-quality drawings and figures for reports. Care in this respect is particularly important for slide copy since the magnification inherent in projection tends to emphasize irregularities, diverting a viewer's attention from the substantive content. It cannot be stressed too often that a slide containing a limited amount of information—one or two formulas or curves— is far more effective than a cluttered one.*

We conclude with a few special remarks concerning *overhead transparencies*. "Oversized slides" of this type are quickly and easily prepared with the aid of standard copying machines. Usually one needs only to replace the paper supply with appropriate transparent sheets. Suitable materials are available from a variety of sources. If one has a choice, it is best to utilize the heaviest sheets possible; thin sheets are difficult to manipulate, a problem that becomes especially apparent when one is in a stressful situation.

Audiences generally anticipate that overhead transparencies will be less formal than conventional slides, so careful free-hand preparation directly on transparent stock is also permissible—provided one's handwriting is legible. Special marking pens are available for the purpose. These differ from ordinary marking pens in that they contain a *transparent* ink, facilitating the preparation of striking multi-colored images.[†] "Black and white" transparencies prepared with a photocopier can also be color-enhanced in this way. Certain transparency pens are even designed to produce "erasable" marks, permitting one to modify the projected image as one talks. Indeed, the "flexibility" of transparencies in this respect—and the fact that projection is completely under the control of the speaker—are special advantages whose exploitation should be considered. Another interesting example of the flexibility of transparencies is the fact that they can be superimposed during projection, making it possible to add new information to a screen image as a talk progresses. The techniques involved in the preparation and use of transparencies are so straightforward that even students can be encouraged to enliven their talks with projected illustrations—although it should be stressed that the traditional blackboard remains an effective medium for supplementing the spoken word.

---

* This advice must not be carried to the opposite extreme, however. Audiences grow suspicious of speakers whose slides are very "slick" and professional—but devoid of content.
† The ink in ordinary marking pens is opaque. Consequently, if these are used to prepare a transparency the projected image will be black regardless of the original ink color.

Most of the recommendations presented above for preparing 35 mm slide copy also apply to transparencies. The usual format of a transparency is A4 (or 8 1/2 in. x 11 in.), and the corresponding original can be of the same size. Simple extrapolation of the standard guidelines reveals that lettering on a transparency should have a height of at least 7 mm. This means that typewritten copy is always inappropriate unless one has access to a special type element (cf. Sec. 5.2.1) or the presentation is to take place in a very small room.

Finally, it should be noted that computers provide an increasingly attractive option for the preparation of slides and transparencies. The combination of word-processing and graphics software together with a high-quality printer/plotter (even better: a laser printer) is capable of producing truly outstanding originals for slides, and transparencies can often be prepared directly.

# A.4    The Lecture Itself

Prior to the beginning of a lecture (or the session of which it is a part) it is wise to make a careful survey of the room, keeping the following points in mind:

- Is there chalk available?
- Is it certain that the projector works?
- Are spare projector bulbs handy?
- What sort of communication is possible with the projectionist?
- How does the microphone operate?
- What about a pointer?

It may also be useful to do some preparatory writing on the blackboard (e.g., an outline, or a few complicated chemical formulas).

Important as such preliminaries may be, it is one's *speaking style* that deserves the most attention. One indispensable prerequisite is careful *enunciation*. Every listener in the room must be able to understand every word clearly without special effort, and in spite of disruptive noises such as those coming from the street. One must speak *loudly enough* and *slowly enough* so that all vowels and consonants are distinct. The lecturer should bear in mind that the audience may contain foreigners for whom spoken English in *any* form can present problems. Speaking volume must also be adjusted to suit the size and acoustics of the auditorium. One experienced lecturer has provided the following advice: "Speak as though you are directing your talk to a particular person in the back row."

If a *microphone* is used it is important that it always be kept at the proper distance. In some cases, moving a microphone 10 cm away from one's mouth is sufficient to destroy all voice projection, while having it too near causes distortion. Unfortunately, many lecture rooms are not equipped with portable or adjustable microphones, so one may be forced simply to "make do" with an awkward situation.

Science is complicated enough without further complicating it through confusing and inappropriate words. Verbal descriptions should be formulated as simply as possible. Moreover, *short sentences* are far preferable to long and complex ones, especially in the case of an oral presentation. Short sentences are easier for the listener to comprehend, and they minimize the likelihood that the *speaker* will get lost in a tangle of verbiage. A style considered ideal for a written report (cf. Sec. 1.4 and Appendix B) is often *not* appropriate for a lecture.

The opening minutes of a lecture should be devoted to introducing the subject matter and to clarifying essential terminology that may not be familiar to many of the listeners. As suggested previously, one should never hesitate to repeat material that is especially important. In particular, *spontaneous* repetition is called for whenever one has the sense that something already said has not been fully or correctly understood. The opportunity to reiterate—certainly one of the major differences between written and oral presentations—should be used sparingly, but it can be turned to good advantage. The following additional techniques available to a speaker are also worth noting:

● The controlled use of *gestures* enlivens a presentation; movement keeps an audience alert.

● Changes in *intonation* and occasional *pauses* can help signal important points in an argument.

● One should constantly seek to maintain direct *contact* with the listeners; that is, one should give the impression of addressing spontaneously (many!) individual members of the audience.

Eye contact with an audience is crucial, and it is almost impossible to achieve if a text is *read*. Most effective speakers shun the reading of a complete text, preferring to *talk* confidently about their topic, although in a disciplined way. Too often a carefully honed written manuscript resembles a *report* rather than a lecture. If this should happen, the audience would be far better served by reading the material *themselves*, perhaps in their hotel rooms. A lecture audience is seeking the opportunity to "be in touch" with a live human being, the originator of the work being described, someone who is able, through a particular personal style, to convey some of the emotion and significance associated with a piece of work.

If one prepares a full manuscript, it should be glanced at only occasionally. A better practice is to rely instead on notes alone. Even the lecturer forced to deliver a talk in a foreign language will make a better impression by speaking freely. The risk of an occasional grammatical error is far less serious than the consequences of a dull, monotone delivery resulting from word-for-word reading of a polished manuscript.

One excellent technique for maintaining a sense of contact with those addressed is to focus one's attention alternately on several individuals in the audience, seeking in the process to sense the way each is reacting. A relaxed atmosphere can often be established by beginning with a few (apparently) spontaneous remarks directed to real individuals in the particular situation of a given auditorium or lecture room on a specific day.

Whenever one is discussing *visual materials* projected behind the podium it is a good idea to turn slightly *sideways*. This stance allows one to maintain eye contact with the audience, but at the same time to direct attention to the visual material. Turning one's back to the audience is a serious mistake. It is impolite, and it breaks the illusion of personal contact. Furthermore, it alters the projection of one's voice, especially if a stationary microphone is used.

If slides are employed to present *data* it is extremely valuable to make frequent use of a *pointer* or a *light pointer*. Properly handled, these devices help focus audience attention on the detail of immediate interest.

The conclusion of a talk should never be too abrupt, but natural and appropriate. It is also unwise to disappoint the listener who thinks the end has arrived by suddenly declaring "We will now examine another aspect of the problem". The most sensible way to conclude is by providing a brief, explicit summary, one that is recognizably "final" and presents a logical encapsulation of what has been discussed. Regardless of the way one concludes, one should do so promptly, and at the expected time.

Often it is expected that a talk will be followed by a *discussion period*, a session that may produce lively intellectual exchange, especially in English-speaking countries. The speaker is well-advised to be prepared for a variety of questions—including critical ones. The situation is in some ways analogous to the oral defense of a thesis, and similar advance planning is called for (cf. Sec. 2.3.2, "One Last Hurdle"). Experienced lecturers commonly lay the groundwork for a discussion period by deliberately "setting up" the audience: answers to certain obvious questions are carefully (but subtly) omitted from the talk, thereby increasing the likelihood that the speaker will be able to maintain some control over what follows.

After the talk, or after an ensuing discussion, the speaker should immediately collect all slides and transparencies and, if appropriate, clear the blackboard for the next speaker. One should not forget, however, that at most conferences speakers are expected to remain in the hall until the conclusion of the session.

## A.5   Poster Sessions

In recent years, so-called *poster sessions*—opportunities to display one's results on a bulletin board—have enjoyed increasing popularity at scientific meetings. Participants find them a time-saving and effective device for *exchanging* information, as opposed to merely "presenting" or "receiving". Those who have information they wish to dispense, as well as those seeking information, appear to appreciate the informality of this "new" medium.

A *poster* has the same assignment as an advertising billboard. The reader is supposed to acquire at a glance at least enough information to become curious and read further, and then to be interested in consulting the "distributor" for more details. At the same time it must be remembered that a poster is intended as a substitute for a talk. It should serve to *announce* significant new results, not *review* previously published work.

Anything appearing on a poster should be readily understandable to the average scientist trained in the broad specialty represented. A good poster is analogous to a mosaic: it provides a reasonably clear overview, but does not pretend to supply detail. A few key literature references (especially one's own) are helpful in leading the interested visitor further into the subject. The complete *address* of the author should certainly be made a part of the poster.

Posters are often displayed only for a few hours, although they sometimes remain in place for the duration of a meeting. They may hang in the halls and corridors of the conference area, or they might be restricted to specific *bulletin boards* in special rooms. Interested participants are encouraged to study them at their leisure. During specific time periods announced in advance the authors of the posters make themselves available for questions and discussion, and these times constitute the actual poster "sessions". Such sessions are generally listed in the official conference program.

The main advantage of this form of scientific information transfer is obvious: one has the opportunity to confront a researcher directly, permitting a true two-way dialogue and exchange of views. Lectures, by contrast, even those accompanied by discussion, suffer from the condition that they, like journal articles, are virtual "one-way streets". Poster session abstracts are often printed together with the abstracts of conference *lectures*, so a poster presentation can be the origin of a recognized "publication" (see below).

The space made available for a poster is typically a wall 2 m wide and 1 to 1.5 m high, often covered with a special surface (wood, cork) designed to accept thumbtacks. Sometimes one has the option of utilizing several such spaces, especially if "authors" will not be available for discussion. In a technical sense, posters share certain common features with slides and transparencies. For example, the size of the writing must be chosen with a concern for readability under a particular set of circumstances. The major letters and symbols on a poster, those designed to attract attention, should be clearly legible from a distance of 2 to 3 m. In other words, they should be between 10 mm and 15 mm in height.

Uniformity is often expected with regard to poster size and form. Typical lettering specifications issued by conference organizers suggest:

- titles comprised of 30 mm letters;
- captions 10 mm to 20 mm high;
- labels at least 10 mm high for figures and tables, as well as for short blocks of text (Results, Discussion, Summary, etc.).

Materials required for hanging the posters (e.g., ladders, tape, thumbtacks, etc.) are generally available from the conference sponsors, and personnel may also be available to help with the job. Nevertheless, it is wise to bring a set of personal supplies so that in an emergency one can arrange for hanging (or perhaps repairing) one's own poster—on whatever surface happens to be available: wood, felt, cork, or even a plaster wall.

The illustrations and text portions of all posters displayed at a given session are frequently reproduced in a set of *Meeting Abstracts* or *Conference Proceedings* (cf. Sec. 11.4.5).* Poster presentations thereby acquire the same status as brief oral reports delivered at scientific meetings. From the perspective of the author they represent a hybrid of the two standard classes of presentations, written and oral.

---

* The clever author will maintain a personal supply of small reproductions of the material for distribution to interested visitors.

# Appendix B
# Aspects of Scientific English

## B.1    The Grace of English

Let us begin with a brief critical appraisal of the English language. Many would argue that such an appraisal can lead only to a chorus of approval for the most practical and versatile means of communication yet developed by man. This judgment is clearly subjective, and it might well prove impossible to sustain on "scientific" grounds, but it is nonetheless shared even by some whose native tongue is other than English, including two authors of the present book. They freely acknowledge that they are more fluent in German and French, yet they enjoy communicating in English and are convinced that it is more than mere historical coincidence—the colonial might of the British— that has led English to become the *lingua franca* of modern science (cf. "Choice of Language" in Sec. 3.2.5).*

We hasten to add that our praise of English presupposes its full and proper use. Unfortunately, what many regard as English—particularly young people, and perhaps especially in the United States—is an impoverished remnant of its progenitor. Much of the power of English is embodied in the breadth of its

---

* The eventual triumph of English as a medium of international communication was predicted at least as early as 1780 by none other than John Adams, second President of the United States, who was referring to American English when he wrote: "English is destined to be in the next and succeeding centuries more generally the language of the world than Latin was in the last or French is in the present age. The reason of this is obvious, because the increasing population in America ... will ... force their language into general use, in spite of all the obstacles that may be thrown in their way, if any such there should be." (Quoted by Mencken, 1936. This source provides valuable and fascinating insights into the fundamental differences between "English" and "American".)

vocabulary, born of a propitious union of two linguistic streams: the Germanic dialects of the Anglo-Saxons and the Romance language of the Norman followers of William the Conqueror. To this has been added a constant influx of new words—brought by immigrants from nearly every language group in the world, and accepted with unusual openness by the recipients. The size alone of a typical "unabridged" English dictionary is an awesome spectacle. For nearly any concept one might envision, English provides a series of near-synonyms, words whose meanings differ from one another only in subtle ways. But these differences permit the expression of equally subtle distinctions in perception, response, or attitude—provided, that is, that both speaker (or writer) and listener (or reader) understand the subtleties. Comprehensive mastery of English vocabulary—with proper adherance to the grammatical principles enabling sets of words to coexist as sentences—is a source of great power, not only for the poet, but also for the scientist concerned with accurate, full reporting of observations and insights.

It is not our intent in this brief appendix to attempt to teach anyone the correct use of the English language, or even to provide a manual of preferred writing style. This would exceed the bounds of our assignment, and were we to attempt such a task we would quickly be recognized as "connoisseurs" playing with the stuff of professionals. We deliberately chose to relegate the subject to an appendix where we could provide the interested reader with guidance to authoritative sources offering further instruction. Nonetheless, we do claim the right to offer a few general but somewhat random observations about the niceties and the pitfalls of English.

Before proceeding, we should justify our contention that English is distinguished in its *practicality* and its *versatility*. Both terms are appropriate for describing a language so well-suited to a wide variety of applications: poetry, conversation, diplomacy, rhetoric, and certainly science, among many others. Part of its success may be due to its relative lack of formal structure. Indeed, foreigners are often amazed to discover how few "rules" seem to apply in English. Unlike nouns in French and German, English nouns are essentially devoid of gender, and articles are unaffected by the case of the nouns to which they apply. Inflected word endings are rare: mastery of "s" (or "es"), "d" (or "ed"), and "ing" will nearly suffice. "Correct" punctuation presents little problem; even the authorities frequently disagree about what is correct (cf., for example, the fourth footnote in Sec. 3.5.4). Acceptable sentences can be built in a remarkable number of ways, though the sequence subject/verb/indirect object/direct object is most common. The "parts of speech" in English often seem quite interchangeable. For example, nouns are sometimes found modifying other nouns as if they were adjectives (e.g., "reaction mechanism"), and many identical forms appear for both nouns and verbs (e.g, "precipitate", "change") and adjectives and nouns (e.g., "integral"). Anyone who has studied other languages—especially Latin—cannot help being struck by the differences.

Flexibility is commendable, but it also carries a risk: a complex sentence can easily become unintelligible, or at least ambiguous, in the absence of obvious "markers" to point out relationships between words, or the sense in which a given word is used. Schoenfeld (1986) quotes the classic example "Flying airplanes can be dangerous" (is "flying" a gerund or a participle?), and then adds a more complex but also more relevant case:

> The aqueous layer was removed, the solvent concentrated, and the residue exploded.

Was the explosion spontaneous, or did it require initiation? Both examples are admittedly somewhat contrived, and in both instances the sentences are so nearly correct that their author can be excused for not noticing the ambiguity. Nevertheless, very real problems arise with sentences that clearly violate those rules the language does possess, for the reader is left completely helpless:

> The latter product was characterized by IR spectroscopy and resulted to form during the chromatographic operation, likely by air oxidation of ....*

It is the purpose of this appendix to assist writers who are concerned that their work not prove as inaccessible as that of the alchemists (cf. Sec. 6.1)

# B.2    A Few Useful Aids

Just as a library is an indispensable aid to successful scientific laboratory work, so certain books can provide valuable help for the expression of

---

* This sentence is based on an excerpt from a manuscript submitted for publication in a major journal.

scientific ideas. Unfortunately, many who write about science appear to devote proportionately too little of their effort to perfecting their mastery of the language. This is a pity, because it is *other* scientists who suffer the consequences. One need consider only the amount of time spent deciphering obscure, confusing prose in the current journals to appreciate the truth of this assertion. In principle, there is no reason why professional reading should be more difficult than recreational reading: both require the same skills. The only sense in which professional reading should be more time-consuming is that the ideas, *once acquired*, must be pondered, probed, and applied. When a skilled reader finds the *acquisition* of ideas or information a problem, it may be the *writer* who is at fault. Even the writers of scientific textbooks are notorious in this regard, and students may be forgiven if they conclude early in their studies —as many do—that scientific writing is somehow not meant to be *understood*, only memorized.

How should one proceed to become a better, more effective writer? A useful first step is to read what others have said on the subject. Schoenfeld (1986) provides an attractive starting point. His book *The Chemist's English* combines a wealth of sober advice, somewhat randomly presented, with a delightfully witty style certain to hold the reader's attention. More systematic and business-like, but remarkably concise and to the point, is the highly praised classic by Strunk and White (1979), *The Elements of Style*, a work that first appeared over fifty years ago but still retains its freshness and impact. Two or three hours spent studying these 92 pages will pay handsome dividends. *The ACS Style Guide* by Dodd (1986) contains a long chapter entitled "Grammar, Usage, and Style". It is directed explicitly to the chemist, and emphasizes technical problems associated with chemical writing. This work owes much to a famous predecessor, Fieser and Fieser's *Style Guide for Chemists* (1972). We recommend that Dodd's book be read (even by non-chemists) and referred to frequently, though it is certainly no substitute for the more general works cited above. Among many others who devote attention to the communication problems of scientists we would single out Farr (1985), and its chapter "Writing Good English"; and Day (1983), especially the chapter "Use and Misuse of English", which contains a number of amusing illustrations of serious points. A book by King (1978), written from the viewpoint of the medical sciences, also deserves special mention for the directness with which it attacks the primary problem: *Why Not Say It Clearly —A Guide to Scientific Writing*. Other recommended general works include those by Turk and Kirkman (1982) and Zinnser (1980).

All the sources identified so far presuppose a basic knowledge of English *grammar*, which permits them to concentrate largely on *style*. However, many scientists could also profit by a "refresher course" on "the basics". Here the list of works for possible study is limitless. Many are designed as texts for general college-level use. A typical example is Crews (1980); it combines a review of formal language rules and conventions with a general treatment of "the composing process". A delightful alternative guide to proper grammar is *The Transitive Vampire—A Handbook of Grammar for the Innocent, the Eager, and the Doomed* by Gordon (1984). This book is unlikely to be surpassed in terms of imaginative examples.* The writer seriously interested in improvement should certainly acquire a copy of *Line by Line—How to Improve Your Own Writing* by Cook (1985). This marvellous work, prepared by an editor, combines many of the best features of several kinds of guides—style manuals, grammars, dictionaries—into a practical self-study manual filled with "incorrect" and "correct" examples.

The non-native English writer seeking formal guidance to the language will probably be served best by more specialized works, such as the two-volume *English as a Foreign Language for Science Students* by Brookes and Ross (1975, 1976) and *Communication Skills for the Foreign-Born Professional* by Barnes (1982). The latter deals with communication in the broadest sense, including its non-verbal components.

Even the writer with an excellent command of the language will need to refer occasionally to an authoritative source regarding details of style and usage. One of the most respected books in the United States for this purpose is *The Chicago Manual of Style* (1982), noteworthy for its extensive classification of the information provided. Every author should take time to become familiar with this standard work, to which we have frequently made reference. A much smaller but otherwise somewhat analogous British work is *Hart's Rules for Compositors and Readers at the University Press Oxford* (1983). Those in search of a more exhaustive volume addressed to the writer with a scholar's interest in the language will turn to *A Dictionary of Modern English Usage* by Fowler (1965/1984), a book with a misleading title, since it more nearly resembles a collection of mini-essays on words and phrases, all arranged alphabetically and in dictionary format. Nicholson (1957) has prepared an abridged version (of an earlier edition) more atuned to American English.

---

* "The behemoth *that* trundled into view was covered with tatoos" (use of a subordinate conjunction); "Trinculo *became* her confidant as the slivovitz disappeared" (a linking—or "copulative"—verb).

One aspect of English almost equally troublesome to both native and non-native writers is correct *spelling*. Anyone forced to learn English as a foreign language quickly discovers the closely related problem of correct *pronunciation*. Indeed, words such as "rough", "dough", "cough", "through", and "bough" raise questions about the extent to which English differs from an unabashedly non-phonetic language like Chinese. With respect to spelling let us simply acknowledge that it is foolhardy for anyone to attempt to write more than a few English sentences without access to a dictionary—which should be opened at the slightest sign of doubt ("Is it '-ible' or '-able'?"; "Is 'principal' or 'principle' the adjective form?"). Unfortunately, different dictionaries often disagree with respect to which of two spellings is preferred (e.g., "judgement" vs. "judgment"; "catalogue" vs. "catalog"). Variation in spelling becomes especially apparent if one compares a standard British dictionary such as the *Concise Oxford Dictionary of Current English* with *Webster's Third New International Dictionary* or some other American work ("colour" vs. "color"; "cheque" vs. "check"). In the absence of an editor to establish a set of binding local rules one's only recourse is to select one correct form and then strive to maintain consistency (not an easy matter, as any book author will testify).

Dictionaries are important for purposes other than the verification of spelling. *Hyphenation* should never be introduced at the end of a line without first ascertaining the "official" location of a break between syllables (cf. Sec. 5.3.1). An analysis of the way a word is pronounced may provide clues to proper division, but it is not an infallible guide. A dictionary should also be consulted to verify the *precise meaning* of any word that causes even the slightest doubt. We spoke earlier of the wealth of the English language and its remarkable ability to convey subtleties, but careless word-usage increasingly threatens to squander this inheritance from our ancestors. Perhaps this is the proper context in which to remind the reader of an invaluable supplement to a dictionary, one developed precisely because of the immensity of the English vocabulary. That supplement is a *thesaurus*, such as the widely acclaimed *St. Martin's Roget's Thesaurus of English Words and Phrases*, edited by Dutch (1965), or the *Roget's Thesaurus* edited by Lloyd (1962 and later editions). A thesaurus is intended to help the writer find the elusive synonym that refuses to come immediately to mind, the dimly perceived "better" way of expressing something. Careless use of a thesaurus can of course produce marvellous—or irritating—gibberish, but the writer who consults a thesaurus regularly and carefully verifies word meanings in a dictionary performs a valuable service for the reader.

# B.3 Notes from a Short Course in Good Writing

## B.3.1 Some Credos

The ultimate goal of any piece of scientific writing is to transmit a precise and coherent message from an author to a reader, from one scientist to another. Scientific training stresses the need for abstracting the important from the unimportant, distinguishing between the specific and the general, and striving for precision. Why then does the *writing* of scientists often lack these very qualities? Why don't they "say it clearly" (King, 1978)? Much scientific writing is littered with idle words, awkward constructions, and inaccurate phrasing simply because few scientists take seriously the importance of good writing. They see the job of reporting their efforts as peripheral to the work itself, and regard writing as little more than a necessary evil. Educators—including scientists!—are also at fault when they place too little stress on the mechanics of writing. Worst of all, virtually no one has insisted that the situation be changed (this comment may be taken as an editor's sigh), although many have expressed alarm at what they see as an accelerating deterioration of linguistic skills.

*Readers* of scientific prose are altogether too tolerant and too willing to shoulder an inappropriate amount of the burden. Perhaps this is a reflection of the scientist's love of puzzle-solving, but it is certainly *not* conducive to effective communication. If that were the aim, all writers would adhere firmly to a basic set of rules such as these:

- Write the first draft as if it were a transcription of a clear, informal, *oral* account, perhaps delivered over the telephone.
- Keep all sentences short and simple.
- Avoid fanciful, ambiguous, and repetitive terminology and phrasing, and eliminate all slang.
- Guard against the use of unnecessary superlatives.
- Group together those phrases or words that belong together.
- Minimize the use of overly specialized "insider" vocabulary.

It may be that such a list of "dos and don'ts" is too general to be of any practical value, yet it inculcates the essence of "saying it clearly". If all writers would take the time and trouble to test each of their sentences against these injunctions, the quality of scientific writing would increase immensely. Indeed, a measurement of the time saving for readers would reveal that the effort has a very remarkable cost/benefit ratio. Science itself is the source of enough inherent complications without clothing it in artificial obscurities born of inconsideration and carelessness.

We are certainly not the first to deplore the state of scientific writing and to plead for reform. In Sec. B.2 we cited numerous sources of advice on ways to improve one's writing, and each of these sources in turn lists others. Some readers might conclude that yet another summary of the most important principles would be superfluous, that there is no need for an "Ebel/Bliefert/ Russey Catechism of Rules for Writing Effective Scientific Prose". The essential points have all been made by others—and more eloquently than we are likely to do. Nonetheless, we have compiled our own set of observations regarding effective writing, and these are summarized in the following sections. In so doing we reaffirm the strength of our belief in the importance of the crusade, and we venture to hope that our modest contribution will lead at least a few readers to explore the more extensive guidance provided by others. We shall consider the function and application of the various parts of speech before examining the ways in which words are combined as sentences.

## B.3.2    Verbs and Nouns

The two types of words we examine first seem to have little in common. "Actions" are certainly different from "things", but in fact the two are very closely related, *too* closely in much scientific prose. Verbs are words of action. Correctly employed they imbue a sentence with life. The activating influence of a verb is most apparent when one employs the *active voice* rather than the *passive* (compare "I poured the solution ..." with "The solution was poured..."). Passiveness is the antithesis of action. Unfortunately, scientists seem to prefer the impersonal when describing their work, thus obligating themselves to frequent use of passive constructions (cf. the second footnote in Sec. 1.4.1).

However there is another "deactivating" convention that *can* be avoided: *nominification*, the conversion of an expression of action into a noun ("*Distillation* was conducted on the residue" instead of "The residue was *distilled*").* The habit may owe something to the fact that alphabetical indexes tend to be based on nouns, and concepts are usually closely linked to "terms" (i.e., nouns) and an associated "terminology". Nevertheless, a competent indexer will have little difficulty discovering the appropriate noun for characterizing a verb (e.g., "extraction" for "to extract"). It is the reverse transformation that often requires a bit of imagination. The sin of nominification is not confined to the ranks of scientists; indeed, Rathbone (1985) has characterized it as symptomatic of *Federal prose*.

Perhaps a more blatant problem within this category is the unnecessary wordiness inherent in many popular *verb-noun combinations*; e.g.:

> take into consideration (better: consider)
> lend support to (support)
> arrive at a conclusion (conclude)
> have a preference for (prefer)
> conduct an investigation (investigate)
> serve as a substitute for (replace)

The list could easily be extended to include dozens of examples. To those who wish to take issue with the significance of the problem we propose going further and instituting the phrase "to effect scission" in the context of requesting a haircut. It should be noted that one important consequence of substituting the verb form in the above cases is that the resulting text becomes shorter.

There are other types of "combination phrases" involving verbs that also contribute to unnecessary wordiness:

> make shorter (shorten)
> become aware of (realize, or sense)
> be devoid of (lack)

The "of" construction in the last two examples is particularly subject to overuse. Here again a noun must be imported—this time in the genitive—to carry out a task more properly given to a verb. Why should an observation that "is suggestive of" something not simply "suggest" it? A little reflection will quickly confirm that scientific writing is exceptionally rife with such

---

* Interestingly, Schoenfeld (1986) appears more concerned about the reverse transformation: noun-to-verb conversion, a sin whose origin he traces to the Elizabethans in his chapter "To Reflux or not to Reflux?".

constructions. "The application of isoelectric focusing was found to lead to ..." says neither more nor less than "Isoelectric focusing gave ...", but it certainly takes longer to read (or type!), and if the remainder of the sentence is equally verbose the reader will have a legitimate excuse for prematurely drifting off to sleep.

Consider another example: "This approach is representative of ..." instead of "... represents...". The verb "to be" has much to commend it, as Hamlet so eloquently demonstrated. Nevertheless, the "verb of existence" was not invented to enable nouns to take over the work of verbs. A review that "is an excellent summary of" something even more excellently "summarizes" it. Some authors who use the verb "to be" in this way display their own discomfort with the "weak" results by yielding to an "is-and construction", a practice rightly criticized by King (1978). The sentence fragment

"The reagent is usually formed in situ and can ..."

is certainly better replaced by

"The reagent, usually formed in situ, can...".

One "is" has been eliminated, emphasis has been shifted to the active word "formed", and the reader's attention has been directed to the more important statement that follows.

## B.3.3   Prepositions

For anyone learning a foreign language, *prepositions* usually create the biggest pitfall. Prepositions are generally used together with verbs, providing the necessary link to a noun functioning as an *object*.* The combination is what conveys meaning, not the preposition alone, and for this reason it is impossible to "translate" a preposition from one language into another.† Indeed, a foreigner should always consult a dictionary when in doubt about the appropriate preposition. (Does one say "in analogy *to*" or "in analogy *with*"? Either may be permissible, although there is a subtle distinction.) Sometimes the simplest approach is to avoid the problem by choosing a different verb;

---

* Prepositions also link nouns to nouns, as in "the day *after* the fire ...".
† Many of the peculiar problems inherent in translation are lucidly explored by King (1978; Chapter 9).

e.g., "to look *at*" and "to look *for*" can be replaced by "to examine" and "to seek".

The proper choice of a preposition is not always as straightforward as one might wish, even for the native speaker of English. For example, one generally "compares" something *with* something else, but if *similarities* are the issue one should compare A *to* B instead. We may discuss a "contrast" *between* two things, and later contrast something *to* something else. Again, a dictionary is a valuable source of guidance.

English, unlike many other languages, often permits two different prepositional constructions to relate simultaneously to the same noun:

> Capillary isotachophoresis has always been related to,
> and influenced by, other electrophoretic methods.

Elegant as such constructions may be, we still must warn *about* (or caution *against*) excessive use of prepositions. Some cannot be avoided. English lacks a true genitive form and thus requires a proliferation of "ofs", but this is all the more reason for avoiding unnecessary prepositional phrases, especially those that merely circumvent verbs (cf. the previous section). Let that which "results in the destruction of ..." simply "destroy" and delete the superfluous "of" from "in all *of* the species examined..."!

There is a general rule lurking behind all this (not "all of this"!). We cannot live *without* prepositions (as shown *by* this sentence). They are required for establishing relationships (e.g., between the *object* of the preposition and the word or element modified by the prepositional phrase). One might even go so far as to mourn the apparent death of certain once-common English prepositions, such as *betwixt* and *amidst*. Nevertheless, technical writing abounds in aesthetically offensive prepositional constructions. King (1978, p 39) proposes the rule that no sentence should contain more than four prepositions, or more than three consecutive prepositional phrases. As a beginning, the reader might practice looking for "in the course ofs" that could be replaced by "durings".

## B.3.4    Adjectives

Adjectives are the words that convert generalized nouns into descriptors of the specific: a *gas*\* chromatograph, a *differential* equation, or a *colorless* liquid. Scientific writing probably suffers more from an absence of adjectives than from their overuse. Adjectives can help confer precision, and can do so concisely, without the need for long qualifying phrases. Perhaps the only note of caution required is a suggestion that adjectives not be chosen lightly. Readers come from very diverse backgrounds, and a description that may seem appropriate to the writer may be meaningless to others. For many years, one of the authors of this book has assigned to students a preparation requiring action when a particular solution acquires a "robin's egg blue color". French and German exchange students have always found this statement rather puzzling— for good reason: the eggs of European robins are not blue!

## B.3.5    Adverbs

Adverbs are so named because they frequently *modify* verbs, but they also serve as the modifiers of adjectives and other adverbs.† Thus, in the phrase "the temperature dropped suddenly..." the last word must be an adverb because it modifies "dropped" (a verb). Many adjectives can be converted into adverbs by introducing the suffix "-ly" (e.g., "sudden" and "suddenly", "true" and "truly"), a distinction careless writers sometimes fail to observe. Prepositional phrases or clauses can also assume an adverbial role, as in "The process starts *with some delay* and *after the color has disappeared*".

Some readers may wonder why we bother here to recite such elementary points of grammar. Our justification is the fact that adverbs are among the more troublesome components of English, especially for the foreigner. The problem is largely one of placement. In the examples above, the adverbial elements *follow* the verb, but this is not always the case; e.g., "*After the color disappears* the reaction begins"; "One should *immediately* lower the temperature". Proper choice of adverbial placement is particularly critical when several adverbial modifiers compete for the reader's attention. An analysis of

---

\* Notice that the noun "gas" is functioning here as an adjective, a possibility mentioned in Sec. B.1.

† Those adverbs that modify adjectives are sometimes referred to as *intensifiers*. Examples include *very* (e.g., "a very thorough investigation"), *slightly, exceedingly* , and *surprisingly*.

adverb placement in English leads to the recognition that adverbs must be classified according to what they specify, such as *place, time, manner, frequency,* or *condition.* For example, adverbs of place generally precede those of time: "I shall be *there tomorrow".* Adverbs of *frequency* generally appear foremost in a sentence: "I *never* go *there".* The latter example is interesting in that one of the two adverbs precedes the modified verb, while the other follows it. The same happens in the sentence "We *normally* use SI units *in scientific work",* where the second adverbial element (a prepositional phrase) is even separated from the verb by a direct object.

In nearly every case cited, alternative arrangements are also permissible. Thus, "In scientific work we normally use SI units" is also correct, as is "Normally we use SI units in scientific work". The placement of adverbs should not be a random process, however; the preferred order in a given situation depends on what the writer wishes to emphasize. The last of the arrangements above would be preferable if the sentence were to continue "... , but sometimes an exception is permissible." "Normally" would then provide an early clue of what follows; *time* is emphasized by placing a "time adverb" first.

There are, however, two general rules that apply to the placement of all adverbs. First, an adverb should not divide a *compound verb.* Thus, correct formulations include

> *usually* is generated
>
> *sometimes* can be isolated
>
> will be acknowledged *in due course*

but one should avoid wordings such as "is usually generated". The major exception to this rule is an adverb that modifies only a portion of a compound verb; e.g.:

> The method is *particularly well* tested.

The second general rule states that an adverbial element should not be placed between a transitive verb and its objects, particularly if the adverbial unit is long or complex. "We chose reluctantly and after exploring every available option distillation as the best alternative" asks too much of the reader's patience —and memory.

## B.3.6     The "Little Words"

While incorrect placement of adverbs is one of the surest signs that a passage in English has been written by someone whose native language is different, another common clue is incorrect usage of *articles*. When does one use the definite article "the", the indefinite article "a", or no article at all? The corresponding differences in meaning may be slight—or seemingly nonexistent —but in most cases only one of the choices "sounds right" to someone who has grown up with the language; e.g., "*The* separation of the product was unexpected"; but "Separation of the product was readily accomplished". The title "Fundamentals of Chemistry" is quite inoccuous; it has been used for dozens of freshman chemistry texts. But "*The* Fundamentals of Chemistry" is used more cautiously, because it implies (although subtly) an authoritative treatment of *all* the fundamentals.

Surely the most abused "little words" are pronouns, especially "it". All too often the reader is left in the position of having no idea what "it" is. "A sample of the product was distilled under vacuum. *It* was then chromatographed." *What* was chromatographed, the distilled material or the crude product? Every use of "it" should be considered suspect—and the same injunction applies to the demonstrative "this": "*This* indicates that ..." commonly poses an interesting challenge to a reader. Indeed, the "this" device is a favorite of authors who are not sure *themselves* precisely what they mean (or not sure how to express it). A little reflection on the part of the author will often lead to the discovery that more is involved in a "this" than meets the eye, and some elaboration may be necessary in order to define "this" more clearly—even to the extent of an additional paragraph. (Several paragraphs in this book had such an origin!).

Awkward "its" can also arise through a tendency to be verbose. "It is essential that the device be modified" says nothing more than "The device must be modified", and "it" wastes precious space. Similarly, there is no point in saying "it is apparent that" instead of "apparently". "It was decided that" is no better, but the obvious substitute ("I—or we—decided") will probably continue to meet with resistance as long as scientific writing remains impersonal (cf. the introductory remarks to Sec. B.3.2).

Other "little words" are candidates for intensive examination. For instance, *conjunctions* are subject to frequent misuse, especially when they are called upon to join elements of unequal standing. "What is and why should one study chemistry?" An interesting question, but one that raises a further question: is "chemistry" here in the nominative or the accusative case? The intransitive verb "is" demands the former, while the transitive "to study" requires the latter. Thus, the original question is incorrect; "and" links two "unlinkables", even though in English the two forms of the beleaguered noun are indistinguishable. A correct formulation need not be significantly longer: "What is chemistry and why should one study it?"

## B.3.7    On the Length and Structure of Sentences

We have frequently condemned the tendency to indulge in meaningless verbosity. The most obvious target is the pompous phrase hiding a simple idea:

> a number of (many)
> a majority of (most)
> at this point in time (now)
> despite the fact that (although)
> due to the fact that (because)
> for the purpose of (for, to).

Others have made the same point and provided longer lists of examples (e.g., Day, 1983, especially in the chapter "Avoiding Jargon"; Dodd, 1986). Awareness is the first step in avoiding the problem. The same is true with respect to superfluous words. An early draft of Appendix A contained a sentence including the phrase "... a type of ink which is transparent ...". It now says "... a transparent ink ...". Nothing has been lost except four useless words. Empty adjectives and adverbs are also candidates for ruthless extermination. What distinguishes a proven result from a "definitely" proven result? Is it really necessary to announce an "extremely high yield"? The elimination of such excess baggage has been compared to the process of increasing a signal-to-noise ratio. Of course there is a limit; a certain number of words is required to convey any information, and a few more may be necessary to ensure clarity. Schoenfeld (1986) suggests that if one were to plot the rate at which a reader acquired a given piece of information as a function of the amount of paper needed to describe it, the resulting curve would show a clear maximum ("the good paper") lying between two regions labeled the "rocks of terseness" and the "swamp of verbosity".

Before leaving the subject of words we should also stress the importance of avoiding slang and "local" expressions. One's benchmate may be accustomed to hearing about the latest "break in the vac line", and the trouble you had "running the IR machine", but scientists expect written reports to display more formality. Along with Webster (or Oxford) they would prefer a description of a "ruptured vacuum system" and the problems entailed in "acquiring data" with the aid (?) of a recalcitrant "infrared spectrophotometer".

But what can we say about the primary "message units", the sentences? Certainly the first rule is that main ideas should appear in *main clauses*—a rule often violated. Consider:

> One can proceed from the assumption that ...
> It is important to note that ...

In both cases, the real meaning of the sentence is relegated to an inferior position. This criticism does not rule out introductory phrases altogether: proper placement is preserved in a sentence beginning

> "As these results show, ...".

The *number* of clauses must also be held within bounds. A complex sentence may be perfectly correct, but it may also be a serious stumbling block to the reader, especially the reader forced to translate it into another language.* This is not to say that all sentences should be of the unembellished "subject/verb/object" type. Variety in sentence form and length helps ensure that readers will remain alert (cf. Sec. 1.4.2).

Enough! We have already said much, but it is still only a beginning. Perhaps the best piece of concluding advice is "be open to learning". There is no writer whose work cannot be improved, given the *will* to improve. Find examples of good writing produced by *others*; study them, identify their characteristics, and attempt to emulate them. Science and your peers will welcome your efforts.

---

* Those whose native language is English may rejoice over the fact that English has become the universal medium of science, but they must accept the related responsibility of writing in a way that *others* will readily understand. The significance of this injunction in the context of sentence structure will be appreciated by anyone not fluent in German who has tried to read German papers.

# Appendix C
# Authors and Their Rights

## C.1    The Essence of the Copyright

A *copyright* is a legal instrument designed to protect an original work of *intellectual creativity*. Whether in the sciences, in literature, in art, or in music, any *original work* that can be defined and identified as the product of one (or several) person's independent thought and labor is eligible for a copyright. It is important to recognize, however, that authors cannot expect copyrights to prevent others from *using* their ideas or data. That which is protected is the *material form* of the author's work, the *expression* of ideas, rather than the ideas themselves.

Even though musical works and works of the poetic or visual arts are also "expressions of ideas", and thus may be copyrighted, it is *literary works* that are thought of most often in conjunction with copyright law; i.e., books, pamphlets, journal articles, and the like. In some cases computer programs are also considered literary works (e.g., as in the May, 1985, amendment to the German Copyright Law), and the term covers material designed for communication to the public either orally or in writing.

Of particular interest to scientists and engineers is the fact that drawings, graphs, diagrams, charts, illustrations, tables, or even three-dimensional models also qualify, in the sense that these are also scientific or technical (and possibly artistic) presentations. Copyright protection must be taken into account any time one proposes to make use in one's own work of a figure taken from an extrinsic, published source. The law makes it mandatory, within the time period of protection, that permission be obtained before copyrighted material is reproduced. Permission can be secured either from the creator or the owner of the copyright (usually a publisher).

It is occasionally a matter of debate whether or not a given piece of work can be considered as an "intellectual creation" eligible for copyright protection. Certainly originality (*novelty*) is demanded; the work in question must lead beyond what was already known or existed previously. Thus, a scholar who only repeats what others have already said, written, or printed will not be regarded as having "created" something. On the other hand, complete originality in the sense of *unrelatedness* is not required, because it is obvious that virtually everyone's work rests to some extent on cultural achievements of the past. As a rule, all one must do to show the presence of a (new) creation is to demonstrate that a new combination of elements exists, either among themselves, or together with new elements. The principal requirement is that a certain degree of uniqueness has been achieved. Nevertheless, the achievement must go beyond the trivial if creativity is to be claimed. For example, the straightforward application of commonplace mathematical rules in the form of a computer program would not result in a new program that could be copyrighted. Novelty in this case would be recognized only if the new program were to demonstrate significantly the personal style and ingenuity of the programmer.

In all other respects the term "creative work" is, for copyright purposes, neutral and devoid of judgment. Works are eligible for *copyright protection* regardless of their form, manner of expression, purpose, or destination, and irrespective of their scientific, literary, musical, or artistic merits.

## C.2    Copyright Law

The creator of any work is referred to in *copyright law* as its "author", regardless of whether the work is a scientific treatise, a novel, a painting, a piece of sculpture, or some other art object. In most instances the author is the original *owner* of the copyright. In many countries—probably a majority—the law decrees that only human beings (*natural persons*) can be the original owners of literary or artistic works, so that the only way a *legal entity* can acquire copyright to a work is through purchase or some similar transaction.

However, the matter is different in the United States and other countries following the Anglo-Saxon legal tradition; here, *authorship*, and hence copyright, in the first instance can be vested in a corporate body or a legal entity (such as an employer, for example, if the work has been created in the course of employment), or in an agency. It can also rest with a person other than the author if the work has been *commissioned*. One consequence of this situation is that scientists in the United States affiliated with government authorities (services, agencies) are usually prohibited from transferring to publishers the publication rights to their writings. All copyrights are retained by the authority, and the works are considered public property.

Copyright law is concerned with rights in *intellectual property*, and the rights covered are of various kinds. Most authors have an economic interest in the publication, presentation, and communication of their works, and from this interest a set of defined *economic rights* of authors has evolved. These are subject to protection by national laws all over the world. Copyright protection goes further than the economic sphere, however. Authors have been described as "living in their works", a way of saying that intellectual property reflects, and is an extension of, an author's personality. In other words, more is involved than mere physical property, and a kind of protection is called for that transcends economic considerations. This second, non-pecuniary sphere has been termed *moral right*. Copyright speaks to the moral as well as the material interests inherent in an author's work. The law protects authors from unauthorized economic exploitation of their work, but it also provides protection for their psychological and personal relationships to that work.

Moral right is complex. Besides safeguarding the integrity and inviolability of a work it deals with such questions as whether a given work should be made public at all. One aspect of moral right is the establishment of the *right of authorship*, under which authors can claim recognition due for their work, ensuring that their names will be permanently associated with it.

Not surprisingly, however, the economic aspects of authorship are usually the focus of copyright attention. The most fundamental right granted to authors by copyright laws is their right to authorize the *making of copies* of their works. It is here, in fact, that the term "copyright" originates, although the scope of the term is obviously far broader than this origin implies.* "Making

---

* The German term "Urheberrecht" and its French equivalent, "droit d'auteur", are both more appropriate than the English word "copyright" in that they more clearly reflect the diversity of the legal considerations involved.

copies" was originally taken to mean *reproduction* of an entire book or other printed work, duplication in a form essentially identical to the original. However, recent technological achievements have broadened the meaning of "reproduction", so that present law applies to a multitude of methods of copying: printing, photographing, photocopying, and the making of phonograms, whether mechanically, photomechanically, electronically, or otherwise.

Closely allied to the *right of reproduction* are the author's *right of distribution* and *right of exhibition,* where distribution includes such nonmaterial forms of transfer as broadcasting.

# C.3    Copyright Ownership and Transfer

The production and distribution of literary works is an involved process entailing a complex "apparatus" not at the average author's disposal. This is why *publishers* are important. Authors not in a position to exploit their own works *transfer* their copyrights to publishers engaged on their behalf. Transfer of copyright is also regulated, and in different ways in different countries. In the United States, for example, copyright of a work can be easily transferred. Only a simple legal agreement is required. Anglo-Saxon tradition regards copyright as a personal piece of movable property, and as such it may be assigned wholly or in part to others. In certain other countries, however, copyright can only be inherited (i.e., by the *heirs* of the author), and then only in its entirety, consistent with the principle that *copyright ownership* in a work belongs first to the person who created it, its author. In particular, the moral rights associated with copyright are often considered inherent in the author, and hence inalienable. During their lifetimes, only the authors may exercise these rights; they cannot transfer their powers to any other person.

The author's *economic* rights, however, are recognized nearly everywhere as transferable. Thus, an author may, through a process of *licensing*, entitle another person or entity (e.g., a publisher) to exercise specific rights to *uses* of a work. Both exclusive and non-exclusive licenses may be granted, and these may be either limited to single modes of usage or they may extend to all applications. An *exclusive license* entitles the licensee to use a work in the manner described to the total exclusion of all other persons—even the author! A non-exclusive license, on the other hand, permits concurrent use by other claimants, including the author. A licensee may or may not be given permission to transfer primary or subsidiary rights to third parties.

# C.4    Duration of Copyright

Copyright protection of economic rights is usually limited to a specific period of time. This *period of protection* (*duration of copyright*) is not the same everywhere. Portugal, Yugoslavia, and Czechoslovakia, for example, recognize unlimited duration of rights, but with the understanding that after certain periods of protection the state becomes the owner of the copyright. In Spain, a work is protected for 80 years after the author's death; in the Federal Republic of Germany, Austria, and Israel, for 70 years; in Brazil and Belgium, for 60 years (plus the life span of the author in each case). The majority of national laws—including those of the United States, the United Kingdom, France, and the German Democratic Republic—provide 50 years of protection *post mortem auctoris*, although in the Soviet Union 25 years is the rule.

In contrast to these time limitations on economic rights, moral rights (where these are recognized as a separate aspect of copyright) are usually considered perpetual.

Of special interest to scientific authors is the situation with respect to publications appearing in periodically issued collective works, e.g. *journals*. Provided no alternative arrangements have been made, the publisher of a periodical normally acquires exclusive rights of reproduction only for a limited period of time; in the Federal Republic of Germany, for example, the rights extend only one year beyond publication.* Thereafter, authors regain the right

---

* This short period of time is considered sufficient to allow the publishers of periodicals to pursue their rightful economic interests.

to use their contributions as they see fit. Thus, a request for reproduction of an illustration from a journal article would correctly be sent to the author once the period of protection has elapsed; alternatively, one might contact both the author and the publisher. For various reasons, however, it is common practice to ask permission only from the publisher of the periodical no matter how "old" the publication is.*

# C.5    Integrity of the Work

The acquisition of economic rights in a work does not imply the right to change at will its contents, title, or, for that matter, anything about it; on the contrary, the author's consent to changes is always required. A publisher or publisher's representative (e.g., the editor of a journal, or a book editor) may correct errors in spelling or grammar, but more sweeping changes cannot be forced upon an author. In the case of a controversial journal article an editor or publisher may simply refuse to publish the manuscript, but with book manuscripts the legal situation is more delicate. The author of a contracted book can, upon delivery of the manuscript to the publisher, insist on its publication, although the publisher does have the right to demand the removal of obvious errors. Publishers of scientific works are particularly likely to enforce such demands, because they can be held responsible for erroneous data or false information contained in their publications. For this reason, publication agreements often contain a clause stating that the submitted work must reflect the current *standard of knowledge* (state of the art), and that the publisher retains the right to introduce necessary improvements or even deny publication if these criteria are not met.

---

* One obvious reason for applying directly to the publisher is that one may not know the address of the author. Another is lack of information about what particular laws apply, or what individual arrangements exist between author and publisher.

# C.6    Limits of Copyright Protection

Authors frequently seek to include in their papers or books *figures* (e.g., graphs, illustrations, or photographs) that have already appeared elsewhere. The direct *reproduction* of a figure from an outside source is considered equivalent to a *citation*. The unrestricted "citation" of *graphic material* in this way is entirely permissible, but only if the figure is crucial for understanding the contents of the new work, as opposed to being merely a convenient way of supplementing or perfecting it. Unfortunately, an author's intent will usually be judged as the latter. For this reason it has become common practice to obtain permission before reproducing figures from alien sources.

Permission is usually sought from the original publisher of the illustration (cf. C.4), and it is generally granted. In the case of purely scientific work the practice is quite routine, and it seldom involves payment of any fees. However, the situation may be quite different if works of art (e.g., paintings) are to be reproduced. A museum or the heirs of an artist may well impose a charge, a *reproduction fee,* before granting such permission. Textbook publishers, too, often require payment for permission to reproduce figures. Having invested a great deal of money in the preparation of first-rate illustrative material they feel entirely justified in demanding compensation from third-party users.

A typical reproduction request letter to a publisher is shown in Fig. C-1. The source of a reproduced illustration must be clearly identified in the new publication. One way is to incorporate the information into the figure caption; e.g.,

> Fig. xy ... (reproduced, with kind permission, from Klyne and Miller [12])

Sometimes, despite the extra effort involved, it is actually preferable to redraw an illustration that has appeared elsewhere, taking the opportunity to adapt it so that it better suits the style of the new publication, and perhaps modifying some of the terminology or symbolism. If a figure has been significantly altered, the author is no longer obligated to obtain permission for reproduction, nor does the problem arise of obtaining for one's publisher the kind of high-quality copy necessary for photomechanical processing.* Instead

---

* A xerographic copy taken from published material is rarely adequate since it leads to inferior reproductions (cf. Sec. 9.1.3).

Dear XXX:

I am preparing a manuscript for a book
tentatively entitled
XXXXX
to be published in 19XX by XXX. I would like to
include the following material which originally
appeared in
XXXXX
published in 19XX.

It is my understanding that you have the rights
to this material. I would be grateful for your
permission to include this material in the original
and all subsequent editions of the above book. I
also request that this permission include
nonexclusive world rights for this edition as well as
for any subsequent translations and editions.

Please indicate your agreement by signing and
returning the enclosed copy of this letter. In the
space provided you are also requested to supply the
appropriate credit and/or copyright notices that
should appear with this material.

If you do not control these rights, I would
appreciate knowing with whom I should communicate.
If the permission of the author is also required,
please provide his/her address.

Thank you for your cooperation.

Sincerely,

We grant the permission on the terms stated in this
letter for non-exclusive world rights.
CREDIT LINE TO BE USED:

Date_____

By_____
                (Signature)

*Fig. C-1.* Sample letter for requesting permission.

the resulting figure may be regarded as one's own. Nevertheless, honesty still requires that one cite the source of the original with a phrase such as

... (adapted from Kolikowski and Trojanovsky, 1980).

*Text citations* are always permissible. No special permission is needed as long as the citation is kept within the limits dictated by the purpose. Thus, a citation must always be subordinate to the new work itself with respect to both size and importance. Nevertheless, direct quotes should be used only where they are truly necessary to clarify or illustrate opinions identical or in contrast to one's own, or where they give important support for one's lines of thought —situations that arise less frequently in the natural sciences than in the humanities.

Works issued by legal entities (laws, government documents, and the like) are usually not copyrighted, and may be freely used by others. This tradition is consistent with the premise that citizens are entitled to have access to information related to the activities of their governments.

# C.7   The Publishing Agreement

The traditional route to wider distribution of a literary work is the transfer of copyright from author to publisher. Even today this is the most common practice, despite the advent of new methods of creating and communicating "works of the mind". The exact nature and extent of the rights to be transferred are typically spelled out in a document known as a *publishing agreement* (*publication contract*). However, even in the absence of such a contract an author-publisher relationship can be subject to copyright law. In the case of journal articles, absence of written contracts is the rule: a contract-like situation is tacitly created by the editor's acceptance of an article submitted for publication.*

---

* Essential elements of the contractual implications of acceptance of an article are often spelled out in an "Acknowledgment of Receipt" form issued by the publisher.

In the absence of a contract, only those rights required for the purpose of publication will be considered to have been transferred. Copyright itself tends to remain with the author (Lat. *in dubio pro auctore*). As a consequence of the limitations imposed by this legal presumption, publishers usually prefer to negotiate contracts, and to stipulate transfer of as many additional rights as possible. Their motivation is clear: if they fail to do so they may find themselves restricted in their activities in unforeseeable and undesirable ways. This attitude on the part of publishers is fortified by the fact that copyright is not compulsory in nature. It rather aims at filling gaps left by incomplete or non-existent agreements (*dispositive right*). Particularly in the case of a book, in which large sums of money are involved and where the potential impact can be significant, the drawing up and signing of a *contract* is highly recommended for the sake of both author and publisher.

Publication agreements between authors and one or more publishers are even more important today than they were in the past, due to the increasingly international dimensions of the modern publishing business and the resulting legal complexities. International licensing of *distribution rights* or *translation rights*, for example, has to be undertaken in conformity with a wide diversity of national laws. Correctly drawn contracts are essential if legal ambiguity is to be avoided.

Every contract incorporates the basic principle that the origin, scope, and termination of copyright to a work are governed by existing law in the country where protection is claimed. In order to maximize the likelihood that a given contract will be effective in several countries, a proviso is often included stating that the terms are subject to all applicable national laws, and that the contract as a whole will not become invalid should any clause prove incompatible with a particular national law. A similar proviso is recommended for publication agreements of all kinds, national or international; its absence could make an entire agreement null and void if one clause turns out to be at variance with some relevant piece of legislation.

# C.8     Book Agreements: Form and Content

Publishing agreements covering books are usually termed *book agreements*. They originate with the publishers, and often are nothing more than preprinted forms containing a series of filled-in blanks, though some are individually typewritten. Agreements of the latter kind are also highly standardized even if they do not look that way. Most are produced on word processors, with custom elements introduced as needed into pre-formulated text units.

Despite their aesthetic shortcomings, printed forms have the advantage of indicating more clearly the unique features of a given agreement. Moreover, the fact that the unique parts are embedded in what is obviously a more general contractual context can be a source of confidence to the author that "his" or "her" contract is not really novel, and therefore not likely to be deceptive or exploitive: other authors will already have subscribed to similar conditions. In some countries, "master contracts" have been developed jointly by publishers' associations and authors' leagues. Where these exist, publishers' agreements usually correspond closely to the recommended terms and wordings. Such documents tend to represent an especially careful balance of the interests of both authors (or editors) and publishers.

The most important particulars in any book agreement are:

- name(s) of author(s);
- provisional title (working title) of the book;
- size of the book (e.g., estimated number of manuscript pages);
- promised date of manuscript delivery;
- projected modes of editing, proofreading, and indexing;
- author's royalty terms; and
- extent and nature of primary and subsidiary rights transferred.

All specifications will be the result of prior negotiations between author and publisher. The *author's royalty* is usually given as a share in either the *sales price* (e.g., 10%) or in the *net proceeds* (e.g., 15%) of every copy of the book sold. For books with multiple authorship it is common for lump sum payments to be made to the individual authors, the amount usually being based on a fixed fee per contribution or per printed page. Reimbursements of the latter kind are known as *author's fees*.

The *mode* of royalty payment (e.g., annually, semiannually, or on the date of publication) should also be clearly specified in the agreement. Occasionally, authors (as well as editors or translators) are promised *advance payments* to be deducted at a later date from the royalties. Particularly with respect to financial matters it is important that the contract be complete in every detail, even to the point of specifying bank account numbers if direct deposit is envisioned, or tax liability provisions. Special care is advised if several authors are involved, in order to minimize the likelihood of subsequent controversies among the parties. Not all disputes are occasioned by money matters; immaterial rights can also be a source of friction. For example, the order in which a book's authors are listed may be of real concern to certain individuals.

Some important parameters related to book production are left to the discretion of the publisher and fixed at a later date, even though they may be of considerable interest to the author. These typically include the printrun and the ultimate selling price of the book.

# C.9    *Contracts with Editors and Translators*

Editors* and translators receive from the publisher contracts similar to those issued to authors. Contracts with *editors* need to be particularly detailed in spelling out what their responsibilities are and the forms of control they will be expected to exercise, with respect to both authors and the publisher. Editors have moral and economic rights comparable to those of authors even though they may not contribute a single word to the volumes they edit (except, perhaps, a preface). The creativity of an editor, which is more hidden than that of an author, manifests itself, for example, in the subject matter, titles, and sequence of chapters, and in the selection of authors. Editors normally claim the right to have their names printed on the title pages of the works they edit, as well as in any related promotional material.

---

* The term "editors" is used here in the context of *outside* editors, persons not normally employed by the publisher. This usage does not, however, exclude the possibility that a publisher might occasionally wish to offer a separate contract to an employee, who would then concurrently hold the rank and status of an outside editor.

Arranging for publication of an edited volume usually requires two kinds of agreements. In addition to an editor's contract there also must be agreements between the publisher and prospective authors, although these often take the form of short *letters of agreement*. Occasionally the latter are replaced by direct agreements between the editor and the authors, in which case it is the editor's responsibility to represent the authors by ensuring that their contributions are published. For example, the several hundred speakers at a conference may submit written versions of their contributions to an editor, who in turn arranges for publication of a *conference proceedings* volume. Upon delivering their manuscripts to the editor the authors implicitly agree to have their papers published, although this tacit approval may be reinforced by their agreement to sign vouchers provided by the editor. In essence, the procedure is comparable to that attending acceptance of a contribution for publication in a journal (cf. C.7).

*Translation* of a previously (or concurrently) published work presents a rather different situation. A translation, like an original manuscript, is considered to be an intellectual creation, a "work" in its own right. Thus, a *translator* enjoys author-like status. Nevertheless, since most translations are commissioned by publishers, translators are less directly identified with "their" works than are original authors. One consequence of this distinction is that translators are usually paid on a flat fee basis, negotiated in advance.

# C.10   The Work Contract

Encyclopedias and other large compendia are usually handled somewhat differently from other books. Typically, a publisher of a compendium will select a large number of authors, each of whom is expected to contribute one or more essays on topics chosen by the publisher. The specific terms applicable to a given contributor are defined in a *work contract*, which replaces the normal publishing agreement. The alternative designation *hire contract* is perhaps more descriptive, in that it stresses the reversal of the traditional author/publisher dependency relationship. This time the publisher is in command, and there is no firm commitment to publish any particular manuscript. An author's fee must be paid if a contribution has been properly and promptly submitted, but that is the end of the matter. This is in sharp contrast to the situation with an ordinary book, where a publisher operating under a normal publishing agreement is obliged to publish any work submitted, provided, of course, it has been prepared and delivered according to agreement (cf. Sec. C.5).

## C.11    The Copyright Notice

In order to be legally protected from piracy or plagiarism a book must contain a *copyright notice*. This information must be on the title or imprint page and preceded by the standard copyright symbol, an encircled letter "c" (©). The notice must contain both the date of publication and the identity of the holder of the copyright. Although this form of copyright notification is now in worldwide use, it originates in United States copyright law, which defines certain *formal requirements* that must be met if copyright guarantees are to be applicable, one being the presence of the copyright notice. A copyright notice must appear in all editions of a work, beginning with the first publication. The obligation to ensure that a legally valid copyright notice is incorporated rests with the publisher, and usually the publisher is identified as the *holder of copyright*, although occasionally the original copyright holder, the author, is named instead.

# Appendix D
# Using the Chemical Literature

## D.1    Acquiring an Overview

That "the literature" plays a central role in science is a recurring theme in this book (e.g., in Sections 1.2.2, 2.2.13, 2.3.3, 3.1, and 4.3.2, as well as Chapter 11). Indeed, the book itself could be viewed as our attempt to exert a positive influence on the quality of certain types of scientific literature. The purpose of this appendix is to provide a brief description of the chemical literature as a whole, and to indicate some of the many ways it would be approached by the chemist in search of information.* The reader with little training in chemistry may find the subject particularly interesting. Analogies to the literature of other disciplines will surely suggest themselves, and this introduction may aid the non-chemist in locating needed bits of chemical information.

The chemical literature comprises works of many types. This diversity is partly a reflection of the fact that chemists—scientists in general—search libraries for answers to various types of questions. For convenience, we might group the objects of these quests into three broad categories:

(1) an *overview* of a broad area within some discipline,
(2) a *status report* reliably summarizing current knowledge and efforts in a specific field, or
(3) a detailed and exhaustive *answer to a specific question.*

---

* The subject is extremely complex, and our treatment here is both selective and cursory. Additional information is available in the excellent works by Bottle (1979), Maizell (1987), Mücke (1982), and Woodburn (1974), to name but a few.

Various components of the chemical literature will be examined to see how they relate to these three types of search, beginning with the problem of surveying a broad landscape.

If one requires a concise but general description of a subject, the best place to look is often outside the technical literature altogether. For example, one might first consult a major *encyclopedia* such as the *Encyclopedia Britannica*.* It is astonishing how useful sources of this type can be—yet they are rarely consulted. Another possibility is the 6th edition of the McGraw-Hill *Encyclopedia of Science and Technology* (1987), encompassing some 13 000 pages of information. More commonly, the chemist relies on a *textbook* to provide a first orientation within a discipline, thereby making the not unreasonable (though optimistic!) assumption that textbook authors are particularly mindful of the problems confronted by the novice. Such a search is most easily carried forward to more technical ground with the aid of the *reference list* that usually accompanies a general introduction.

It is also possible to access the chemical literature directly, since several broad information sources are designed to provide a direct entry into chemical information. These include the *Encyclopedia of Chemistry* (van Nostrand, 4th ed., 1984), *Römpps Chemie-Lexikon* (8th ed., 1979- ),[†] the series *Techniques of Chemistry* (Wiley, 1971- ), the *CRC Handbook of Chemistry and Physics* (CRC Press), and *Lange's Handbook of Chemistry*, the last two subject to frequent revision (e.g., 67th ed., 1986, and 13th ed., 1985, respectively).

The next level of comprehensive information sources comprises somewhat more specialized encyclopedic works whose content is restricted to particular subdisciplines. Examples include *Ullmann's Encyclopedia of Industrial Chemistry* (VCH Verlagsgesellschaft, 5th ed., 1985- ), *Kirk-Othmer Encyclopedia of Chemical Technology* (Wiley, 3rd ed., 1978-1984), *Encyclopedia of Polymer Science and Engineering* (Wiley, 2nd ed., 1985- ), and *Biotechnology* (VCH Verlagsgesellschaft, 1981- ), sources that could be described as retaining a measure of "general" appeal despite their specialized nature. Other major treatises in this category are likely to be of more interest to the specialist. One might assign to this category such works as *Comprehensive Inorganic Chemistry*, *Comprehensive Organic Chemistry*, *Comprehensive*

---

\* We have deliberately omitted full references to standard works familiar to most readers.—It may be noted that editors' names are often utilized when referring to certain major work (e.g., "Weissberger" as a synonym for *Techniques of Chemistry*).

† Unfortunately, this extremely valuable reference work, edited by O.-A. Neumüller and published by Franckh'sche Verlagshandlung in Stuttgart, is available only in German.

*Organometallic Chemistry* and *Comprehensive Biotechnology* (all from Pergamon, Oxford; 1973, 1979, 1982, and 1985, respectively), *Rodd's Chemistry of Carbon Compounds* (Elsevier, 2nd ed., 1964- ), *Houben-Weyl Methoden der Organischen Chemie* (Thieme, 4th ed., 1958- ), *Treatise on Analytical Chemistry* (Wiley, 1959- ), and *Methods in Enzymology* (Academic Press, 1955- ). A recent addition to the ranks is *Reactions and Methods in Inorganic Chemistry* (Zuckermann, Ed.; VCH Publishers,* 1986- ), in which chemical reactions are classified according to the nature of the elements involved in bond formation.

Here we should digress briefly to point out that comprehensive works of the type cited above usually fall into one of two distinct categories depending on how they are structured. Some can be described as *closed publications*, for which a complete organizational framework is established prior to the publication of any volumes. Thorough treatment of a given subject is thus assured within the confines of a defined and limited number of volumes, and a strictly logical structure is feasible. *Comprehensive Organic Chemistry* is an example of such a work; it is divided into discrete sections assigned to particular topics, including stereochemistry, hydrocarbons, and halogen-, oxygen-, and nitrogen-containing compounds, to name but a few. The entire work is confined to five volumes, accessible by means of a single index.

A closed publication has the distinct advantage that the information contained is relatively easy to locate. Indeed, an information search can often be conducted in such a work solely on the basis of its organizational scheme, allowing one to bypass the index. Nevertheless, there is also a major disadvantage: literature coverage is necessarily restricted to information available at the time of publication, either of the entire work or of a particular volume within it.[†]

The alternative is an *open-ended series*. Works of this type are continuously under development and are—in theory—published somewhat more promptly, which adds to their usefulness. However, they lack a systematic structure, a disadvantage that results in their being difficult to access. Many of the "Advances" series listed in Table G-2 are ongoing publications in this sense. As we shall see below, some publications (e.g., *Beilstein* and *Gmelin*) are best described as hybrids of the two basic types of encyclopedic resource.

---

* VCH Publishers, New York, is the U.S. subsidiary of VCH Verlagsgesellschaft, Weinheim.
[†] Actually the restriction is more severe; time must also be allowed for writing, editing and printing.

The list of major sources of chemical information provided above is to some extent random, but it contains many titles familiar to every chemist (at least in the Western world). With but two exceptions (*Römpp* and *Houben-Weyl*) they are all in English, further testimony to the assertion that English has become the dominant language of chemistry (cf. Sec. 3.2.5, "Choice of Language", and Sec. B.1). Two other major—and nearly indispensable—encyclopedic works, *Gmelins Handbuch der Anorganischen Chemie* and *Beilsteins Handbuch der Organischen Chemie*, were for decades available only in German. Indeed, they constitute one of the principal reasons why generations of chemists throughout the world felt compelled to learn at least a smattering of German. Times have changed, however: both "Gmelin" and "Beilstein" have now begun the process of converting to English, the language in which all future volumes will appear.*

We shall not attempt to describe the structures of these giant works in any detail, or the ways in which they should be approached. For our purposes it is sufficient to note that both attempt to summarize systematically everything that is known about the materials they treat: either inorganic or organic substances. Both rely on a strict organizational scheme based on a fixed number of volumes, to which supplements are regularly issued. Chemistry students are expected to become familiar with both, and both are heavily utilized. Nevertheless, and despite their size, neither can be regarded as a practical source of *general* information. Therefore we shall return to them briefly in Section D.3.

Mention must also be made of one other major work of German origin: *Zahlenwerte und Funktionen aus Naturwissenschaften und Technik*, known generally as *Landolt-Börnstein*. This enormous collection of physical data has long been relied upon by natural scientists of every kind despite the difficulties posed by the absence of a general index (a problem the publisher has finally addressed). One attractive feature of *Landolt-Börnstein* is its emphasis on graphic presentations. The latest (sixth) edition of the work was completed in 1980. According to current plans, more recent data will be incorporated into a set of supplementary volumes known as the "New Series". Other more

---

* *Ullmann's Encyclopedia of Industrial Chemistry* has taken the same step for its 5th edition. A July, 1985, issue of *Nature* announced the latest developments in a note headed "Chemical References in English" which began: "Two of the oldest and best known reference works for chemists, *Beilstein Handbook of Organic Chemistry* and *Ullmann's Encyclopedia of Industrial Chemistry*, published in German for 104 years and 71 years, respectively, are now appearing in English ...".

specialized sources of physical data (e.g., thermodynamic values, spectra, etc.) are described in Maizell (1987) and other works on the literature.

Some general questions are best answered not by a comprehensive information source but rather by a work providing an in-depth treatment of a narrower subject, supported by extensive references to relevant primary literature. The ideal resource in such a case is an appropriate *monograph.* Unfortunately, sources of this type are not always easy to find, and ascertaining whether such a source even exists can be a challenging assignment. A possible starting point in a search for recent monographs published in English is the latest edition of *Books in Print* and its companion volume, *Subject Guide to Books in Print,* both published by Bowker (New York).* These are readily accessible (at least in the United States) through most libraries and book dealers. *Books in Print* currently includes approximately 700 000 titles issued by some 7000 publishers. As the title of the work indicates, each listed book is available for purchase at the time of publication (*Books in Print* is published annually). Four volumes are required to index the listings. Volumes 1 and 2 are arranged alphabetically by author, Volumes 3 and 4 by title. Each entry specifies the author, coauthor(s), editor(s), title, price, publisher, year of publication, number of volumes, Library of Congress (LC) number, and ISBN (cf. Appendix E). Bowker Co. also prepares the more limited *Scientific and Technical Books and Serials in Print,* the content of which is extracted largely from *Books in Print.* This work (listing approximately 100 000 books and 17 000 serials) is popular among science librarians because of its smaller format and correspondingly lower price. Other more specialized *directories* (guides) to published resources include various lists of serials,† government documents, microform editions, and library holdings (cf. Herner, 1980).

Despite the existence of resources such as these, finding suitable monographs remains a challenge. Even if promising titles do appear in *Books in Print,* a chemist will rarely order, sight unseen, several books of uncertain

---

* Equivalent publications in the United Kingdom and the Federal Republic of Germany, respectively, are *British Books in Print* (Whitaker, London) and *Verzeichnis lieferbarer Bücher* (Verlag der Buchhändler-Vereinigung, Frankfurt). *Books in Print* has recently become available in compact disk form. No source of this type is exhaustive: a given title is included only upon the request of the corresponding publisher, who must pay for the service. Most publishers are prepared to cooperate, however, because it is in their interest to inform the public of their offerings.

† Certainly the most important of these is *Ulrich's International Periodicals Directory* (Bowker), listing over 68 000 periodical titles.

relevance. More commonly one would first attempt to examine the books themselves, usually in an appropriate library—at a university,* research institute, or corporate facility. This will entail searching a *card catalog,*[†] which may reveal additional potentially interesting books (monographs) no longer included in *Books in Print.*

Today, thanks to computerized data banks, many libraries are equipped to identify not only the books on their own shelves, but also the resources available in a host of other libraries. Such unified catalogs are usually linked to interlibrary loan services, allowing one to obtain materials of interest from remote institutions in a matter of days.[‡] It is to be hoped that similar facilities will soon become available on an international basis.

Following the monograph, the next level of information is the *review* (or *review article*). Unlike the sources described above, reviews are *dependent* publications (cf. Sec. 4.1.1 and Sec. 11.4.1, "Document Classification"), so they are not listed in the catalogs of publishers or libraries. A review is a particular type of journal article, hence the search for such material usually begins with the most important index to the *periodical literature* of chemistry, *Chemical Abstracts (CA)*. Each weekly issue of *CA* is divided into 80 subject sections, and each of these begins with references to any newly published review articles. The *CA* indexes (both the twice-yearly "Volume Indexes" and the "Cumulative Indexes", currently appearing at five-year intervals) clearly distinguish references to review listings by appending the letter "R" to the corresponding abstract number.

An important guide specifically devoted to review publications is the *Index to Scientific Reviews*, published semiannually by the Institute for Scientific Information (ISI). Other more specialized resources include the *Index of Reviews in Organic Chemistry* of the Royal Chemical Society. If a pertinent review has appeared in a journal to which one lacks immediate access, *Chemical Abstracts Service Source Index (CASSI,* cf. Sec. 11.4.3) can be used to obtain a list of potential source libraries.

---

* The libraries of major universities are usually subdivided into disciplinary branches. One's search is therefore most likely to begin in a specialized collection housed in or near the relevant academic department.

[†] "Card catalog" is rapidly becoming a misnomer as more and more institutions establish computerized cataloging systems. These not only provide the information traditionally found in a card catalog but also indicate whether or not a book is currently available.

[‡] Telephone transmission from one library to another of *facsimile copies* of short documents is a relatively new and increasingly popular option.

Major technical information services like *Chemical Abstracts* can also prove useful in calling attention to monographs. The titles of many newly published books appear in *CA* immediately following the listing of reviews. Each entry contains complete bibliographic data, including the number of pages in the book. *BioSciences Information Service* (*BIOSIS*) is a similar resource, and it also lists books of interest to chemists. *BIOSIS* listings have the advantage of providing brief content summaries along with bibliographic information. However, it is important to note that coverage of book titles by both *CA* and *BIOSIS* is quite fragmentary because it depends on voluntary cooperation between book publishers and the respective information service.

Locating a key monograph or review is an important achievement, because it provides access (through *citations*) to hundreds or even thousands of further sources of information, mainly articles in scientific journals—the *primary literature* (cf. Sec. 3.1.2). At this point the problem becomes one of finding time to *evaluate* and *study* a set of sources rather than of identifying and locating them.

# D.2    Keeping Current

Nearly every scientist has experienced the frustration of trying to keep abreast of new developments in a discipline—of maintaining a comfortable level of *current awareness*—and each develops a unique set of personal reading habits in an attempt to reach this ever-elusive goal. One obligatory step is regularly reading or at least browsing through several specialized primary and secondary journals. A chemist's reading list should also include such journals as *Science*, *Nature*, *Accounts of Chemical Research*, *Angewandte Chemie*, and *Journal of the American Chemical Society*—even *Chemical and Engineering News* or *Journal of Chemical Education*. These publications broaden one's horizons and provide insight into progress on all the major research frontiers.

Nevertheless, generalized reading represents a more or less random approach to the acquisition of knowledge, and it includes the strong probability that some crucial publication will be overlooked—perhaps because it has been published in an obscure place. Many feel it essential to encounter promptly *all* significant new results in their field even though they cannot possibly read every scientific journal. This has resulted in the development of a number of specialized information services designed to meet precisely these needs. An example is *Chemischer Informationsdienst* (*ChemInform*), sponsored by the German Chemical Society (Gesellschaft Deutscher Chemiker, GDCh) and published weekly by VCH Verlagsgesellschaft. *ChemInform* is directed mainly toward the organic chemist, and its most distinctive feature is extensive use of structural formulas and reaction schemes, permitting the reader to assimilate the key features of a newly published paper at a single glance. Brief abstracts (in German) are also provided. *ChemInform* is selective in its coverage, and it is limited to results published in what the editors regard as "important" primary journals. The same holds true for the analogous ISI publication *Index Chemicus* (*IC*). This is also addressed primarily to the organic chemist, but unlike *ChemInform* it is supported by a set of comprehensive indexes, which facilitates *information retrieval*.

One of the most ambitious recent developments in the realm of current awareness is the concept called *CA* SELECTS, introduced by Chemical Abstracts Service (CAS). By early 1987, CAS was distributing, on a bi-weekly basis, individualized listings of all current *CA* abstracts considered relevant to each of 180 specific chemical and chemical engineering topics. The list of topics continues to expand, and it is carefully tailored to meet a wide variety of specific needs. Most of the topics cut across the boundaries of the traditional *CA* subject sections. Thus, the interested chemist has only to *select* the listed topic or topics of greatest interest (e.g., atomic spectroscopy, steroids, antitumor agents, pollution monitoring), pay a relatively modest fee, and then await the arrival twice-monthly of a highly specific (and correspondingly small and manageable) edition of *Chemical Abstracts*. *CA* SELECTS provides the opportunity to review at leisure *all* the recent (relevant!) results appearing in the chemical literature. The concept has been warmly received by a surprisingly large number of individual subscribers.*

---

* When *CA* was introduced early in this century it was assumed that most serious chemists would skim through each issue in its entirety, and for several decades this was indeed the case. However, the explosive growth of the chemical literature since the 1950s has made this ideal utterly impractical. Subscriptions peaked in 1949. ACS members received special (subsidized) rates for individual subscription, but these increased in 1963 from $40 to $500 annually and were eliminated entirely in 1966. Maizell (1987) provides interesting background information on *CA* (see also Schulz, 1985).

Publishers have shown considerable ingenuity in devising new ways of supplying timely information to researchers in chemistry. Two additional examples worthy of special mention are *Current Contents* and *Chemical Titles*, publications of ISI and CAS, respectively. These sources are designed to provide the reader with organized lists of the *titles* of recently published articles. The former, available for various disciplines, consists of miniaturized reproductions of the contents pages of selected journals, accompanied by a very limited set of indexes. *Chemical Titles* features consolidated title lists, the most important being a *Keyword in Context* (*KWIC*) index. Here, all significant words found in the titles of all current articles are alphabetized and displayed in context (i.e., surrounded by several adjacent words from the same title). This permits one to scan the entire contents rapidly by looking only for relevant terms.*

Specialized current-awareness services are becoming available in increasing numbers, not only for research scientists, but also for marketing managers, librarians, and others located at the fringes of science. One example of a publication designed for a wide readership is *CAS BioTech Updates*, available since January, 1986, from CAS. This bi-weekly bulletin includes abstracts from *Chemical Abstracts*, but it also contains other information related to biotechnology, such as reports of newly developed processes, news related to people in the industry, and significant government activities. Information is broken down into four sections as a function of source: patents, papers, books and reviews, and industry notes.

Finally, mention should be made of the growing popularity of "customized" current awareness schemes based on computerized database searches. These may be performed either according to a fixed schedule or only on demand. Searches of this type generally rely on keyword analysis of the titles of published articles, and they are offered by both CAS and private database proprietors (*hosts*), including *DIALOG* Information Services (a division of Lockheed), *InfoLine* (Pergamon, London), STN International (FIZ, Karlsruhe, West Germany), and *DARC*-Questel (Paris). Anyone wishing to take advantage of such a system first establishes a *search profile* in order to define a field of interest. On the basis of this profile the proprietor or host prepares a comprehensive list of bibliographic data for all newly recorded and

---

* The existence and importance of publications such as these emphasizes rather sharply the need for authors to avoid ill-defined or trivial titles. An important set of findings can easily be overlooked by potential readers because of the poor choice of a title (cf. Sec. 3.3.1).

(apparently) relevant documents. Lists may be provided at intervals as short as two weeks. This particular route to current awareness is often referred to under the heading *selective dissemination of information* (SDI). In essence it could be described as a personally tailored alternative to *CA Selects*, and it is probably the most elegant approach currently available. Details concerning the costs and advantages of various SDI services can be obtained from the appropriate database managers. Maizell (1987) provides extensive discussion of the subject (see also Pichler, 1986).

# D.3   Finding Answers to Specific Questions

It is much more difficult to generalize about how one should approach the literature when one's need is for a *specific* piece of information. Too much depends on the precise nature of the question. If it relates to a specific *compound*, however, the first step is straightforward: one turns to *Gmelin* if the substance is inorganic, or *Beilstein* if it is organic (cf. Sec. D.1). Sometimes both may prove useful, particularly since there is an increasing degree of overlap between the two works. For example, although *Gmelin* began to treat *organometallic compounds* only in the early 1970s, the extent of this coverage has expanded to such a degree that by 1986 it encompassed 25 volumes.

These two sources have the special advantages that they are exhaustive but at the same time systematically organized. Moreover, all data are subjected to critical evaluation prior to their inclusion, a most unusual feature for a secondary source. The fact that earlier volumes are in German in no way diminishes their importance to the English-speaking chemist. Most of the information contained can be acquired with minimal linguistic skills, and the effort is usually more than compensated. Both works are organized according to compound structure, and a person familiar with the systems can proceed almost immediately to the appropriate volume and page without recourse to an index (although partial indexes to both do exist).* Compound entries in both works are divided into subsections, including natural occurrence, published syntheses, physical and spectroscopic properties, chemical behavior and reactions, and applications. The only limitation is one of timeliness: the extent to which a comprehensive answer can be found to a given question depends on how recently a relevant volume was published. Further information regarding these sources can be obtained from the excellent descriptive brochures prepared by the publisher (Springer-Verlag: Berlin, Heidelberg, New York).

If more recent information is required about a compound, the only practical solution is to conduct a search through *Chemical Abstracts*, and this requires the use of an appropriate index. The time required for such a search can be minimized by making use of the *CAS Registry Number* (cf. Sec. 6.4) of the compound in question, although a search by molecular formula or by name is also feasible. In the latter case it is important that one first consult the *CA Index Guide* to determine what name one can expect to find used in the current index. For additional information on efficient use of *CA* the interested reader is advised to consult Schulz (1985),† Maizell (1987), or the various descriptive publications available from CAS.

The organic chemist in search of information regarding a *class* of compounds will probably turn first to works such as *Houben-Weyl Methoden der Organischen Chemie* (another remarkable German work for which no English equivalent exists), the volumes in the set *Chemistry of Functional Groups* (Patai, Ed.; Wiley), and *Rodd's Chemistry of Carbon Compounds* (Elsevier, 2nd ed., 1964- ). These sources provide information on general *reactivity* and *synthetic methods*, as well as discussions of *reaction types*. Another important tool in the same category, but with a periodic publication schedule, is *Synthetic Methods in Organic Chemistry* (formerly edited by Theilheimer—and still known by that name—but now produced by Derwent under the editorship of Finch; ongoing). The latter is organized according to types of reactions, types of atoms involved in bond making and bond breaking, and types of reagents. Similar organizational features have also been

---

* The Beilstein Institute now offers a search system—known as *SANDRA* (Structure and Reference Analyzer)—for the purpose of locating compounds within the *Handbuch*. The software is adapted to personal computers and permits graphic entry of structures with the aid of a "mouse".

† This book is in German; at the time of publication of the present book an English edition is being prepared. The chemist familiar with German is also well-served, not only with respect to *CA* but to *Beilstein* and *Gmelin* as well, by Mücke (1982).

incorporated into the publication *Chemical Reactions Documentation Service* available from Derwent. The service consists of two parts, *Journal of Synthetic Methods*, designed to meet current awareness needs, and a *database* (including the Theilheimer volumes) for conducting retrospective retrieval.

*Chemical Abstracts* can also be used for performing an exhaustive search for a particular reaction or class of reactions, but the task is one requiring far more skill than is needed for locating references to a compound. Success depends heavily on establishing an appropriate set of sufficiently narrow search terms (e.g., "hydrogenation", "methanation", "deoxymercuration", etc.), which in turn requires considerable study of the *CA General Subject Index* in conjunction with the *CA Index Guide*. As many relevant synonyms as possible should be utilized in such a search, since concepts—unlike compounds—defy unique specification. Moreover, even a conscientious and time-consuming search is likely to yield only fragmentary results.

More and more frequently, searches of the latter kind are carried out with the aid of *online* techniques; i.e., by appealing to *CAS OnLine* or some similar computerized database, accessed either directly or through the intermediacy of a *host*. In some cases the entire process is carried out locally (provided sufficient computer capacity is available) by utilizing magnetic tapes supplied by CAS. More often a *terminal* is used to communicate by telephone with a remote database. However accomplished, a computer-aided literature search can provide astounding efficiency (cf. Sec. 11.2.3). This efficiency manifests itself in two ways: time is saved and the quality of the search is enhanced, resulting in considerably more relevant information than could possibly be obtained manually. One of the great strengths of *online retrieval* is the fact that the questioner can maintain a *dialog* with the database, permitting the evolution of complex, multi-dimensional searches of a type inconceivable to the researcher forced to rely on the resources of a traditional library.

The current state of search technology is dramatically illustrated by the fact that one can now retrieve information concerning molecules on the basis of *topology* through the specification of key *substructure* elements. Not surprisingly, a video screen plays a major role in the process. Such searches first became possible with the advent of the DARC service in France (accessed through the host agency Télésystèmes-Questel) and *CAS Online*, the latter available to the public since 1980.

Another important and more recent computer-based search system is *ACS Journals Online*, a service of the American Chemical Society. This service provides access to the full text of articles published in recent ACS journals, and is designed primarily to supply answers to analytical questions. The area of computer-assisted literature searching is characterized presently by extremely rapid progress, and scientists are well-advised to follow the developments closely and take advantage of opportunities to experiment with any available facilities.

Returning to the subject of literature searches in general it should be noted that information in *CA* can also be retrieved on the basis of authors' names and patent numbers. This applies to both conventional and computer-aided searches. Patents pose a number of unique problems to the chemist seeking information, and their nature and use are discussed at length by both Mellon (1982) and Maizell (1987). One important source of patent information is the *Chemical Patents Index* (*CPI*), prepared by Derwent Publications (London, 1970- ). Each weekly issue of *CPI* treats more than 12 000 documents, about 5500 of which (so-called "basic patents") relate to inventions never described previously. *CPI* is both an alerting and retrieval service, and it is capable of rapidly yielding extensive information. The retrieval function is supported by cumulated indexes, coded cards, magnetic tapes, and online search capability. One important feature of the *CPI* is its classification system, under which patents are categorized into twelve major sections and 135 well-defined "Derwent Classes".

One final documentation system deserves special mention because it is so unlike all the others: ISI's *Science Citation Index* (*SCI*). This source provides the unique opportunity to carry a search *forward* in time by pointing to publications that have *cited* an earlier publication. In other words, one can proceed from some known piece of definitive literature to more current publications—from past to present. This also makes it possible to define a topic in terms of an entire *article* rather than simply a set of keywords or index terms. Brochures available from the Institute for Scientific Information (Philadelphia, Pennsylvania) provide a complete description of *SCI* and the ways in which its organizational principles can best be utilized (see also Woodburn, 1974; Maizell, 1987; and Mücke, 1982). *SCI* is an excellent example of a tool made substantially more valuable through a computerized database. The need to consult two types of index (the *Citation Index* and the *Source Index*) and the lack of a single set of cumulative indexes makes a conventional search of *SCI* a tedious undertaking tolerated only because the results are so valuable.

# Appendix E
## International Standard Serial Number and International Standard Book Number

## E.1    The ISSN

The concept of the *International Standard Serial Number* (ISSN) originated in 1968 at a joint meeting of committees representing the *American National Standards Institute* (ANSI) and the *International Organization for Standardization* (ISO). The purpose of the meeting was to seek some means of keeping track of the serial publications appearing in ever-increasing number throughout the world. The committees' recommendations were accepted by an ISO assembly in 1971 and led to the creation of the International Serials Data System (ISDS), with headquarters in Paris and subsidiary national centers worldwide.

Representatives of the sciences were particularly interested in the development of a universal serials cataloging system because of the internationally interdependent character of scientific research and the literature supporting it. The ISSN finally provides a means of unmistakably identifying any serial published, regardless of its place of origin or language. In a sense, an ISSN plays the same role within the world of information and documentation that a CAS registry number (cf. Sec. 6.4) plays within the field of chemistry.

An ISSN always consists of eight digits. In the interest of readability and recognition these are written as two four-digit sets, connected by a hyphen and preceded by the letters "ISSN". A particular sequence of digits corresponds to a particular serial, but implies nothing more. The numbers have no intrinsic meaning and provide no clues to a serial's subject area, country, language, or any other characteristic. The ISSN is simply an administrative instrument for use in the maintenance of records dealing with serials and their publishers: new serials, old serials, even serials whose publication has been terminated. A given ISSN can be assigned only once, and it is withdrawn from active service upon the death of the serial with which it is associated. Even though its primary purpose is to assist publishers, librarians, and book dealers, the ISSN is also important for scientists because it facilitates literature searches.

The definitive publication on ISSNs, Standard *ANSI Z39.9-1979 (R 1984)*, defines the word "serial" rather broadly. Journals are obviously included, but so are book series, yearbooks, government bulletins, corporate magazines, the "proceedings" volumes issued in conjunction with regularly recurring conferences, and routine reports of various organizations and institutes. In short, any publication issued either regularly or as an indefinite number of successive parts is eligible for an ISSN. Sometimes the individual parts of a serial publication also qualify as books; when this is the case each is also given its own ISBN (cf. Appendix E.2). The principal restriction upon ISSN assignment is that no publication will be regarded as a serial if a pre-planned limit exists on the number of issues. Most systematic reference works are therefore excluded. Newspapers and popular magazines are covered by the definition, but with these ISSNs play only a minor role.

ISSNs are assigned by the ISDS central office in Paris. Publishers can apply for a number by writing to the appropriate ISDS national office.\* A request should contain basic information on the serial (including the frequency with which it is published, its editor's name and address, the holder of the copyright, and the price) and should be accompanied by a copy of the publication.† The information submitted is stored by ISDS in both national and central databanks along with what is called a *key title* for the publication. A key title is a kind of abbreviation formed according to a fixed set of international rules (cf. Sec. 11.4.3 and Appendix G), and its form is unaffected by other recommended nomenclature or referencing guidelines. Once issued, an ISSN continues to be valid as long as the recorded key title exists. In the event that a

---

\* Alternatively, application may be addressed directly to the ISDS International Center, 20 rue Bachaumont, F-75002 Paris, France.
† Actually, two issues of the serial should be sent: a copy of the first or *premier* issue (or part), and the most recent issue. However, publishers may also apply for an ISSN prior to a serial's initial publication.

serial's title changes in a way that also requires a change in its official key title, a new ISSN is assigned.

Publishers are requested to print a serial's ISSN in a specified format on every issue, and to inform ISDS promptly of any changes with respect to title, ownership, publication mode, etc. Cooperation is usually assured, because it is in every publisher's interest to help international information services keep abreast of the current status of their publications.

Despite the almost universal acceptance of the ISSN system, Chemical Abstracts Service designates scientific journals by a second computer-oriented identifier, the "CODEN". A CODEN consists of a unique combination of six letters* chosen to bear some obvious relationship to the corresponding journal title. It is beyond our scope to delve further into the functions and virtues of the CODEN system apart from noting that a given CODEN, like an ISSN, is specific to one particular serial.[†] Most scientific journals possess (and display) both ISSN and CODEN identifications.

# E.2    The ISBN

An *International Standard Book Number* (ISBN)[‡] is similar to an ISSN in that it also permanently identifies an individual publication, but the subject is a *book*, a non-serial work. Just as the term "serial" was used in Appendix E.1 to cover a wide variety of publications, so the word "book" can be broadly understood to encompass brochures, pamphlets, and the individual parts (issues, installments) of collective works (including loose-leaf editions). Every copy of a book produced in a single printing carries the same ISBN. Different versions of the "same" book, however, such as hardcover and softcover, have

---

* Or five letters and a digit; the sixth character performs a control function.
† CODENs are discussed briefly in Sec. 11.4.5, however.
‡ The ISBN system is supervised by the International Standard Book Number Agency, Staatsbibliothek Preußischer Kulturbesitz, Reichpletschufer 72/76, D-1000 Berlin 30. So-called "group agencies" exist in various countries to deal with routine local management of the system.

different ISBNs. A *reprint* of a book by the original publisher retains the original ISBN, but a *new edition* of a work is always assigned a new ISBN even if the changes are minor, because in the strictest sense it is a new publication.

As with the ISSN, an ISBN provides unambiguous identification for a publication. The work is thereby made more accessible not only to prospective purchasers but also to someone wishing merely to examine it, perhaps after seeing it cited in a journal article. The similarities between the two systems end with their common purpose, however. Unlike an ISSN, an ISBN possesses a definite internal structure, and it conveys certain intrinsic information (cf. Standard *ISO 2108-1972*). Each ISBN consists of ten digits,* arranged in four sets. The sets are called *identifiers*. In a printed ISBN the four identifiers are separated by either hyphens or spaces, and the whole is preceded by the letters "ISBN". A typical example is:

ISBN 3-527-25944-9

The first "set" in the above example, the numeral "3" alone, is the *group identifer*. It defines the *language domain* to which the book belongs. For example, group "0" (together with group "1") encompasses the English-speaking countries, group "2" covers France and other francophone territories, "3" includes all countries using the German language, "4" is Japan, and so on. It should be noted, however, that this identification relates solely to the place of publication and does not necessarily indicate the language of the text. The principal domains, those expected to have the most publications, are assigned single-digit numbers.†

The next set of digits, which may range from two to seven in number, is the *publisher's identifier*. A large publishing house with an extensive publication program is assigned a low number containing only a few digits, while smaller publishers receive longer numbers. The fewer the digits consumed by the publisher's identifier, the more that remain available for the third set, specifying a particular publication. Thus, a publisher from a single-digit language group who is assigned a three-digit publisher's identifier has five

---

* Or nine digits and the letter "x", as explained in a subsequent footnote.
† Only the numbers 0-7 are made available as single-digit group identifiers. Two-digit combinations are restricted to the range 80-94, three-digit groups to 950-997, etc. Similar limitations are placed upon permissible publisher identifiers. The reason for omitting so many apparent possibilities is to permit computer recognition of the end of one identifier and the beginning of another without the need for processing either hyphens or spaces.

digits available for titles. This is the case with our publisher, VCH Verlagsgesellschaft, Weinheim, Federal Republic of Germany, which has the number 527 in group 3.*

Next appears the *title identifier*, that part of the ISBN which specifies a particular title within a given publisher's program. As noted above, a "3-digit-publisher" is entitled to a maximum of five digits for title numbers, and can therefore publish up to one hundred thousand registered items (numbered from 0 to 99 999). So-called "5-digit-publishers", on the other hand, have at most three additional digits at their disposal, permitting the registration of no more than a thousand different titles.

The fourth and last part of the ISBN is a check number. This is derived from the nine numbers preceding it[†] by a rather sophisticated mathematical process.[‡] The computer systems employed by most information services, book dealers, publishers' accounting departments, etc., are equipped to verify the accuracy of any ISBN by independently determining what check number should be present, and making sure the result matches the number listed. If a mismatch occurs, it can be assumed that an error has been introduced somewhere into the listed ISBN.

---

* 3-527 was originally assigned to Verlag Chemie GmbH of Weinheim. Even though the firm changed its name on 1 January 1985 to VCH Verlagsgesellschaft mbH, its publisher's number was unaffected. However, the same publisher's U.S. subsidiary, VCH Publishers Inc. (formerly Verlag Chemie International), located in New York, is assigned an entirely different publisher's number and even belongs to a different group (language domain): its titles bear ISBNs beginning with 0-89573-.
[†] Every ISBN must contain exactly ten characters, not counting the three hyphens. Any spaces in the title identifier not otherwise occupied by digits are filled with zeros.
[‡] The actual procedure for obtaining the check digit is as follows: the first digit of the ISBN (which is from the group number) is multiplied by 10. The second digit is then multiplied by 9, the third by 8, and so forth. Multiplication proceeds according to this pattern until the last digit of the title identification has been multiplied by 2. Next, all the resulting products are added to one another, after which the sum is subtracted from the *next largest multiple of 11*. The remainder derived from this subtraction, which obviously must lie between 0 and 10, becomes the check digit. A remainder of exactly 10 is designated as "x." (Quoting Standard *ISO 2108-1972*, "the check digit is calculated on a modulus 11 with weights 10-2, using x in lieu of 10 where ten would occur as a check digit.") For our previously cited example, ISBN 3-527-25944-9, the calculation leads to:
$(3 \times 10) + (5 \times 9) + (2 \times 8) + (7 \times 7) + (2 \times 6) + (5 \times 5) + (9 \times 4) + (4 \times 3) + (4 \times 2) = 233$;
$22 \times 11 = 242$, and $242 - 233 = 9$, the check digit that has indeed been assigned. Thus, the complete ISBN is far from an arbitrary number. The one we have chosen for our example has, for us, yet another (and very special) meaning: it was assigned to a precursor to the present book (Ebel and Bliefert, 1982).

Unlike ISSNs, most ISBNs are issued and administered directly by publishers. The publishers are expected to inform their national libraries and central book catalog services each time a new title appears, and simultaneously to report the corresponding ISBN. Publishers are also responsible for printing the correct ISBN on each book's imprint page, back cover, and dust cover, and for including ISBNs in all promotional materials and book catalogs.

ISBNs also become a part of a so-called *CIP entry* on the imprint page of a book if the publisher participates in the *Cataloging-in-Publication* (CIP) system (cf. Sec. 4.5.2, "Imprint Page"). This system ensures timely release of key bibliographic data for all forthcoming publications, and it is designed to provide early notification to book dealers, librarians, and others in the publishing business (cf. *Guidelines for Cataloguing-in-Publication*, 1986).*

The attention of the central ISBN Agency has recently been focused on the increasingly important role of so-called "new media", including video and software products. International cooperation has been urged in accepting such materials into the ISBN system, a practice already common in several countries, including the United States.

---

* *CIP records* may be obtained by submitting all necessary data on a forthcoming publication (including a copy of the title pages) to the appropriate national library; e.g., to the Library of Congress Cataloging in Publication Division, Washington DC 20540, in the US; to the British Library, Great Russell Street, London, WC1B 3DG, in the UK; and to CIP-Zentrale der Deutschen Bibliothek, Zeppelinallee 8, D-6000 Frankfurt 1, in the Federal Republic of Germany. The appropriate CIP data slip is then returned to the publisher, and it is to be incorporated on the imprint page of the publication. At the same time, the data are released for inclusion in the national library's bibliographic computer service and in the (weekly) lists of new publications circulated by the library to subscribers. A book entry in such a list often predates actual publication by as much as eight weeks.

# Appendix F
# Preparing an Index

## F.1    General Considerations

In this appendix we provide the details required to complete the general discussion of index preparation begun in Sec. 4.5.1. Proper indexing of a book is a difficult and extremely time-consuming task. Nevertheless, most readers of technical books, and certainly those who consult such books for reference purposes, will agree that the quality of an index can be decisive in determining a book's value. The following suggestions and guidelines are drawn mainly from a pamphlet prepared by VCH Verlagsgesellschaft for the use of its authors.

The most challenging and difficult part of preparing an index is selection of the terms to be included. Considerations of space and cost usually dictate a limited number. In general, an average of five to eight index terms per printed page is regarded as reasonable.*

In selecting terms one should always keep the *reader* in mind: given the title and content, what will a reader of the book most likely seek? In principle, compilation of a list of index terms can begin at any time during the preparation of a manuscript, but completion of the index itself must await the arrival of page proofs, because only then will pages have their proper numbering.

One prerequisite for an effective index is that each referenced entity or concept is associated with a single unique term. The reader must be assured that references to *all* relevant passages will be collected in one place and not

---

* This particular book represents an exception. It is designed for use as a *handbook*, and therefore it has an unusually high density of index terms.

scattered among two or three index terms. However, it is equally important to ensure that concepts which can be designated in any of several ways (i.e., for which *synonyms* exist) will be located even if the reader happens to look under a term other than the one the author prefers. *Page numbers* for a given concept should appear only once in an index, but a serious attempt must be made to provide extensive *cross-references*. Such cross-references typically take the form

mesomerism *see* resonance

Terms that are *related* but not *identical* call for index cross-references as well, but these are distinguished by the label *"see also"*. In this case, independent page number entries are provided for *both* terms. Examples of the two types of linkage appear in the index of this book.

Most readers are especially interested to find passages in which a term is treated in depth. For this reason, page number references to discussions involving more than a single page are followed by the letter "f" (or "ff" if the reference extends over more than two successive pages).* If the reference is to a footnote, many publishers recommend placing an "n" after the page number, or printing the number in italics.

Certain index terms will probably figure prominently in the discussion more than once in a book, thereby requiring several different page references. It is especially important to provide a reference to the *first* point at which a given term is introduced, used, or defined. All subsequent relevant passages should also be noted, especially any in which the term is discussed in a different context.

Nevertheless, we recommend that no term be followed by more than five page references. If more are necessary it is likely that the word has been used in various contexts. Readers will be grateful if this fact is made clear; otherwise they may waste time examining material irrelevant for their purposes. The simplest solution is to *subdivide* the term, introducing *qualifiers* for added precision. For example, the term "acidity" might be expanded to:

Acidity constant
Acidity, definition
Acidity, measurement of
Acidity scales

---

* Some publishers use the italicized forms *"f"* and *"ff"*, while others request authors to supply page *ranges* (e.g., "219-223").

An alternative is breaking the original term into a series of *subentries*. Subentry qualifiers are set below a main index term, and are indented and preceded by a symbol such as "-" or "~" in order to indicate repetition of the preceding principal term. Adjectives are generally separated from the connecting symbol by a comma; i.e.:

> Ionization
> ~ constant
> ~, hydrolytic
> ~ of metal complexes
> ~, non-aqueous

Should it be found that a given subentry still requires more than five page references, further breakdown is permissible, resulting in *second order subdivisions*. If this fails to solve the problem, the chosen index entry term is almost certainly too broad to be useful.

# F.2    *Index Card Files*

## F.2.1    *Card Preparation*

The traditional method of preparing an index involves creating a separate *index card* for each index term, to which appropriate page numbers will be added. The process entails systematically reviewing the entire book in a search for passages worth referencing. Each time such a passage is found, the corresponding page number is entered on a card labeled with the appropriate index term. Ideally, one should begin at the front of the book so that all numbers on a given card will be in ascending order. Even titles, headings, and captions (i.e., of text divisions, figures, and tables) should be examined with care, because these often contain terms worth indexing. The completed cards are arranged alphabetically to give an organized set, which then constitutes the *index manuscript*.

Publishers usually supply their authors with the necessary cards. VCH, for example, provides perforated sheets that divide into eight cards. The sheets are initially left intact to simplify typing, and later separated for easy sorting. Each card has a hole punched near the bottom. This allows the author to bind the completed set together, in correct order, by simply threading the cards on a string. The procedure is straightforward:

● Begin by typing in the upper left-hand corner of a card each index term as it is encountered, each time inserting the appropriate page number.*

● Subsequent page number entries for a given card may be carefully added by hand to save constantly returning the sheets to the typewriter.

● As sheets are completed, the cards are separated from one another and placed alphabetically (see below) in a file box.

● If alphabetization reveals more than one card for a given term, all entries are consolidated onto a single card and duplicate cards are destroyed.†

● Any card found to contain too many page entries is subdivided as described in Sec. F.1. Extra cards can be used as necessary, but each card must contain an appropriate *principal* index entry in the upper left-hand corner. The term should be enclosed in parentheses on all but the first card in a series.

● The complete, alphabetized card set is then carefully and thoroughly checked for errors and lack of uniformity with respect to terminology and style.

● Finally, the entire set of alphabetized cards is firmly bound with a sturdy string. If the set proves too large and unwieldy it can be subdivided into manageable portions.

The index manuscript is now ready for delivery to the publisher.

## F.2.2   Alphabetization

Rules for alphabetization can be rather complex, and they vary somewhat among publishers. The following suggestions are typical, however, and they should provide a useful starting point. Details should be clarified with one's own publisher at an early date.

● In general, alphabetization proceeds as one would expect based on such precedents as a dictionary or telephone directory.

---

* Cards can also be filled in by hand, but only if one's handwriting is extremely clear. Typesetters cannot be expected to identify technical terminology written carelessly by hand.

† It is actually preferable to enter all page numbers relevant to a particular index term on *one* card from the beginning, the main reason being that excessive accumulation of page entries is immediately revealed and can be counteracted by forming subentries.

• Most *hyphenated prefix letters* (as in "*d*-glyceraldehyde" and "α-amino acids") are ignored for alphabetization purposes. The same is true for numbers (e.g., 3-chlorobutane) and structural prefixes such as *iso-*, *tert-*, etc. On the other hand, if multiple entries of this kind occur together the prefixes are used for further ordering (e.g., "1-chlorobutane" precedes "2-chlorobutane" and "α-elimination" precedes "ß-elimination").

• Free-standing *Greek letter symbols* (e.g., "Ω" or "σ") should be alphabetized either according to their English names ("omega", "sigma") or placed at the beginning of the entries for the corresponding letters in the Latin alphabet.

• Publishers of books in English generally treat *accented letters* as if the accent were not present. The principal exceptions are German "ä", "ö", and "ü", which are often more correctly alphabetized as "ae", "oe", and "ue". However, even in this case the supplemental marking is sometimes ignored and the letters are treated—or even printed—simply as "a", "e", and "u".

• Index *subentries* are also alphabetized, but "unimportant" words such as conjunctions and prepositions are usually not taken into account (some guidelines suggest otherwise, but the rule stated here seems most appropriate, and it has been followed in the sample subentry list in Sec. F.1).

# F.3    Computerized Indexing

A major disadvantage of the traditional method of index preparation—using file cards—is that much time is required for typing, accumulation of page numbers, and alphabetization. Moreover, mistakes are often made in alphabetization. In principle, both problems can be alleviated with the aid of a *computer*, and such an index would be indistinguishable from its counterpart prepared in the traditional manner. Nevertheless, the process is a problematic one. At least for the present, the following discussion must be regarded as a description of an ideal case.

In general, and from the point of view of the author, little more is involved in preparing a computerized index than simply *underlining* index terms wherever they appear in a complete set of page proofs. The typesetter at the printing house then takes the responsibility for introducing all marked words into a computer. Appropriate software attends to cumulating the entries, alphabetizing them according to the publisher's guidelines, and printing the results.*

The principal prerequisite for successful use of this technique is a high degree of concentration on the part of the author, who must also have a good command of the contents of the book to be indexed. In particular, great care is required to assure proper treatment of index subentries. Otherwise, much time will be required for study and reorganization of the computer printout, using the page proofs as a guide. If, for example, a single index term becomes inundated with page references, each reference must be re-examined in a search for suitable qualifiers. Any potential time-saving relative to manual index preparation might thereby disappear—indeed, there could even be a net loss of time.

Care is also required to counter the inherent "mindlessness" of computers. For example, the average computer is incapable of recognizing plural or adjectival forms of a noun, but it *would* be sensitive to the presence or absence of a hyphen. Each form encountered is treated as a separate word, and a computer assigns to each a separate index entry.

Publishers are quite willing to offer advice concerning computerization of the indexing process, and any suggestions should be followed meticulously. However, it should be noted that such a service is not likely to be available unless an index will be relatively large—more than 500 entries, perhaps—because implementation of the procedure is rather costly.

The following guidelines are typical of those supplied to authors whose books will be indexed by computer.

● Index markings are to be made in a *separate copy* of the page proofs, *not* the one used for proof correction.

● Index terms should be *marked* by *underscoring* with colored felt-tip pens (red, blue, and green) having fine points (ca. 0.5 mm).

● *Principal* index terms are identified by red underlines, *first-level* subentries by blue, and *second-level* subentries by green.

● The underline for a *subentry* should be connected by an arrow to the corresponding principal index term (located at the point of the arrow). If necessary, several arrows may point to the same principal index term.

---

* Recent software developments permit automated preparation of an index—including cumulation and alphabetization—as a part of the actual writing process. Indexing is attended to by a subroutine of the word processing system residing in the author's personal computer.

● Words that appear in a form inappropriate for indexing purposes (plurals, for example, or adjectives) should be restored to their *principal forms* with the same colored pen used for underlining.

● If a term to be indexed consists of *several* words in sequence, all should be joined by a single underline.

● If the term consists of several words, and these appear *separated* from one another, only the words actually constituting the index entry should be underlined. Intervening words should be bridged by a curved line. A curved line should also be used to connect parts of an index term falling on two successive text lines.

● If the actual index term fails to appear on a page where it is nonetheless relevant (e.g., if a discussion of phenols is to be referenced in the index under the term "acids, organic"), it should be written by hand in the *margin* using the same pen that would be employed for a corresponding underline. This includes words like "of" and "in" that may be required to relate a subentry to the corresponding principal term. Principal terms that fail to appear in a discussion of a subentry term should be similarly written in the margin.

● If a word is to be indexed in *multiple ways* (e.g., as a principal entry and in connection with one or several qualifiers), each use should be treated independently with respect to pen color and association with qualifying subentry terms.

● One should append an "f" or "ff" (as appropriate) to a term whose discussion continues on successive pages of text. Such a term would then *not* be indicated a second time in the course of the same referenced passage.

The annotated page proofs should be submitted promptly to the editor. In due course, the marked proofs will be returned to the author, together with a computer printout. The printout takes the form of a list recording all entries in the order in which they were introduced into the computer by the typesetter. The author is expected to proofread the list for completeness, spelling, proper identification of principal and subentries, and correctness of page numbers. Proofreading is best accomplished by two people working together, one reading aloud the material marked in the proof and the other carefully checking the printout. Carelessness at this stage will result not only in increased costs, but also in errors that will probably survive the remaining steps in the publication process.

The corrected printout and its accompanying marked page proofs should be returned as quickly as possible to the publisher, who will then order a computer printout of the actual index. This will contain all entries arranged in alphabetical order, with all their page references. The material should be examined as soon as it is received, with attention directed first to finding entries that contain too many references. These should be subdivided as necessary, a step that entails fresh consultation of the page proofs of the book. At this stage duplications also become apparent, often arising as a result of spelling variations or inflections in number or tense. The extent of the problem will be inversely related to the care given to earlier stages in the process. If the number of errors is too large, all benefits of computerization will be lost due to the time and expense involved in correction.

Some publishers distribute simultaneously the sorted and unsorted computer printouts. In this case corrections should be applied *only* to the unsorted printout. The sorted index printout is intended merely as an aid to the detection of spelling inconsistencies, or entries with too many page references. If this is the procedure to be followed, final proofreading of the index itself must be carried out with special care.

The last step in preparation of any computer-generated index is proofreading the page proofs of the finished index, which are prepared from the corrected second-stage computer printout. It should be noted that computerized index preparation entails three stages of correction, while only two are necessary when file cards are used.

The optimal role of the computer in index preparation remains a subject of active investigation; other approaches have also been tested, as noted briefly in Sec. 4.5.1. Improved procedures are certain to be introduced, but for the present—and the foreseeable future—authors of most complex works must continue to bear the burden of manual indexing.

# Appendix G
## Serial Citation Titles

The titles of scientific and technical serials (periodicals) are almost invariably abbreviated for reference purposes. Appropriate *serial abbreviated titles* (also called *short titles* or *citation titles*) are derived by applying rules developed by various national (e.g., Standard *ANSI Z39.5-1985*) and international (Standard *ISO 4-1984(E)*, Standard *ISO 833-1974*) organizations. Familiarity with the rules and their application makes it possible at least to approximate an acceptable short title from the official full title (or *key title*, cf. Sec. E.1) of nearly any serial.

The basic premise underlying the abbreviation rules is the converse of the claim just made; that is, a scientist reading an *abbreviated* title should be able without difficulty to ascertain the corresponding full title, or at least closely approach it. An attempt has also been made to ensure that the *language* of the title will be apparent from its abbreviation, since this is usually an indication of the language of the contents. (It is assumed that no one would be disappointed to find that articles in *Theoretica Chimica Acta* are not in Latin!) The principal rules are as follows:

● Serial titles comprised of only one word (*Nature, Synthesis*) are spared all abbreviation.

● All short, "unimportant" words are omitted from serial short titles. This includes conjunctions and articles, such as "and", "of", "the", "in", "de", and "der".

● Most remaining words are replaced by consistent abbreviations.

A selection of words commonly found in the titles of scientific serials appears in Table G-1, together with the corresponding ISO abbreviations. Table G-2 contains a selection of the short titles of some of the major journals

*Table G-1.* Selected journal title word abbreviations.

| Title word | Abbreviation | Title word | Abbreviation |
|---|---|---|---|
| Abstracts | Abstr. | Experimental | Exp. |
| Accounts | Acc. | Forschung | Forsch. |
| Academy | Acad. | Fortschritte | Fortschr. |
| Advances | Adv. | Gesellschaft | Ges. |
| American | Am. | Industrial, Industry | Ind. |
| Anales | An. | International | Int. |
| Analytical | Anal. | Journal | J. |
| Annalen, Annales, Annals | Ann. | Medical, Medicine, Medizin | Med. |
| Annual | Annu. | National | Natl. |
| Applied | Appl. | Organic | Org. |
| Berichte | Ber. | Physical, Physics, Physik, Physique | Phys. |
| Biochemical | Biochem. | | |
| Biology, Biologie, Biological | Biol. | Proceedings | Proc. |
| | | Quarterly | Quart. |
| Bulletin | Bull. | Recueil | Rec. |
| Chemical, Chemie, Chemistry | Chem. | Report | Rep. |
| | | Research | Res. |
| Chimie | Chim. | Review | Rev. |
| Comptes | C. | Science, Scientific | Sci. |
| Communications | Commun. | Society | Soc. |
| Engineering | Eng. | Wissenschaft | Wiss. |
| European | Eur. | Zeitschrift | Z. |

in chemistry and related fields. This tabulation is an extract* from the comprehensive list published in *Chemical Abstracts Service Source Index* (*CASSI*), which provides bibliographic data for more than 60 000[†] serials.

---

* Abbreviated titles are not listed explicitly in *CASSI*, but boldface type is used for those elements (words, word fractions, letters) of each full serial title that constitute its abbreviated title. Alphabetization in *CASSI* is based on bold–faced characters only (cf. Sec. 11.4.3).

[†] This number represents the status as of the *CASSI 1907-1984 Cumulative*. Supplements are issued quarterly, typically revealing the addition of about 5000 entries per year. A somewhat less extensive and now outdated compilation was published in 1974 by ACS and CAS under the title *Bibliographic Guide for Editors and Authors*. The latter contained "only" about 27 700 entries and enjoyed a larger circulation than the former due to its more compact size.

*Table G-2.* Selected journal title abbreviations.

Acc. Chem. Res.
Acta Chem. Scand.
Acta Chim. Hung.
Acta Crystallogr.
Acta Endocrinol.
Acta Metall.
Acta Physiol. Scand.
Adv. Biol.
Adv. Biol. Med. Phys.
Adv. Biophys.
Adv. Carbohydr. Chem.
Adv. Chem. Eng.
Adv. Chem. Phys.
Adv. Chromatogr.
Adv. Comp. Physiol. Biochem.
Adv. Enzymol. Relat. Areas Mol. Biol.
Adv. Heterocycl. Chem.
Adv. Inorg. Chem. Radiochem.
Adv. Mass Spectrom.
Adv. Organomet. Chem.
Adv. Org. Chem.
Adv. Phys.
Adv. Polym. Sci.
Adv. Protein Chem.
AIChE J.
AIChE Symp. Ser.
Am. J. Bot.
Am. J. Physiol.
Am. J. Sci.
Am. Nat.
Anal. Abstr.
Anal. Biochem.
Anal. Chem.
Anal. Chim. Acta
Anal. Lett.
Analyst (London)
An. Bot. (London)
Angew. Chem.
Angew. Chem. Int. Ed. Engl.
Angew. Makromol. Chem.
Anim. Behav.
Ann. Chim. (Paris)
Ann. Chim. (Rome)
Ann. N.Y. Acad. Sci.
Ann. Phys. (Leipzig)

Ann. Phys. (Paris)
Annu. Rev. Biochem.
Antimicrob. Agents Chemother.
Appl. Opt.
Appl. Phys. Lett.
Appl. Spectrosc.
Arch. Biochem. Biophys.
Arch. Pharm. Ber. Dtsch. Pharm. Ges.
Arch. Pharm. (Weinheim)
Arch. Tech. Mess. Ind. Meßtech.
Arch. Tech. Mess. Meßtech. Prax.
Ark. Kemi
Arzneim.-Forsch.
Aust. J. Chem.
Ber. Bunsenges. Phys. Chem.
Ber. Dtsch. Chem. Ges.
Ber. Dtsch. Keram. Ges.
Biochem. Biophys. Res. Commun.
Biochemistry
Biochem. J.
Biochem. Pharmacol.
Biochim. Biophys. Acta
Biochimie
Biofizika
Biokhimya
Biol. Cyber.
Biol. Rev.
Biol. Rundschau
Biopolymers
Bull. Chem. Soc. Japan
Bull. Soc. Chim. Belg.
Bull. Soc. Chim. Fr.
Cancer Res.
Can. J. Biochem.
Can. J. Bot.
Can. J. Chem.
Can. J. Zool.
Carbohydr. Res.
Chem. Abstr.
Chem. Ber.
Chem. Biol. Interact.
Chem. Br.
Chem. Commun.
Chem. Eng. News
Chem. Eng. (N.Y.)

*Table G-2.* (Continued)

| | |
|---|---|
| Chem. Eng. Sci. | Helv. Chim. Acta. |
| Chem. Erde | Helv. Phys. Acta |
| Chem. Ind. (London) | Hoppe-Seylers Z. Physiol. Chem. |
| Chem.-Ing.-Tech. | IEEE J. Quantum Electron. |
| Chem. Lett. | Ind. Eng. Chem. |
| Chem. Listy | Indian J. Chem. |
| Chemotherapy (Tokyo) | Ind. Quim. |
| Chem. Pharm. Bull. | Inorg. Chem. |
| Chem. Phys. | Inorg. Chim. Acta |
| Chem. Phys. Lett. | Inorg. Nucl. Chem. Lett. |
| Chem. Process. Eng. (N.Y.) | Int. J. Chem. Kinet. |
| Chem. Rev. | Int. J. Mass Spectrom. Ion Phys. |
| Chem. Soc. Rev. | Int. J. Quantum Chem. |
| CHEMTECH | Isr. J. Chem. |
| Chem.-Tech. (Heidelberg) | Izv. Akad. Nauk SSSR Ser. Khim. |
| Chem. Technol. | J. Agric. Food Chem. |
| Chem. Unserer Zeit | J. Am. Chem. Soc. |
| Chem. Zentralbl. | J. Appl. Chem. |
| Chem.-Ztg. | J. Appl. Chem. Biotechnol. |
| Chimia | J. Appl. Electrochem. |
| Chim. Ind. (Paris) | J. Appl. Phys. |
| Clin. Chem. (Winston-Salem, N.C.) | J. Assoc. Off. Anal. Chem. |
| Cold Spring Harbor Symp. Quant. Biol. | J. Bacteriol. |
| Collect. Czech. Chem. Commun. | J. Biochem. (Tokyo) |
| Comments Solid State Phys. | J. Biol. Chem. |
| C. R. Séances Acad. Sci. | J. Catal. |
| Croat. Chem. Acta | J. Chem. Educ. |
| C. R. Seances Soc. Biol. | J. Chem. Eng. Data |
| Curr. Mod. Biol. | J. Chem. Inf. Comput. Sci. |
| Dokl. Akad. Nauk SSSR | J. Chem. Phys. |
| Drug Metab. Dispos. | J. Chem. Res. |
| Electrochim. Acta | J. Chem. Soc. A, B, C, or D |
| Endocrinology (Baltimore) | J. Chem. Soc., Chem. Commun. |
| Envir. Sci. Technol. | J. Chem. Soc., Dalton Trans. |
| Eur. J. Biochem. | J. Chem. Soc., Faraday Trans. |
| Eur. Polym. J. | J. Chem. Soc., Perkin Trans. |
| Exp. Cell Res. | J. Chim. Phys. Phys.-Chim. Biol. |
| Experientia | J. Chromatogr. |
| FEBS Lett. | J. Chromatogr. Sci. |
| Fed. Proc., Fed. Am. Soc. Exp. Biol. | J. Comput. Phys. |
| Fluorine Chem. Rev. | J. Educ. Res. |
| Fortschr. Chem. Forsch. | J. Elastomers Plast. |
| Fortschr. Phys. | J. Electroanalyt. Chem. |
| Fresenius Z. Anal. Chem. | J. Electrochem. Soc. |
| Gazz. Chim. Ital. | J. Endocrinol. |

*Table G-2.* (Continued)

| | |
|---|---|
| J. Eng. Math. | Macromol. Chem. |
| J. Exp. Biol. | Macromolecules |
| J. Fluorine Chem. | Med. Biol. Eng. |
| J. Genet. Physiol. | Mess., Steuern, Regeln |
| J. Heterocycl. Chem. | Microbiol. Rev. |
| J. Immunol. | Microchem. J. |
| J. Indian. Chem. Soc | Mol. Pharmacol. |
| J. Inorg. Nucl. Chem. | Mol. Phys. |
| J. Insect Physiol. | Monatsh. Chem. |
| J. Lipid Res. | Nachr. Chem. Techn. |
| J. Macromol. Sci., Chem. | Nachr. Chem. Techn. Lab. |
| J. Magn. Magn. Mater. | Nature (London) |
| J. Magn. Res. | Naturwissenschaften |
| J. Mater. | Organometallics |
| J. Mater. Sci. | Org. Magn. Reson. |
| J. Math. Phys. | Org. Mass Spectrom. |
| J. Math. Phys. (N.Y.) | Org. React. |
| J. Med. Chem. | Org. Synth. |
| J. Mol. Biol. | Pharm. J. |
| J. Mol. Spectrosc. | Phys. Abstr. |
| J. Mol. Struct. | Phys. Ber. |
| J. Non-Cryst. Solids | Phys. Bl. |
| J. Organomet. Chem. | Phys. Chem. Space |
| J. Org. Chem. | Phys. Chim. |
| J. Ornithol. | Physica (Utrecht) |
| J. Pharmacol. Exp. Ther. | Physics (N.Y.) |
| J. Pharm. Pharmacol. | Phys. Inorg. Chem. |
| J. Pharm. Sci. | Physiol. Rev. |
| J. Phys. | Physiol. Zool. |
| J. Phys. Chem. | Phys. Lett. |
| J. Phys. Chem. Ref. Data | Phys. Rev. |
| J. Phys. Chem. Solids | Phys. Unserer Zeit |
| J. Physiol. (London) | Polymer J. (Tokyo) |
| J. Phys. (Paris) | Power Eng. |
| J. Plant Physiol. | Proc. Chem. Soc. (London) |
| J. Polymer Sci. | Proc. Jpn. Acad. |
| J. Prakt. Chem. | Proc. Natl. Acad. Sci. U.S.A. |
| J. Sci. Instr. | Proc. R. Soc. (London) |
| J. Sci. Technol. India | Proc. Soc. Exp. Biol. Med. |
| J. Vac. Sci. Technol. | Pure Appl. Chem. |
| Kolloid-Z. Z. Polym. | Q. Rev., Chem. Soc. |
| Kunststoffe | Rec. Chem. Progr. |
| Langmuir | Recl. J. R. Natl. Chem. Soc. |
| Liebigs Ann. Chem. | Recl. Trav. Chim. Pays-Bas |
| Lipids | Rev. Anal. Chem. |

*Table G-2.* (Continued)

| | |
|---|---|
| Rev. Sci. Instrum. | Z. Anorg. Allg. Chem. |
| Rocz. Chem. | Z. Chem. |
| Science (Washington, D.C.) | Z. Elektrochem. Angew. Phys. Chem. |
| Spectrochim. Acta | Z. Elektrochem., Ber. Bunsenges. |
| Spectrosc. Lett. | Z. Phys. Chem. |
| Steroids | Zh. Fiz. Khim. |
| Sven. Kem. Tidskr. | Zh. Neorg. Khim. |
| Synth. Commun. | Zh. Obshch. Khim. |
| Synthesis | Zh. Org. Khim. |
| Tetrahedron | Zh. Struct. Khim. |
| Tetrahedron Lett. | Z. Klin. Chem. Klin. Biochem. |
| Theor. Chim. Acta | Z. Metallkd. |
| Trans. Faraday Soc. | Z. Naturforsch. |
| Trans. Inst. Chem. Eng. | Z. Pflanzenphysiol. |
| Usp. Khim. | Z. Phys. |
| Uzb. Anal. Chem. | Z. Phys. Chem. (Wiesbaden) |
| Z. Anal. Chem. | Z. Phys. Chem. (Leipzig) |
| Z. Angew. Phys. | |

# Appendix H
# Abbreviations

*Abbreviations* are shortened forms of names, terms, or other words. Judicious use of abbreviations is a service to the reader because it leads to more concise text. Nevertheless, only abbreviations that are widely accepted and understood are permissible in a scientific manuscript.*

Occasionally a distinction is made between a "true" abbreviation, in which the *end* of a word has been excised (i.e., "truncation" has occurred), and a form that is a "suspension" or "contraction" (where something in the *middle* of the word has been removed). In Britain, it is customary to write suspensions without periods (Mr, Dr), but this practice seems to have few advocates elsewhere. The usual rule is that an abbreviation always ends with a period regardless of its origin. There are exceptions, however, such as the occasional use of "p" (or "pp") and "f" (or "ff") for "page(s)" and "following", respectively. Other exceptions common in scientific and technical writing include the two-letter country codes employed in patent document references (cf. Table 11-1 in Sec. 11.4.4) and the familiar abbreviations for the chemical elements. However, these can be rationalized by arguing that they are not abbreviations at all, but rather *symbols* (cf. Sec. 7.3.1).

*Acronyms* (Greek: *akros*, topmost, outermost; and *onyma*, name) are also commonly employed to replace long expressions. An acronym is a combination of letters formed from the initial letter(s) of each of the successive or major parts of a compound term ("US" for United States, or "NMR" for nuclear magnetic resonance). Acronyms are normally written entirely in capital letters and without periods, although there are other conventions (e.g., n.m.r., N.M.R., nmr). If an acronym happens to lend itself to pronunciation, it may

---

* This rule also applies to such special text components as tables and figures. Table key lines (cf. Sec. 10.2.1) littered with (unidentified!) abbreviations coined for the purpose of saving space are particularly distasteful.

*Table H-1.* Standard English abbreviations occuring frequently in scientific text.

| Abbreviation | Meaning | Abbreviation | Meaning |
|---|---|---|---|
| approx. | approximately | instr. | instrumental |
| bk. | black | It. | Italian |
| bl. | blue | Lat. | Latin |
| biol. | biology, -ical, -ist | lb., lb | pound |
| br. | brown | lit. | literature |
| ca. | *circa*, about, approximately | loc. cit. | *loco citato*, in the place cited |
| cf. | *confer*, compare | lt. | light |
| Chap., ch., chap. | chapter | math. | mathematics, -ical |
| | | max | maximum |
| Co. | Company | min | minimum |
| Corp. | Corporation | misc. | miscellaneous |
| Dept. | Department | nat. | national, natural |
| deriv. | derivative | neg. | negative |
| dk. | dark | No., no. | number |
| Div. | Division | obs | observed |
| Ed. | editor (*pl.* Eds.); edition; edited by | op. cit. | *opere citato*, in the work cited |
| | | oz. | ounce |
| ed. | edition | p | page (*pl.* pp) |
| e.g. | *exempli gratia*, example given, for example | pers. | person, personal |
| | | pos. | positive |
| Eng. | English | publ. | publication, publisher; published by |
| Eq., eq. | equation | | |
| esp. | especially | ref. | refer, reference |
| et al. | *et alii*, and others | repr. | reprint, reprinted |
| etc. | *et cetera*, and so forth | resp. | respectively |
| f | and following (*pl.* ff) | rev. | review; revised, revision |
| Fig., fig. | figure | Sec., sec., sect. | section |
| ft., ft | foot | | |
| Fr. | French | ser. | series |
| G., Ger. | German | sl. | slightly |
| gal. | gallon | Sp. | Spanish |
| geog. | geography, -ical, -er | sq. | square |
| geol. | geology, -ical, -ist | subj. | subject, subjective |
| geom. | geometry, -ical | Suppl., supp. | supplement |
| Gk. | Greek | syn. | synonym, -ous |
| grn. | green | univ. | university |
| ibid. | *ibidem*, in the same place | vol., Vol. | volume |
| i.e. | *id est*, that is | vs. | versus |
| in., in | inch | wh. | white |
| incl. | inclusive; including; includes | wt. | weight |
| | | yel. | yellow |
| Inst. | Institute | yd., yd | yard |

come to be regarded as a true word. New meanings may even be added as the original derivation of the term is forgotten, and eventually the language finds itself enriched by the presence of a new word, although one with a peculiar origin. Two familiar contemporary examples are "BASIC", from beginners all-purpose symbolic instruction code, and "laser", from light amplification by stimulated emission of radiation. The second example is an acronym so far along the path of becoming an ordinary word that it is nearly always written lowercase.

Thousands of words are subject to abbreviation in general writing: first names, components of company names, geographical terms, academic titles—the list is endless. Compilations of standard abbreviations are to be found in most dictionaries and style guides, although it is not uncommon for two "authoritative" sources to disagree about the "preferred" abreviation for a given word (e.g., "ch" vs. "Chap." for "chapter", or "Sec." vs. "Sect." for "section"). Table H-1 contains a selection of general English-language abbreviations. Many of the entries come from the realm of bibliographers and writers, and thus relate to publications. In several cases, two (or more) alternative abbreviations have been provided. Either may be used, but consistency should be maintained.

Writers in science also make heavy use of abbreviations for technical terms, many of which have become standard. A list of some of the more important chemical abbreviations is presented in Table H-2. Dodd (1986) provides a collection of standard acronyms and abbreviations for structural units occurring within chemical compounds and for the ligands most frequently encountered in chemical, biochemical, or biological work. The Aldrich Chemical Company has distributed an extremely valuable collection of acronyms related to organic chemistry.*

Spectroscopists seem particularly prone to introducing abbreviations for characterizing and specifying intensities and locations of spectral bands or signals. Table H-3 contains a representative selection, with emphasis on IR and NMR spectroscopy.

The abbreviation of *serial titles* is the subject of separate treatment in Appendix G.

---

* The list originally appeared as "Table II" in an article on "The Use of Acronyms in Organic Chemistry" by G.H. Daub, A.A. Leon, and I.R. Silverman, published in the corporate publication "*Aldrichimica Acta* **1984**, *17* (1), 13-23". It is available separately in Europe through Aldrich-Chemie (Steinheim, West Germany) as *Technical Information Bulletin A74*.

*Table H-2.* Abbreviations of common scientific terms and institutions, especially ones of interest to chemists.

| Abbreviation | Explanation |
|---|---|
| abs., abs | absolute |
| AAS | atomic absorption spectroscopy |
| Ac | acetyl substituent ($-COCH_3$) |
| a.c. | alternating current |
| ACS | American Chemical Society |
| al., alc., al | alcohol (generally ethanol) |
| Alk | alkyl substituent (or alkali!) |
| alt. | altitude |
| anh, anhyd. | anhydrous |
| ANSI | American National Standards Institute |
| aq., aq, aqu | aqua; aqueous; water |
| Ar | aryl substituent |
| ASTM | American Society for Testing and Materials |
| atm | atmosphere |
| b.p., bp | boiling point |
| Bu | butyl substituent ($-C_4H_9$) |
| Bz | benzyl substituent ($-CH_2C_6H_5$), or benzene |
| calcd. | calculated |
| CAS | Chemical Abstracts Service |
| cat. | catalyst; catalyzed |
| con, conc., conc | concentrated |
| const. | constant |
| cor, corr. | corrected |
| Cp | cyclopentadienyl substituent ($-C_5H_5$) |
| cps, CPS | cycles per second |
| CW | continous wave |
| d.c. | direct current |
| d, decomp., dec | decompose, decomposition |
| dil., dil | dilute |
| DIN | Deutsches Institut für Normung (Federal Republic of Germany) |
| DMF, dmf | N,N-dimethylformamide |
| DMSO[a] | dimethyl sulfoxide |
| DNA | deoxyribonucleic acid |
| DSC | differential scanning calorimetry |
| DTA | differential thermal analysis |
| EDTA, edta | ethylenediaminetetraacetic acid |
| EMF, emf, e.m.f. | electromotive force |
| en | ethylenediamine |
| EPR | electron paramagnetic resonance |
| ESCA | electron spectroscopy for chemical analysis |
| ESR | electron spin resonance |
| Et | ethyl substituent ($-C_2H_5$) |

*Table H-2.* (Continued)

| Abbreviation | Explanation |
|---|---|
| f.p., fp | freezing point |
| FT | Fourier transform |
| (g) | gas, gaseous |
| GC | gas chromatography |
| gem | geminal |
| GLC | gas liquid chromatography |
| Hal | halogen substituent (–F, –Cl, –Br, –I) |
| HMO | Hückel molecular orbital |
| HPLC | high performance (*or* pressure) liquid chromatography |
| hyd. | hydrate |
| hygr. | hygroscopic |
| *i*Bu, Bu$^i$ | isobutyl substituent [$-CH_2CH(CH_3)_2$] |
| i.m. | intramuscular |
| *i*Pr, Pr$^i$ | isopropyl substituent [$-CH(CH_3)_2$] |
| IR | infrared |
| ISI | Institute for Scientific Information |
| ISO | International Organization for Standardization |
| IUPAC | International Union of Pure and Applied Chemistry |
| IUPAP | International Union of Pure and Applied Physics |
| (l) | liquid |
| L, L', L$^1$, (L$^2$, etc.) | ligand |
| LCAO | linear combination of atomic orbitals |
| LC$_{50}$ | lethal concentration (at 50% level) |
| LD$_{50}$ | lethal dose (at 50% level) |
| liq., liq | liquid |
| *m* | meta |
| M | metal atom in a chemical formula |
| max | maximum |
| Me | methyl substituent ($-CH_3$) |
| min | minimum |
| MO | molecular orbital |
| m.p., mp | melting point |
| MS | mass spectrum, mass spectrometry |
| NBS | National Bureau of Standards |
| *n*Bu | *n*-butyl substituent [$-(CH_2)_3CH_3$] |
| NMR | nuclear magnetic resonance |
| *o* | ortho |
| *p* | para |
| PE | polyethylene |
| PEG | polyethyleneglycol |
| PFK | perfluorokerosene |
| Ph | phenyl substituent ($-C_6H_5$) |
| ppb | parts per billion |

*Table H-2.* (Continued)

| Abbreviation | Explanation |
|---|---|
| ppm | parts per million |
| Pr | propyl substituent ($-C_3H_7$) |
| p.s.i. | pounds per square inch |
| PTFE | poly(tetrafluoroethylene) |
| PVC | poly(vinyl chloride) |
| R, R', $R^1$ ($R^2$, etc.) | substituent (usually implies alkyl) |
| R.M.S., rms | root mean square |
| r.p.m. | revolutions per minute |
| (s) | solid |
| $s$Bu, Bu$^s$ | *sec*-butyl substituent [$-CH(CH_3)(C_2H_5)$] |
| SD, s.d. | standard deviation |
| SI | International System of Units (Système International d'Unités) |
| sol. | soluble |
| soln. | solution |
| $t$Bu, Bu$^t$ | *tert*-butyl substituent [$-C(CH_3)_3$] |
| temp. | temperature |
| TG | thermogravimetry |
| THF | tetrahydrofuran |
| TLC | thin layer chromatography |
| TMS | tetramethylsilane |
| UV | ultraviolet |
| v, vic | vicinal |
| VIS, Vis, vis | visible |
| VPC | vapor phase chromatography |
| w. | water |
| X | heteroatom (halogen or pseudohalogen atom) in chemical formulas |

[a] Many journals prefer the abbreviation "Me$_2$SO" for dimethyl sulfoxide

*Table H-3.* Abbreviations used in IR and NMR spectroscopy.

| Abbreviation | Meaning | Abbreviation | Meaning |
|---|---|---|---|
| vs | very strong | d | doublet |
| w | weak | dd | doublet of doublets |
| m | medium | t | triplet |
| s | strong | q | quartet |
| vw | very weak | quint | quintet |
| sh | shoulder | sext | sextet |
| b, br | broad | sept | septet |
| s | singlet | m, mlt | multiplet |

# Appendix I
# Proofreader's Marks and
# Examples of Error Corrections
# in Typeset Text

*Table Footnotes*

[a] cf. Sec. 3.5.6. Examples have been taken and adapted from various sources (often inconsistent), e.g. Standard *ANSI Z39.22-1981*, *The Chicago Manual of Style* (1982).
[b] Standard *DIN 16 511*.

| American system[a] | | Alternative system[b] | | Explanation |
|---|---|---|---|---|
| Error with correction mark | Marginal correction | Error with correction mark | Marginal correction | |
| crystallisation<br>hibrydize<br>sulphur | z<br>y i<br>(f) f | crystallisation<br>hibrydize<br>sulphur | z<br>by Li<br>Lf | *Change character(s)* |
| particle<br>rotatory | ip<br>con ( ) | particle<br>rotatory | cip<br>conr | *Add character(s)* |
| vapour<br>hyphenation<br>min | (f)<br>g<br>g | vapour<br>hyphenation<br>min | | *Delete character(s)* |
| sizable fraction | large | sizable fraction | large | *Change word(s)* |
| was assigned | originally | was assigned | originally | *Add word(s)* |
| very large<br>or or distilled | g<br>g | very large<br>or or distilled | | *Delete word(s)* |
| 1.76 | / 5 | 1.76 | ×1.75 | *Correct number(s)* |
| cnocentric<br>Fig. 2) | tr.<br>tr.   (2). | cnocentric<br>Fig. 2) | | *Transpose characters* |

| American system[a] | | Alternative system[b] | | Explanation |
|---|---|---|---|---|
| Error with correction mark | Marginal correction | Error with correction mark | Marginal correction | |
| we have already / Smith and Miller | tr. | we have already / Smith and Miller | (symbol) | Transpose words |
| results (1985) by Snyder | tr. [Miller and Smith] / tr. | results (1985) by Snyder | (symbol) | Rearrange words |
| ...have so far been only / qualitatively evaluated | so far have been / evaluated only / qualitatively | ...have so far been only / qualitatively evaluated | 1-7 | |
| Fisher (1950). The | insert from page X | Fisher (1950). The | (( insert from page X )) | Insert phrase (or paragraph) |
| sp...ed | X | sp...ed | O | Reset damaged character(s) |
| $W_{o}od^{w}{}_{a}rd$ | \|\| | $W_{o}od^{w}{}_{a}rd$ | \| \| | Adjust line |
| but never / 70kg | # # | but never / 70kg | (symbol) | Insert space (between words) |
| proof reading | ( ) | proof reading | ( ) | Delete space (between words) |

| American system a) | | Alternative system b) | | |
|---|---|---|---|---|
| Error with correction mark | Marginal correction | Error with correction mark | Marginal correction | Explanation |
| was ∨ crystallized | eq. # | was ⌢ crystallized | | Reduce space (between words) |
| ...the proton chemical shift ...correlations proved useful | ld. | ...the proton chemical shift ...correlations proved useful | | Insert space (between lines) |
| ...to lowering the | g ld. | ...to lowering the | | Reduce space (between lines) |
| ...and solubility of | | ...and solubility of | | |
| a) by automatic titration | fl. r. | a) by automatic titration | | Move to the right (indent) |
| as that of glucose | fl. l. | — as that of glucose | | Move to left (cancel indent) |
| is a point of view of economy. The alkali sulfide | ¶ | is a point of view of economy. The alkali sulfide | | Start new paragraph |
| ...increased. This result is... | no ¶ | ...increased. This result is... | | Run on (no new paragraph) |
| distrotatory | stet. | distrotatory | | Leave as it stands |
| the trans isomer | ital. | the trans isomer | ～ «italics» | Set in (or change to) italic type |

| American system[a] | | Alternative system[b] | | Explanation |
|---|---|---|---|---|
| Error with correction mark | Marginal correction | Error with correction mark | Marginal correction | |
| compound 13 a has | bf. | compound 13 a has | ((bold)) | Set in (or change to) boldface type |
| dacron | Cap. | dacron | ⊥D | Set in upper case (capitals), capitalize |
| following Chapter | lc. | following Chapter | ⊤c | Set in lower case |
| [ ⊃ or | | | | Follow instruction (as specified) |
| | | | | Re-set as indicated here |

# Appendix J
# Standard Reference Formats

*Table Footnotes*

a The examples presented here have been adapted, with modifications, from the current literature. The style of references closely follows that in ACS journals, with minor deviations. For example, ACS journal style does not introduce a comma before a journal title, and publishers' names and locations are supplied in the reverse order.—The examples are shown in typeset form, with special typefaces used where appropriate (e.g. for journal and book titles and years of publication). Depending on the circumstances it may or may not be advisable to simulate these type styles in a manuscript by underscoring (e.g., a straight line for italics and a double line for bold face; see also Sec. 4.3.7, "Choice of Type"). In references following the name-date citation system the year of publication should be moved forward until it follows the name(s) of the author(s); for examples see the literature references in the present book.

b Comma preceding book title is often omitted *(The Chicago Manual of Style, 1982)*, but sometimes replaced by a colon (see also "Comment" column) or by double spacing, in which case a double space should also follow the period closing the book title (e.g., Campbell, Ballou and Slade, 1986). References to books may be supplemented by ISBNs.

c The reference is framed in compliance with examples in Sec. 11.4.2, "Series and Multi-volume Works". The title of the volume is placed first in the reference because neither the volume as a whole nor the particular segment in question can be attributed to any editors or authors. Alternatively, *Gmelin Handbook of Inorganic Chemistry* could introduce the reference statement.—In the name-date-system of citation, the in-text reference (cf. Sec. 11.3.1) would read *Silicon, 1986*.

d Patent document country is often given in adjectival form, e.g. "British", "German (Fed. Rep.)". Alternatively, the official *CASSI* notation (cf. Sec. 11.4.4, Table 11-1, center column "Abbreviation") could have been used to identify the type and country of the patent document; i.e. "*U.S.*" instead of "*U.S. Patent*".

e The German word "Schutzrecht" is meant to indicate any kind of claim, application, or patent.—See Table 11-1 with respect to "Country code".

f The reference may be extended by a statement such as "Available from: University Microfilms International, Ann Arbor, Michigan; Publication Number 65-00 000".

g Example taken from Dodd (1986), with place of publication and publisher reversed.

h EPA or its office could have taken the role of author or editor. This identification would then have been placed first (see preceding entry).

*Table.* Recommended reference formats for numeric references.—The reader should be aware that there are variations in referencing practice within the scientific literature, and is advised to consult Sec. 11.4 for a critical discussion of the subject.[a]

| Type of reference | Example | Comment |
|---|---|---|
| Journal article, one author | Jensen, D.E., *Biochemistry* **1978**, *17*, 5105-5113. | Modern sequence and type style, numbers of first and last pages given; comma after initials often omitted |
| Journal article, several authors | Fenn, R.H., Graham, A.J., Gillard, A.D., *Nature (London)* **1967**, *213*, 1012. | Commas between names often replaced by semicolons |
| Journal without volume number | Hart, T.W., *J. Chem. Soc. Chem. Commun.* **1979**, 156. | |
| Journal with new pagination in each issue | Haggin, J., *Chem. Eng. News* **1985**, *63* (42), 23-25. | Issue number (42) in parentheses |
| Journal reference with title of article | Chrambach, A., "A device for the sectioning of cylindrical polyacrylamide gels", *Anal. Biochem.* **1966**, *15*, 544-548. | Quotation marks can be omitted and title of article closed with a period |
| Translation journal | Volpin, M.E., *J. Gen. Chem. USSR (Engl. Transl.)* **1960**, *30*, 1207; *Zh. Obshch. Khim.* **1960**, *30*, 1187. | Translation journal data followed by original journal data |
| Reference to abstract of an article | Mirnov, V.F., *Izv. Akad. Nauk SSSR*, **1966**, 1177; *Chem. Abstr.* **1966**, *65*, 16 997. | Journal article data followed by CA abstract reference |

*Table.* (Continued)

| Type of reference | Example | Comment |
|---|---|---|
| Paper not yet published | Giovanni, A.B., *Spectrochim. Acta*, in press. | Journal title given, "in press" or "accepted for publication" added |
| Monograph[b] | Soo, S.L., *Fluid Dynamics of Multi-phase Systems.* New York: Blaisdell, 1967. | Title in italics, main words capitalized; place of publication preceding name of publisher. The latter sequence is often reversed and colon often used instead of comma after author's initials |
| Chapter in monograph | Ugi, I., *Isonitrile Chemistry.* New York: Academic Press, 1971; Chapter 2. | Work-unit fraction ("Chapter 2") stands last |
| Edited book | Clearfield, A. (Ed.), *Inorganic Ion Exchange Materials.* Boca Raton: CRC Press, 1982. | (Ed.) added after editor's name, title of book follows |
| Chapter in edited book | Stoeppler, M., Nürnberg, W., in: *Metalle in der Umwelt*; Merian, E. (Ed.). Weinheim: Verlag  Chemie, 1984; Chapter I.4a. | Editor's name placed after title of book; chapter notation can be replaced by page numbers |
| Chapter in edited book with title of contribution | Porterfield, J.S., "Interference and Interferon", in: Harris, R.J.C. (Ed.), *Techniques in Experimental Virology.* London: Academic Press, 1980; pp 305-326. | Title of contribution in quotation marks, editor's name placed before title of book |
| Article in multivolume work, article location given | Lindsell, W.E., in: *Comprehensive Organometallic Chemistry*; Wilkinson, G., Stone, F.G.A., Abel, E.W. (Eds.); Oxford: Pergamon Press, 1982; Vol. 1, pp 156-187. | Volume identification and location (page numbers) data stand last, entire work appeared in year given |

*Table.* (Continued)

| Type of reference | Example | Comment |
|---|---|---|
| Specified edition of a book, article location given | Dean, J.A., in: *Lange's Handbook of Chemistry*. 12th ed. New York: McGraw Hill, 1979; pp 5-17 to 5-41. | Edition statement follows title of work, location data for article stand last |
| Contribution to a reference work, work-unit fraction identified | Sifniades, S., "Acetone", in: *Ullmann's Encyclopedia of Industrial Chemistry*. 5th ed. Vol. A1. Weinheim: VCH Verlagsgesellschaft, 1985; pp 79-96 (p 85, Fig. 2). | Volume identification data—"Vol. A1"—placed prior to publication data; year of publication refers to volume only |
| Section within a chapter of a large reference work[c] | *Silicon*. Sec. 2.2.1 "Preparation and Manufacturing Chemistry", Chapter 2, "Silicon Carbide". Berlin: Springer, 1986 (*Gmelin Handbook of Inorganic Chemistry*, Suppl. Vol. B3, System Number 15); pp 36-144. | Section/chapter headings and—at the end—pages given; authors not quoted since segment not signed; reference begins with title of volume |
| Abstract of conference paper | Baisden, P.A., *Abstracts of Papers*. 188th National Meeting of the American Chemical Society, Philadelphia, PA; American Chemical Society: Washington, DC, 1984; NUCL 9. | Abstracts only are published, available from the organizing body (society); issuing authority (publisher) named first, location second (ACS style); location of work-unit given last |
| Conference paper fully published | Bruning, D.A., "Review Literature and the Chemist", in: *Proceedings of the International Conference on Scientific Information*. Washington, DC: National Academy of Sciences–National Research Council, 1959; pp 545-570. | Full-length article in published volume; corporate body serves as publisher of the proceedings |

*Table.* (Continued)

| Type of reference | Example | Comment |
|---|---|---|
| Conference paper, treated as contribution to multi-authored book | Gaylord, N.G, "Effects of polymers on grafting", in: Burke, J.J., Weiss, V. (Eds.), *Block and Graft Copolymers.* Syracuse, NY: Syracuse University Press, 1973; 19-35 (p 20). | Edited conference proceedings; extent of work and exact location of work-unit given; conference origin not visible from book title |
| Series of conferences covered by *CASSI* | Salomons, W., in: *Proc. Intern. Conf. Heavy Metals in the Environment.* Edinburgh: CEP Consultants, 1983. | *CASSI* notation supplemented by publication data |
| Conference proceedings in a journal | Singhal, G.N., "Temperature measurements in industries", Proceedings of the Seminar on Instruments and Control in Process Industries. Bombay, India; in: *Ind. Chem. J.* **1974**, 8 (10) Suppl. 17-35. | Location in journal noted |
| Other congress series | Lullmann, H., "Metabolism of glycosides", in: Szekeres, L., Papp, J.G. (Eds.), *Symposium on Drug and Heart Metabolism, Budapest 1971.* Budapest: Akademiai Kiado, 1973. (First Congress of the Hungarian Pharmacological Society.); pp 107-110. | Title of contribution given; nature and setting of symposium explained |
| Patent[d] | Maldonado, P., Nougier, R., *U.S. Patent 197 205,* 1983. | Briefest form |

*Table.* (Continued)

| Type of reference | Example | Comment |
| --- | --- | --- |
| Patent reference with abstracting source | Baizer, M.M., *U.S. Patent 4 293 393*, 1981; *Chem. Abstr.* **1981**, *95*, 228 075. | CA abstract number given |
| Patent reference including assignee and inventors | Hodgson, G.L., Harfrist, M., *Eur. Pat. Appl. 106 988*, 1983, The Wellcome Foundation Ltd.; *Chem. Abstr.* **1984**, *101*, 110 897. | The name of the company to which the patent was assigned usually follows the patent number or year of publication |
| Patent reference using country codes (*EP, DE*)e) | Schutzrecht *EP 2013-B1*, 1980-08-06, Bayer. Pr.: *DE 275 1782*, 1977-11-19. | *DIN 1505* notation, date of publication in year-month-day sequence; "Bayer" is the assignee, "Pr." signals priority |
| Patent reference including title of patent document and international patent classification (IPC) number | Kysilka, J.O., Sawicki, C.A., "Apparatus and method for measuring optically active materials", *Eur. Pat. Appl. EP 87 535* (G01N21/21), 07 Sep 1983; *U.S. Pat. Appl. 352 321*, 25 Feb 1982, American Crystal Sugar Co; *Chem. Abstr.* **1984**, *100*, 8910. | IPC number in parentheses after patent number |
| Thesisf) | Cairns, R.B., *Infrared Spectroscopic Studies of Solid Oxygen*. Berkeley, CA: University of California, 1965; 156 pp. Dissertation. | Degree-granting university treated as publisher, thesis itself as independent publication; "dissertation" added as supplementary note |
| Thesis, shorter version | Holton, T.C., Ph.D. Dissertation, The State University of New York at Buffalo, 1980. | Title of thesis not given |

*Table.* (Continued)

| Type of reference | Example | Comment |
|---|---|---|
| Report | Roberts, D.A., *Review of Recent Developments in the Technology of Nickel-base and Cobalt-base Alloys.* Columbus, OH: Battelle Memorial Institute, Defense Metals Information Center; 1961-08-04; DMIC memorandum 122. 2 pp. Available from: NTIS, Springfield, VA; AD 261 292. | Comprehensive reference with report identifier (DMIC....) and acquisition number (AD...) |
| Report, shorter version | Johnson, C.K., Report ORNL-3794, Oak Ridge National Laboratory, Oak Ridge, TN, 1976. | Title of report not given |
| Report, treated as serial identified by a CODEN | Stallkamp, J.A., Herriman, A.G., *Mariner Mars 1969 Final Project Report.* Technical Report, Jet Propulsion Laboratory, 1971; 32-1460. CODEN: JPLRA6; available from: California Institute of Technology, Jet Propulsion Laboratory, Pasadena, CA. | Availability included |
| Government brochure[g] | Interdepartmental Task Force on PCBs. *PCBs and the Environment.* Washington, DC: U.S. Government Printing Office, 1972; COM 72.10419. | A committee acting as author, the government's printing house as publisher; publication identifier (COM...) added |
| Government brochure, alternative version[h] | *Energy from Renewable Resources. 1985.* Environmental Protection Agency, Office of Environmental Engineering and Technology. Washington, DC: U.S. Government Printing Office, 1985; EPA-000/0-00-000 | Title of independent publication placed first |

*Table.* (Continued)

| Type of reference | Example | Comment |
|---|---|---|
| Standard | Standard ANSI Z 39.5-1979. *American National Standard for the abbreviation of titles of periodicals.* | Reference introduced by the word "Standard" followed by shorthand notation for type of standard (designation of issuing authority), standard document number, year of publication, and title of standard (in italics) |
| Law | "Copyright Act" (GB), 1959, Sec. 1a. | Title of law in roman face and quotes, country code given. |
| Trade brochure | DEGUSSA, *Aerosol.* Frankfurt (Germany, FRG), 1969 (RAG-3-8-369 H). Trade Brochure. | Company treated as author; "trade brochure" added as supplementary note |

# Appendix K
# Selected Quantities, Units, and Constants

| Name of quantity[a] | Symbol[b] | SI unit[c] | Name of unit | Other units |
|---|---|---|---|---|
| *Space and time* | | | | |
| length* | $l$ | m | meter[d] | |
| breadth, width | $b$ | m | meter | |
| height | $h$ | m | meter | |
| radius | $r$ | m | meter | |
| thickness | $d$ | m | meter | |
| area | $A, S$ | m² | square meter | a (are) |
| | | | | h (hectare) |
| volume | $V$ | m³ | cubic meter | L, l (liter)[e] |
| plane angle | $\alpha, \beta, \gamma,$ | 1, rad | radian | ° (degree) |
| | $\vartheta, \varphi$ | | | ' (minute) |
| | | | | " (second) |
| solid angle | $\omega, \Omega$ | 1, sr | steradian | |
| wavelength | $\lambda$ | m | meter | |
| wavenumber | $\sigma, \nu$ | m⁻¹ | reciprocal meter | |
| time* | $t$ | s | second | min (minute) |
| | | | | h (hour) |
| | | | | d (day) |

[a] SI base quantities are marked by asterisks (*)
[b] Recommended by IUPAC
[c] SI base units as well as derived and supplementary units are listed; all are to be used with prefixes as needed
[d] Also spelled "metre"
[e] Also spelled "litre"; $1 \text{ L} = 10^{-3} \text{ m}^3$

| Name of quantity[a] | Symbol[b] | SI unit[c] | Name of unit | Other units |
|---|---|---|---|---|
| frequency | $v, f$ | $s^{-1}$ | reciprocal second | Hz (hertz)[f] |
| relaxation time | $\tau$ | s | second | |
| velocity | $\boldsymbol{u}, \boldsymbol{v}$ | m s$^{-1}$ | meter per second | km/h |
| acceleration | $\boldsymbol{a}$ | m s$^{-2}$ | meter per second squared | |

*Mechanics*

| | | | | |
|---|---|---|---|---|
| mass* | $m$ | kg | kilogram | g (gram) |
| | | | | t (tonne)[g] |
| (mass) density | $\rho$ | kg m$^{-3}$ | | g/cm$^3$ |
| momentum | $p$ | kg m s$^{-1}$ | | |
| angular momentum | $L$ | kg m$^2$ s$^{-1}$ | | |
| force | $F$ | N | newton[f] | |
| moment of force | $M$ | N m | newton meter | |
| weight | $G, W$ | N | newton | |
| pressure | $p$ | Pa | pascal | bar (bar)[h] |
| energy | $E, W$ | J | joule | W h (watt hour)[i] |
| | | | | eV (electronvolt)[j] |
| work | $W, A$ | J | joule | |
| power | $P$ | W | watt | J/s, V A[k] |

*Molecular Physics and Thermodynamics*

| | | | | |
|---|---|---|---|---|
| thermodynamic temperature* | $T$ | K | kelvin | °C (degree celsius)[l] |
| Celsius Temperature | $\vartheta, t$ | | | °C |
| number of entities | $N$ | | | |
| Avogadro constant[m] | $N_A, L$ | | | |
| Boltzmann constant[n] | $k$ | | | |
| Planck constant[o] | $h$ | | | |
| (molar) gas constant[p] | $R$ | | | |
| (quantity of) heat | $Q$ | J | joule | |

[f] 1 Hz = 1 s$^{-1}$; 1 N = 1 kg m s$^{-2}$
[g] Formerly metric ton; 1 t = $10^3$ kg
[h] 1 bar = $10^5$ Pa
[i] 1 W h = 3.6 x $10^3$ J
[j] 1 eV = 1.602 189 x $10^{-19}$ J
[k] 1 W = 1 J/s = 1 V A
[l] cf. eq. (7-3) in Sec. 7.3.1
[m] $N_A$ = 6.022 136 7 x $10^{23}$ mol$^{-1}$
[n] $k$ = 1.380 658 x $10^{-23}$ J K$^{-1}$
[o] $h$ = 6.626 075 5 x $10^{-34}$ J s
[p] $R$ = 8.314 510 J mol$^{-1}$ K$^{-1}$

| Name of quantity[a] | Symbol[b] | SI unit[c] | Name of unit | Other units |
|---|---|---|---|---|
| entropy[q] | $S$ | J K$^{-1}$ | joule per kelvin | |
| internal energy[q] | $U$ | J | joule | |
| Helmholtz function,[q] (Helmholtz) free energy, Helmholtz energy | $F, A$ | J | joule | |
| enthalpy[q] | $H$ | J | joule | |
| Gibbs function,[q] (Gibbs) free energy, Gibbs energy | $G$ | J | joule | |
| heat capacity[q] | $C_p, C_V$ | J K$^{-1}$ | joule per kelvin | |

*Chemical Physics*

| | | | | |
|---|---|---|---|---|
| amount of substance* | $n$ | mol | mole | |
| relative atomic mass | $A_r$ | 1 | | |
| relative molecular mass | $M_r$ | 1 | | |
| atomic mass constant | $m_u$ | kg | kilogram | u (atomic mass unit)[r] |
| mass of a portion (of substance B) | $m_B, m(B)$ | kg | kilogram | g (gram) |
| molar mass (of substance B) | $M_B, M(B)$ | kg mol$^{-1}$ | kilogram per mole | |
| concentration (of substance B) | $c_B, c(B)$ | mol m$^{-3}$ | mole per cubic meter | mol/L |
| mole fraction[s] (of substance B) | $\kappa_B, \kappa(B)$ | 1 | | |
| mass fraction (of substance B) | $\omega_B, \omega(B)$ | 1 | | %, ‰, ppm, ppb |
| volume fraction (of substance B) | $\varphi_B, \varphi(B)$ $\phi_B, \phi(B)$ | 1 | | %, ‰, ppm, ppb |
| mass concentration | $\rho$ | kg m$^{-3}$ | | g/L |
| molality | | mol kg$^{-1}$ | mole per kilogram | mmol/kg |
| volume concentration[t] | $\sigma$ | 1 | | |
| molar volume | $V_m$ | m$^3$ mol$^{-1}$ | cubic meter per mole | L/mol |

q Molar quantities can be distinguished from the quantity of a sytem by adding the subscript m; e.g., molar internal energy $U_m$, in J mol$^{-1}$

r 1 u = 1.660 565 5 x 10$^{-27}$ kg

s A more accurate, but rather uncommon name is "amount-of-substance fraction" (cf. Sec. 7.4.3)

t $\sigma$ refers to the total volume of a mixture, whereas the volume fraction $\varphi$ relates the volume of a substance to the volume of several components before mixing

| Name of quantity[a) | Symbol[b) | SI unit[c) | Name of unit | Other units |
|---|---|---|---|---|
| molar heat capacity | $C_m$ | J mol$^{-1}$ K$^{-1}$ | | |
| molar conductivity | $\Lambda_m$ | S m$^2$ mol$^{-1}$ | | |
| Faraday constant[u) | $F$ | | | |
| *Electricity, magnetism, light* | | | | |
| quantity of electricity[v) | $Q$ | C | coulomb | |
| electric potential | $\vartheta, V$ | V | volt | |
| potential difference | $U, \Delta V$ | V | volt | |
| electric dipole moment | $p$ | C m | coulomb meter | |
| electric current* | $I$ | A | ampere | |
| electric field strength | $E$ | V m$^{-1}$ | volt per meter | |
| magnetic field strength | $H$ | A m$^{-1}$ | ampere per meter | |
| (electrical) resistance | $R$ | | ohm | |
| radiant energy | $Q$ | J | joule | |
| luminous intensity* | $I$ | cd | candela | |

[u] $F = 9.648\ 530\ 9 \times 10^4$ C mol$^{-1}$
[v] Also called electric charge

An *Abbreviated List of Quantities, Units and Symbols in Physical Chemistry*, prepared for publication by K.H. Homann, has been made available by IUPAC. It can be obtained from Blackwell Scientific Publications, Osney Mead, Oxford OX2 OEL, UK. The data presented here have been carefully compared with this list.

# Literature

## References

A Note to Standards: Standards ANSI can be obtained from American National Standard Institute, Inc., 1430 Broadway, New York NY 10018, USA; Standards BS are provided by British Standard Institution (BSI), 2 Park Street, London W1A 2 BS, UK; DIN Deutsches Institut für Normung publishes its standards through Beuth Verlag GmbH, P.O.B. 1145, D-1000 Berlin 30. ISO, the International Organization for Standardization, has its headquarters in CH-1211 Genève (P.O.B. 56), Switzerland.

American Chemical Society (Ed.) (1978), *Handbook for Authors of Papers in American Chemical Society Publications*. Washington, DC: American Chemical Society; 122 pp.

Ash, J., Chubb, P., Ward, S., Welford, S., Wilett, P. (1985), *Communication, Storage and Retrieval of Chemical Information*. Chichester: Wiley; 296 pp.

Barnes, G.A. (1982), *Communication Skills for the Foreign-Born Professional*; Philadelphia: ISI Press; 198 pp.

Bennett, J.B. (1970), *Editing for Engineers*. New York: Wiley; 126 pp.

Biedermann, H. (1984), *Microcomputer und Publication*. Stuttgart: Fischer; 290 pp.

Booth, V. (1985), *Communicating in Science: Writing and Speaking*. Cambridge (UK): Cambridge University Press; 68 pp.

Bottle, R.T. (Ed.) (1979), *The Use of Chemical Literature*. 3rd ed. London: Butterworths; 294 pp.

Bove, T., Rhodes, C., Thomas, W. (1987), *The Art of Desktop Publishing: Using Personal Computers to Publish It Yourself.* 2nd ed. New York: Bantam; 296 pp.

Brookes, H.F., Ross, H. (1975, 1976), *English as a Foreign Language for Science Students.* London: Heinemann Educational Books. (Vol. 1, 210 pp; Vol. 2, 174 pp)

Butcher, J. (1981), *Copy-Editing—The Cambridge Handbook.* 2nd ed. Cambridge (UK): Cambridge University Press; 332 pp.

Cahn, R.S., Dermer, O.C. (1979), *An Introduction to Chemical Nomenclature.* 5th ed. London: Butterworths; 128 pp.

Campbell, W.G., Ballou, S.V., Slade, C. (1986), *Form and Style—Theses, Reports, Term Papers.* 7th ed. Boston: Houghton Mifflin; 228 pp.

*Chicago Guide to Preparing Electronic Manuscripts* (1986). Chicago: University of Chicago; 160 pp.

Cook, C.K. (1985), *Line by Line—How to Improve Your Own Writing.* Boston: Houghton Mifflin; 220 pp.

Crews, F. (1980), *The Random House Handbook.* 3rd ed. New York: Random House; 436 pp.

Day, R.A. (1983), *How to Write and Publish Scientific Papers.* 2nd ed. Philadelphia: ISI Press; 182 pp.

Davis, G.B., Parker, C.A. (1979), *Writing the Doctoral Dissertation. A Systematic Approach.* Woodbury, NY (USA): Barron.

Dodd, J.S. (Ed.) (1986), *The ACS Style Guide—A Manual for Authors and Editors.* Washington, DC: American Chemical Society; 264 pp.

Drazil, J.V. (1983), *Quantities and Units of Measurement—A Dictionary and Handbook.* London: Mansell; 314 pp.

Dutch, R.A. (Ed.) (1965), *The St. Martin's Roget's Thesaurus of English Words and Phrases.* New York: St. Martin's Press; 1406 pp.

Ebel, H.F., Bliefert, C. (1982), *Das naturwissenschaftliche Manuskript—Ein Leitfaden für seine Gestaltung und Niederschrift.* Weinheim: Verlag Chemie; 216 pp.

Farr, A.D. (1985), *Science Writing for Beginners*. Oxford: Blackwell Scientific Publications; 118 pp.

Fieser, F.L., Fieser, M. (1972), *Style Guide for Chemists*. Huntington, NY: Robert E. Krieger; 116 pp.

Fowler, H.W. (1965 and later reprints), *A Dictionary of Modern English Usage*. 2nd ed. Revised by Sir Ernest Gowers. Oxford: Oxford University Press; 726 pp.

Freedman, G., Freedman, D.A. (1985), *The Technical Editor's and Secretary's Desk Guide*. New York: McGraw-Hill; 584 pp.

Gastel, B. (1983), *Presenting Science to the Public*. Philadelphia: ISI Press; 146 pp.

Gordon, K.E. (1984), *The Transitive Vampire—A Handbook of Grammar for the Innocent, the Eager, and the Doomed*. New York: Times (Random House); 150 pp.

*Guidelines for Cataloguing-in-Publication* (1986), prepared by D. Anderson for the International Federation of Library Associations and Institutions (IFLA) and the United Nations Educational, Scientific and Cultural Organization (UNESCO). Paris: UNESCO; 84 pp.

*Hart's Rules for Compositors and Readers at the University Press Oxford* (1983). 39th ed. Oxford: Oxford University Press; 182 pp.

Hawkins, C., Sorgi, M. (Eds.) (1985), *Research—How to Plan, Speak and Write about it*. New York: Springer; 184 pp.

Hellwinkel, D. (1974), *Die systematische Nomenklatur der Organischen Chemie—Eine Gebrauchsanweisung*. Berlin: Springer; 170 pp.

Herner, S. (1980), *A Brief Guide to Sources of Scientific and Technical Information*. 2nd. ed. Arlington, VA (USA): Information Resources Press; 160 pp.

IUB (1979), *Enzyme Nomenclature 1978*. New York: Academic Press.

IUPAC (Ed.) (1971), *Nomenclature of Inorganic Chemistry* ("The Red Book"). 2nd ed. Oxford: Pergamon Press; 110 pp; reprint 1981.

IUPAP (Ed.) (1978), *Symbols, Units and Nomenclature in Physics*. International Union of Pure and Applied Physics [Document UIP 20 (SUN 65-3)];

also published in *Physica* **1978**, *93A*, 1.—The headquarters of IUPAP are always located at the General Secretary's affiliation; at the time of present writing: Physics Department, Imperial College, London SW7, UK.

IUPAC (Ed.) (1979a), *Manual of Symbols and Terminology for Physico-chemical Quantities and Units* ("The Green Book"). Adopted by the IUPAC Council on 7 July 1969. Prepared for publication by D.H. Whiffen. Oxford: Pergamon Press; 41 pp.

IUPAC (Ed.) (1979b), *Nomenclature of Organic Chemistry. Sections A, B, C, D, E, F and H* ("The Blue Book"). Oxford: Pergamon Press; 560 pp.

IUPAC (Ed.) (1987), *Compendium of Macromolecular Nomenclature* ("The Purple Book"). Oxford: Blackwell; 200 pp; in preparation.

Judd, K. (1982), *Copyediting: A Practical Guide.* Los Altos, CA (USA): William Kaufmann; 304 pp.

Kanare, H.M. (1985), *Writing the Laboratory Notebook.* Washington: American Chemical Society; 146 pp.

King, L.S. (1978), *Why Not Say It Clearly—A Guide to Scientific Writing.* Boston: Little, Brown; 186 pp.

Leaver, R.H., Thomas, T.R. (1974), *Analysis and Presentation of Experimental Results.* London: The Macmillan Press; 128 pp.

Lees, R., Smith, A. (Eds.) (1983), *Chemical Nomenclature Usage.* Chichester: Ellis Horwood; 172 pp.

Lloyd, S.M. (Ed.) (1984), *Roget's Thesaurus of English Words and Phrases.* Middlesex (UK): Penguin Books; 776 pp.

MacGregor, A.J. (1979), *Graphics Simplified. How to Plan and Prepare Effective Charts, Graphs, Illustrations, and other Visual Aids.* Toronto: University of Toronto Press; 64 pp.

Maizell, R.E. (1987), *How to Find Chemical Information—A Guide for Practicing Chemists, Teachers, and Students.* 2nd ed. New York: Wiley; 402 pp.

Meador, R. (1985), *Guidelines for Preparing Proposals.* Chelsea, MI (USA): Lewis Publishers; 120 pp.

Mellon, M.G. (1982), *Chemical Publications—Their Nature and Use.* 5th ed.

New York: McGraw-Hill; 352 pp.

Mencken, H.L. (1936), *The American Language—An Inquiry into the Development of English in the United States*, 4th ed. New York: Alfred A. Knopf; 770 pp.

Michaelson, H.B. (1982), *How to Write and Publish Engineering Papers and Reports*. Philadelphia: ISI Press; 158 pp. [See also the article by H.B. Michaelson "Strategy for the Engineer Author" in *Chem. Eng. Int. Ed.* **1986**, *93* (15), 50-60.]

Mücke, M. (1982), *Die chemische Literatur*. Weinheim: Verlag Chemie; 272 pp.

Neuhoff, V. (1986), Der Kongreß. Weinheim: VCH Verlagsgesellschaft; 232 pp.—English-language version in preparation (to appear 1987).

Nicholson, M. (1957), *A Dictionary of American-English Usage*; New York: Oxford University Press; 672 pp.

O'Connor, M., Woodford, F.P. (1976), *Writing Scientific Papers in English—An ELSE-Ciba Foundation Guide for Authors*. Amsterdam: Elsevier; 108 pp. (ELSE European Life Science Editors)

Pichler, H.R. (1986), *Online-Recherchen für Chemiker*. Weinheim: VCH Verlagsgesellschaft; 262 pp.

Rathbone, R.R. (1985), *Communicating Technical Information*. 2nd ed. Reading, MA (USA): Addison Wesley; 136 pp.

Rico, G.L. (1983), *Writing the Natural Way—Using Right-Brain Techniques to Release Your Expressive Powers*. Los Angeles: Tarcher; 288 pp.

Roth, S.F. (1984), *Microcomputers in Publishing*. New York: Bowker; 160 pp.

Schoenfeld, R. (1986), *The Chemist's English*. 2nd ed. Weinheim: VCH Verlagsgesellschaft. 174 pp.

Seybold, J., Dressler, F. (1987), *Publishing from the Desktop*. New York: Bantam; 300 pp.

Schulz, H. (1985), *Von CA bis CAS ONLINE—Die Datensammlungen des Chemical Abstracts Service und deren Nutzung*. Weinheim: VCH Verlagsgesellschaft; English-language version to appear 1987.

Sindermann, C.J. (1982), *Winning the Games Scientists Play*. New York: Plenum Press; 304 pp.

Smith, E.G., Baker, P.A. (1976), *The Wiswesser Line-Formula Chemical Notation (WLN)*. 3rd. ed.Cherry Hill, NJ (USA): Chemical Information Management..

Souther, J.W., White, M.L. (1984), *Technical Report Writing*. Melbourne, FL (USA): Krieger; 104 pp.

Sowan, F., Horwood, E. (1983), *Publishing with Ellis Horwood—An Authors' Guide to the Publication of Works in Science and Technology*. Chichester: Ellis Horwood; 92 pp.

Standard *ANSI Z39.5-1985. American National Standard for the abbreviation of titles of periodicals.*

Standard *ANSI Z39.9-1974 (R 1984). American National Standard for international standard serial numbers.*

Standard *ANSI Z39.14-1979. American National Standard for writing abstracts.*

Standard *ANSI Z39.18-1974. Guidelines for format and production of scientific and technical reports.*

Standard *ANSI Z39.22-1981. American National Standard for proof corrections.*

Standard *ANSI Z39.29-1977. American National Standard for bibliographic references.*

Standard *ANSI/IEEE 268-1982. American National Standard metric practice.*

Standard *BS 3763: 1976. The International System of units (SI).*

Standard *BS 4821: 1972. Recommendations for representation of theses.*

Standard *DIN 108, Blatt 2, 1975. Diaprojektion—Technische Dias, Vorlagen, Ausführung, Prüfung, Vorführbedingungen.*

Standard *DIN 461, 1973. Graphische Darstellung in Koordinatensystemen.*

Standard *DIN 1301, Teil 1, 1985. Einheiten—Einheitennamen, Einheitenzeichen.*

Standard *DIN 1338, 1977. Formelschreibweise und Formelsatz.*

Standard *DIN 1338, Beiblatt 1, 1980. Formelschreibweise und Formelsatz— Form der Schriftzeichen.*

Standard *DIN 1338, Beiblatt 2, 1983. Formelschreibweise und Formelsatz— Ausschluß von Formeln.*

Standard *DIN 1338, Beiblatt 3, 1980. Formelschreibweise und Formelsatz— Formeln in maschinenschriftlichen Veröffentlichungen.*

Standard *DIN 1505, Teil 2, 1984. Titelangaben von Dokumenten. Titelaufnahme von Schrifttum.*

Standard *DIN 16 511, 1966. Korrekturzeichen.*

Standard *ISO 4-1984(E). Documentation—Rules for the abbreviation of title words and titles of publications.*

Standard *ISO 31/0-1981. General principles concerning quantities, units and symbols.*

Standard *ISO 31/1-1978. Quantities and units of space and time.*

Standard *ISO 31/2-1978. Quantities of periodic and related phenomena.*

Standard *ISO 31/3-1978. Quantities and units of mechanics.*

Standard *ISO 31/4-1978. Quantities and units of heat.*

Standard *ISO 31/5-1979. Quantities and units of electricity and magnetism.*

Standard *ISO 31/6-1980. Quantities and units of light and related electromagnetic radiation.*

Standard *ISO 31/7-1978. Quantities and units of acoustics.*

Standard *ISO 31/8-1980. Quantities and units of physical chemistry and molecular physics.*

Standard *ISO 31/9-1980. Quantities and units of atomic and nuclear physics.*

Standard *ISO 31/10-1980. Quantities and units of nuclear reactions and ionizing radiation.*

Standard *ISO 31/11-1978(E/F). Mathematical signs and symbols for use in the physical sciences and technology—Bilingual edition.*

Standard *ISO 31/12-1981. Dimensionless parameters.*

Standard *ISO 31/13-1981. Quantities and units of of solid state physics.*

Standard *ISO/DIS 690-1979. Documentation—Bibliographic references—Content, form, and structure.*

Standard *ISO 833-1974. Abbreviations of generic names in titles of periodicals.*

Standard *ISO 1000-1981. SI units and recommendations for the use of their multiples and of certain other units.*

Standard *ISO 2108-1972. Documentation—International standard book numbering (ISBN).*

Standard *ISO 3166-1974. Codes for the representation of names of countries.*

Stewart, R., Stewart, A.L. (1984), *Proposal Preparation.* New York: Wiley; 319 pp.

Strunk, W. Jr., White, E.B. (1979), *The Elements of Style.* 3rd ed. New York: Macmillan; 92 pp.

*The Chicago Manual of Style* (1982). 13th ed. Chicago: University of Chicago; 738 pp.

Turk, C., Kirkman, J. (1982), *Effective Writing—Improving Scientific, Technical and Business Communication.* London: Spon (paperback edition available from Methuen); 258 pp.

Woodburn, H.M. (1974), *Using the Chemical Literature.* New York: Marcel Dekker; 312 pp.

Zinsser, W. (1980), *On Writing Well—An Informal Guide to Writing Nonfiction.* 2nd ed. New York: Harper and Row.

Zinsser, W. (1983), *Writing with a Word Processor.* New York: Harper and Row; 128 pp.

# *Additional Reading*

American Chemical Society (Ed.) (1969), *Handbook for Speakers.* Washington, DC: American Chemical Society; 16 pp.

American Intitute of Physics (Ed.) (1978), *Style Manual for Guidance in the Preparation of Papers for Journals Published by the American Institute of Physics and its Member Societies.* 3rd ed. New York: American Institute of Physics (Address: 335 East 45 Street, New York NY 10017, USA.); 56 pp.

American Medical Association (Ed.) (1981), *Manual for Authors and Editors—Editorial Style and Manuscript Preparation.* 7th ed. Los Altos, CA (USA): Lange Medical Publications; 184 pp.

Campbell, M. (1977), *The Use of the Chemical Abstracts.* Nathan: Griffith University Library; 38 pp.

Cochran, W., Fenner, P., Hill, M. (1984), *Geowriting—A Guide to Writing, Editing, and Printing in Earth Science.* Alexandria, VA (USA): American Geological Institute; 80 pp.

Council of Biology Editors (Ed.) (1983), *CBE Style Manual.* 5th ed. Bethesda, MD (USA): Council of Biology Editors.

Hampel, C.A., Hawley (1976), *Glossary of Chemical Terms.* New York: Van Nostrand Reinhold; 282 pp.

Hawley, G.G. (Ed.) (1981), *The Condensed Chemical Dictionary.* 10th ed. New York: Van Nostrand Reinhold; 1136 pp.

Huth, E.J. (1982), *How to Write and Publish Papers in the Medical Sciences.* Philadelphia: ISI Press; 204 pp.

ISO International Organization for Standardization (Ed.), *List of Serial Title Word Abbreviations.* Published jointly with the ISDS International Centre, 20 rue Bachaumont, F-75002 Paris, France. (ISDS: International Serials Data System)

Jeffries, J.R., Bates, J.D. (1983), *The Executive's Guide to Meetings, Conferences, and Audovisual Presentations*. New York: McGraw-Hill; 228 pp.

McGraw-Hill Book Company (Ed.) (1968), *The McGraw-Hill Author's Book*. New York: McGraw-Hill; 74 pp.

Standard *ANSI Z39.16-1979. American National Standard for the preparation of scientific papers for written or oral presentation.*

Standard *BS 5261: Part 1: 1975. Copy preparation and proof correction— Recommendations for preparing of typescript copy for printing.*

Standard *BS 5261: Part 2: 1976. Copy preparation and proof correction— Specification for typographic requirements, marks for copy preparation and proof correction, proofing procedure.*

Standard *ISO 832-1975(E/F) Documentation—Bibliographic references— Abbreviations of typical words.*

Standard *ISO 999-1975(E). Documentation—Index of publication.*

Standard *ISO 1086-1975(E). Documentation—Title-leaves of a book.*

Standard *ISO 2145-1978(E). Documentation—Numbering of divisions and subdivisions in written documents.*

The Institute of Physics (Ed.) (1983), *Notes for Authors*. London: Institute of Physics; 36 pp.

Ulick, Z. (1986), Personal Publishing with the Macintosh. Hasbronck Heights, NJ (USA): Hayden Book Company.

Woodford, F.P. (Ed.) (1968), *Scientific Writing for Graduate Students—A Manual on the Teaching of Scientific Writing*. A Council of Biological Editors Manual. New York: The Rockefeller University Press.

# Index